Statistics for Social and Behavioral Sciences

Advisors:

S.E. Fienberg
W.J. van der Linden

For other titles published in this series, go to
http://www.springer.com/3463

Herbert Hoijtink · Irene Klugkist · Paul A. Boelen
Editors

Bayesian Evaluation of Informative Hypotheses

Springer

Editors
Herbert Hoijtink
Department of Methods and Statistics
Faculty of Social Sciences
University of Utrecht
P.O. Box 80140
3508 TC Utrecht
The Netherlands
h.hoijtink@uu.nl

Irene Klugkist
Department of Methods and Statistics
Faculty of Social Sciences
University of Utrecht
P.O. Box 80140
3508 TC Utrecht
The Netherlands
i.klugkist@uu.nl

Paul A. Boelen
Department of Clinical
and Health Psychology
Faculty of Social Sciences
University of Utrecht
P.O. Box 80140
3508 TC Utrecht
The Netherlands
p.a.boelen@uu.nl

ISBN: 978-1-4419-1874-1 e-ISBN: 978-0-387-09612-4

Softcover reprint of the hardcover 1st edition 2008

Printed on acid-free paper

springer.com

Preface

This book provides an overview of the developments in the area of Bayesian evaluation of informative hypotheses that took place since the publication of the first paper on this topic in 2001 [Hoijtink, H. Confirmatory latent class analysis, model selection using Bayes factors and (pseudo) likelihood ratio statistics. Multivariate Behavioral Research, **36**, 563–588]. The current state of affairs was presented and discussed by the authors of this book during a workshop in Utrecht in June 2007. Here we would like to thank all authors for their participation, ideas, and contributions. We would also like to thank Sophie van der Zee for her editorial efforts during the construction of this book. Another word of thanks is due to John Kimmel of Springer for his confidence in the editors and authors. Finally, we would like to thank the Netherlands Organization for Scientific Research (NWO) whose VICI grant (453-05-002) awarded to the first author enabled the organization of the workshop, the writing of this book, and continuation of the research with respect to Bayesian evaluation of informative hypotheses.

Utrecht,

May 2008

Herbert Hoijtink
Irene Klugkist
Paul A. Boelen

Contents

List of Contributors

Veerabhadran Baladandayuthapani
Department of Biostatistics
The University of Texas
M.D. Anderson Cancer Center
1515 Holcombe Blvd.
Houston, 77030 TX, USA
veera@mdanderson.org

Paul A. Boelen
Department of Clinical and Health Psychology
Utrecht University
P.O. Box 80140
3508 TC, Utrecht, the Netherlands
p.a.boelen@uu.nl

Jan Boom
Department of Developmental Psychology
Utrecht University
P.O. Box 80140
3508 TC, Utrecht, the Netherlands
j.boom@uu.nl

Ming-Hui Chen
Department of Statistics
University of Connecticut
215 Glenbrood Road
U-4120, Storrs, CT 06269, USA
mhchen@stat.uconn.edu

Herbert Hoijtink
Department of Methodology and Statistics
Utrecht University
P.O. Box 80140
3508 TC, Utrecht, the Netherlands
h.hoijtink@uu.nl

Marcel van den Hout
Department of Clinical and Health Psychology
Utrecht University
P.O. Box 80140
3508 TC, Utrecht, the Netherlands
m.vandenhout@uu.nl

Rafaele Huntjens
Department of Experimental Psychopathology
Groningen University
Grote Kruisstraat 2/1
9712 TS, Groningen, the Netherlands
r.j.c.huntjens@rug.nl

Geoffrey J. Iverson
Department of Cognitive Sciences
University of California at Irvine
3151 Social Science Plaza
Irvine, CA 92697, USA
giverson@uci.edu

Valen E. Johnson
Department of Biostatistics
The University of Texas
M.D. Anderson Cancer Center
1515 Holcombe Blvd.
Houston, 77030 TX, USA
vejohnson@mdanderson.org

George Karabatsos
College of Education
University of Illinois-Chicago
1040 W. Harrison Street
Chicago, IL 60607, USA
georgek@uic.edu

Bernet Sekasanvu Kato
Twin Research and Genetic Epidemiology Unit
St. Thomas' Hospital Campus
King's College London
Westminster Bridge Road
London, SE1 7EH, United Kingdom
bernet.kato@kcl.ac.uk

SungDuk Kim
Division of Epidemiology
Statistics and Prevention Research
National Institute of Child Health and Human Development NIH
Rockville, MD 20852,USA
kims2@mail.nih.gov

Irene Klugkist
Department of Methodology and Statistics
Utrecht University
P.O. Box 80140
3508 TC, Utrecht, the Netherlands
i.klugkist@uu.nl

Annemarie Kolk
Clinical Psychology
University of Amsterdam
Roeterstraat 15
1018 WB, Amsterdam
the Netherlands
a.m.m.kolk@uva.nl

Rebecca Kuiper
Department of Methodology and Statistics
Utrecht University
P.O. Box 80140
3508 TC, Utrecht, the Netherlands
r.m.kuiper@uu.nl

Olav Laudy
Department of Methodology and Statistics
Utrecht University
P.O. Box 80140
3508 TC, Utrecht, the Netherlands
o.laudy@uu.nl

Michael Lee
Department of Cognitive Sciences
University of California at Irvine
3151 Social Science Plaza
Irvine, CA 92697, USA
mdlee@uci.edu

Guimei Liu
Department of Mathematics
University of Puerto Rico at Rio Piedras Campus
P.O. Box 23355
San Juan, 00931-3355, Puerto Rico
liugm@zucc.edu.cn

Tom Lodewyckx
Department of Quantitative and Personality Psychology
University of Leuven
Tiensestraat 102
3000 Leuven, Belgium
tom.lodewyckx@student.kuleuven.be

Joris Mulder
Department of Methodology and Statistics
Utrecht University
P.O. Box 80140
3508 TC, Utrecht, the Netherlands
j.mulder3@uu.nl

Jay I. Myung
Department of Psychology
Ohio State University
1835 Neil Avenu
Columbus, OH, 43210, USA
myung.1@osu.edu

Carel F.W. Peeters
Department of Methodology and Statistics
Utrecht University
P.O. Box 80140
3508 TC, Utrecht, the Netherlands
c.f.w.peeters@uu.nl

Luis Raúl Pericchi Guerra
Department of Mathematics
University of Puerto Rico at Rio Piedras Campus
P.O. Box 23355
San Juan 00931-3355, Puerto Rico
lrpericchi@uprrp.edu and luarpr@gmail.com

Albert Reijntjes
Department of Pedagogical and Educational Sciences
Utrecht University
P.O. Box 80140
3508 TC, Utrecht, the Netherlands
a.h.a.reijntjes@uu.nl

Marleen Rijkeboer
Department of Clinical and Health Psychology
Utrecht University
P.O. Box 80140
3508 TC, Utrecht, the Netherlands
m.m.rijkeboer@uu.nl

Jan-Willem Romeijn
Department of Theoretical Philosophy
Groningen University
Oude Boteringestraat 52
9712 GL, Groningen, the Netherlands
j.w.romeijn@rug.nl

David Rossel
Bioinformatics and Biostatistics Unit
Institute for Research in Biomedicine of Barcelona
Josep Samitier 1-5
08028 Barcelona, Spain
rosselldavid@gmail.com

Rens van de Schoot
Department of Methodology and Statistics
Utrecht University
P.O. Box 80140
3508 TC, Utrecht, the Netherlands
a.g.j.vandeschoot@uu.nl

David Torres Nunez
Department of Mathematics
University of Puerto Rico at Rio Piedras Campus
P.O. Box 23355
San Juan 00931-3355, Puerto Rico
david.torres.math@gmail.com

Eric-Jan Wagenmakers
Department of Psychology
University of Amsterdam
Roetersstraat 15
1018 WB, Amsterdam
the Netherlands
ej.wagenmakers@gmail.com

Sonja van Well
Clinical Psychology
University of Amsterdam
Roeterstraat 15
1018 WB Amsterdam
the Netherlands
s.m.vanwell@uva.nl

Floryt van Wesel
Department of Methodology and Statistics
Utrecht University
P.O. Box 80140
3508 TC, Utrecht, the Netherlands
f.vanwesel@uu.nl

1

An Introduction to Bayesian Evaluation of Informative Hypotheses

Herbert Hoijtink[1], Irene Klugkist[1], and Paul A. Boelen[2]

[1] Department of Methodology and Statistics, Utrecht University, P.O. Box 80140, 3508 TC Utrecht, the Netherlands h.hoijtink@uu.nl and i.klugkist@uu.nl
[2] Department of Clinical and Health Psychology, Utrecht University, P.O. Box 80140, 3508 TC Utrecht, the Netherlands p.a.boelen@uu.nl

1.1 Bayesian Evaluation of Informative Hypotheses

Null hypothesis significance testing (NHST) is one of the main research tools in social and behavioral research. It requires the specification of a null hypothesis, an alternative hypothesis, and data in order to test the null hypothesis. The main result of a NHST is a p-value [3]. An example of a null hypothesis and a corresponding alternative hypothesis for a one-way analysis of variance is:

$$H_0 : \mu_1 = \mu_2 = \mu_3$$

and

$$H_a : \mu_1, \mu_2, \text{ and } \mu_3, \text{are not all equal},$$

where μ_1, μ_2, and μ_3 represent the average score on the dependent variable of interest in three independent groups. The implication of the null hypothesis has been often criticized. Cohen [1] calls it the "nil hypothesis" because he finds it hard to imagine situations, especially in psychological research, where "nothing is going on," and three means are exactly equal to each other. The meaning of the alternative hypothesis can also be criticized. If H_0 is rejected, and thus H_a is implicitly accepted, we find ourselves in a situation that can be labelled "something is going on but we don't know what." Knowing that three means are not all equal (H_a) does not tell us which means are different or what the order of the means is. Stated otherwise, the null hypothesis describes the population of interest in an unrealistic manner, and the alternative hypothesis describes the population of interest in an uninformative manner.

In this book we will introduce and exemplify the use of informative hypotheses. An informative hypothesis can be constructed using inequality ($<$ denotes smaller than and $>$ denotes larger than) and about equality ($\approx$) constraints. Two examples of informative hypotheses are

$$H_{1a} : \mu_1 > \mu_2 > \mu_3$$

and

$$H_{1b}: \{\mu_1 \approx \mu_2\} > \mu_3.$$

The first hypothesis states that μ_1 is larger than μ_2 and that μ_2 is larger than μ_3. The second hypothesis states that μ_1 is about equal to μ_2 and that both are larger than μ_3. The inequality constraints $<$ and $>$ can be used to add theoretical expectations to the traditional alternative hypothesis H_a, thus making it more informative. The about equality constraint $\approx$ has two advantages. First of all, if, like Cohen [1], researchers consider the traditional null hypothesis H_0 to be a "nil hypothesis," they can replace it by

$$H_{1c}: \mu_1 \approx \mu_2 \approx \mu_3.$$

Of course it has to be specified what is meant by $\approx$. In words, $\mu_1 \approx \mu_2$ means that μ_1 is not substantially different from μ_2. In a simple formula this means that

$$|\mu_1 - \mu_2| < \delta,$$

where δ is the smallest difference between two means that is considered to be relevant by the researchers formulating the hypotheses. This immediately leads to the second advantage of using $\approx$ constraints. If the traditional null hypothesis is rejected by a significance test, it still has to be determined whether the effect found is relevant or not. If $H_{1c}: \mu_1 \approx \mu_2 \approx \mu_3$ is rejected, this is not necessary, because the relevance of an effect is already included in H_{1c}.

Classical and informative hypotheses differ not only in the manner in which they are formulated but also in the manner in which they are evaluated. The classical null and alternative hypotheses can be evaluated using a p-value. For a one-way analysis of variance the p-value is, loosely speaking, the probability that the differences in means observed in the data or larger differences come from a population where the null hypothesis is true. According to a popular rule, the null hypothesis is rejected if the p-value is smaller than .05. Informative hypotheses can be evaluated using Bayesian model selection. The main result of Bayesian model selection is the posterior probability [2]. The posterior probability represents the support in the data for each hypothesis under investigation. For H_{1a} and H_{1b} these probabilities could, for example, be .90 and .10, respectively. This implies that after observing the data, H_{1a} is nine times as probable as H_{1b}.

1.2 Overview of the Book

The book consists of four parts. The first part, "Bayesian Evaluation of Informative Hypotheses," consists of Chapters 2 through 5. Subsequently, informative hypotheses, Bayesian estimation, Bayesian model selection, and the usefulness of the traditional null, alternative and informative hypotheses will be discussed in the context of analysis of variance models. The first part is an introduction and tutorial in which all the steps involved in the formulation and evaluation of informative hypotheses are subsequently discussed. This part of the book is suited for both social scientists and statisticians.

The second part, "A Further Study of Prior Distributions and the Bayes Factor," which consists of Chapters 6 through 9, contains a further elaboration of the use of Bayesian model selection for the evaluation of informative hypotheses. The second part of the book is rather technical and aimed at statisticians. Subsequently, different specifications of prior distributions, Bayesian alternatives for the use of posterior model probabilities, and the contrast between classical and Bayesian analysis will be discussed.

The third part of the book, "Beyond Analysis of Variance," discusses the application of informative hypotheses beyond the context of analysis of variance. It consists of Chapters 10 through 13, in which the application of informative hypotheses will subsequently be discussed for analysis of covariance, latent class models, models for contingency tables, and multilevel models. The third part of the book is suited for both social scientists and statisticians.

The fourth part of the book, "Evaluations," consists of Chapters 14 through 16. The concept of informative hypotheses will be discussed from the perspectives of psychologists, statisticians, and philosophers of science. This part of the book is also suited for both social scientists and statisticians.

1.3 Software

For many of the models and approaches discussed in this book software, and manuals are available. Software for inequality constrained analysis of variance and covariance, inequality constrained latent class models, and models for contingency tables as discussed in Chapters 3, 4, 10, 11, and 12, respectively, can be found at `http://www.fss.uu.nl/ms/informativehypotheses`. Software for inequality constrained analysis of variance as discussed in Chapter 6 can be found at `http://rosselldavid.googlepages.com`. Readers interested in software for the other approaches/models discussed should contact the authors of the respective chapters.

References

[1] Cohen, J.: The earth is round (p < .05). American Psychologist, **49**, 997–1003 (1994)
[2] Kass, R.E., Raftery, A.E.: Bayes factors. Journal of the American Statistical Association, **90**, 773–795 (1995)
[3] Schervish, M.J: P values: what they are and what they are not. The American Statistician, **50**, 203–206 (1996)

Part I

Bayesian Evaluation of Informative Hypotheses

2

Illustrative Psychological Data and Hypotheses for Bayesian Inequality Constrained Analysis of Variance

Paul A. Boelen[1] and Herbert Hoijtink[2]

[1] Department of Clinical and Health Psychology, Utrecht University, P.O. Box 80140, 3508 TC Utrecht, the Netherlands p.a.Boelen@uu.nl

[2] Department of Methodology and Statistics, Utrecht University, P.O. Box 80140, 3508 TC Utrecht, the Netherlands h.hoijtink@uu.nl

2.1 Introduction

In this chapter, three datasets from existing psychological research programs will be introduced that allow for an investigation of differences between groups on a single outcome variable. The first dataset was gathered to study amnesia in people with Dissociative Identity Disorder. The second dataset was originally generated to study emotional reactivity and emotional regulation in children subjected to different kinds of social evaluation by peers. The third dataset was obtained from research on coping with loss that, among other purposes, was used to study gender differences in coping. These three datasets will be used in subsequent chapters to illustrate Bayesian inequality constrained analysis of variance. To set the stage for these illustrations, in the current chapter, we will provide background information for the three datasets and will introduce theories and corresponding hypotheses that can be tested with these datasets. Some of these hypotheses were (implicitly) formulated by the researchers who gathered the data. However, since the approach introduced in this book allows more flexibility (viz. construction of hypotheses using inequality constraints), additional hypotheses will be formulated. We will describe how traditional hypothesis testing could be used to evaluate these hypotheses. Limitations of more conventional approaches to hypothesis testing will also be addressed. In the chapters that follow, these hypotheses will be evaluated using Bayesian inequality constrained analysis of variance.

2.2 Amnesia in Dissociative Identity Disorder: The Investigation of a Controversy

In psychiatry and clinical psychology, the Diagnostic and Statistical Manual of Mental Disorders (DSM [1]) is probably the most frequently used system to classify mental disorders. It includes specific criteria for dozens of disorders, subdivided into several categories among which are the categories of mood disorders, anxiety disorders, and personality disorders. There is a lot of controversy surrounding the system. Among other things, critics have noted that many of the disorders in DSM lack reliability and construct validity [40]. One of the most controversial disorders in the DSM is the Dissociative Identity Disorder (DID) – among the lay public also known as Multiple Personality Disorder. According to the last edition of the DSM, DID is defined as present when the person has at least two distinct identities or personality states that recurrently take control of the person's behaviour and has an inability to remember important personal information that cannot be explained by ordinary forgetfulness [1].

The controversy surrounding DID basically comes down to the question if it is indeed possible that people can have two or more separable identities (so called "alters"), with currently dominant identities being amnesic for events experienced by the other identity. Some say that this is indeed possible, whereas others have questioned if this is indeed so (for a review see [20]). A lot of research has been conducted to study this topic. Yet, as with the disorder itself, much of this research has been criticized. For instance, some studies have simply asked one alter of a DID-patient if he/she remembered what was experienced by the other alter. Potentially problematic is that, say, subjective experience of amnesia does not necessarily reflects the objective presence of this phenomenon as present in people with, for instance, dementia or other organic mental disorders. To curb this problem, researchers have used implicit measures of amnesia that allow for an examination of amnesia without study participants being aware that amnesia is tested. Elegant examples of this approach are represented in several studies by Huntjens [16], who used experimental designs to answer the question if inter-identity amnesia reported by DID-patients represents true, objectively verifiable amnesia or is perhaps attributable to other processes. In the present book, one of the studies by Huntjens et al. [17] will be used for illustrative purposes in several chapters.

As a starting point of their study, Huntjens et al. [17] observed that it is still uncertain whether or not the amnesia of DID-patients is "real" amnesia, if it is iatrogenic (i.e., induced by therapists), or if it is caused by suggestive influence of media and cultures on suggestible individuals. Noticing limitations of extant studies on this topic, they felt it was timely to further examine the issue of symptom simulation in DID-patients. To this end, a group of DID-patients ($N_{pat} = 19$) and three controlgroups were subjected to a recognition task. In the first phase of this task, which was the learning episode, patients were subjected to part of the Wechsler Memory Scale-revised (i.e., the Logical Memory-story A and the Visual Reproduction subtests [39]); patients were told a brief story and they were shown several drawn figures after which they performed a recall test of both the story and the figures. Then, after a delay, patients were asked to switch to another alter that was subjected to the second phase of the task. In this phase, these other alters were subjected to a recall test and a 15-item multiple-choice recognition test. Ten multiple-choice questions asked participants about particular story details, offering three possible answers for each question (e.g., "Was the story about a man, a woman, or an animal?"). Five questions asked participants to pick out the previously seen figure among five alternatives including four foils. The number of correct answers were summed to obtain a total "Recognition Score." This score ranged from 0 to 15 and represented an index of remembering.

As noted, apart from the DID-patients ($N_{pat} = 19$), three control groups were included. The first control group was a normal control group ($N_{con} = 25$). The second group consisted of normal people who were asked to deliberately simulate inter-identity amnesia ($N_{sim} = 25$). One week before the experiment took place, participants in this group were extensively informed about DID and its possible iatrogenic nature. They were asked to make up an imaginary, amnesic identity and to practice switching from one identity to the other. The third control group consisted of normal people who only underwent the second phase of the experimental task and had to guess the right answers to the recognition questions ($N_{amn} = 25$). As such, they represented a "truly amnesic control group" in that they were truly unable to remember anything about the story because they didn't get a chance to hear the story in the first place.

The design allowed the authors to compare the overall memory performance (recognition scores) among true DID-patients, Controls, Simulators, and True amnesiacs. There were at least two hypotheses implicated in the study. A first hypothesis was based upon the viewpoint that DID-patients actually suffer from "real amnesia." From this viewpoint, it could be hypothesized that, in terms of differences in memory performance between groups, Controls would remember the story details best and that both DID-patients and True amnesiacs would perform worse than normals but would not differ from each other. Finally, Simulators could be expected to perform worst because they knew the correct answer and thus could deliberately choose a wrong answer. Their score could be expected to be worse than the score of the True amnesiacs who had to guess the right answer and obviously occasionally guessed right. Using equality and inequality constraints, this hypothesis could be represented as

$$H_{1a} : \mu_{con} > \{\mu_{amn} = \mu_{pat}\} > \mu_{sim}, \tag{2.1}$$

where μ denotes the mean recognition score in the group indicated and $>$ means larger than. A second hypothesis was based upon the viewpoint that DID-patients actually feign their amnesia. From this viewpoint, it could be hypothesized that the memory performance of DID-patients would be similar to that of Simulators and that both groups would have a poorer memory performance than both normals and True amnesiacs. This hypothesis could be depicted as

$$H_{1b} : \mu_{con} > \mu_{amn} > \{\mu_{pat} = \mu_{sim}\}. \tag{2.2}$$

Table 2.1. Recognition scores in the DID data

	M	SD	N
1. DID-patients	3.11	1.59	19
2. Controls	13.28	1.46	25
3. Simulators	1.88	1.59	25
4. True amnesiacs	4.56	1.83	25

As a means to enhance clarity on the validity of both hypotheses, Huntjens et al. [17] compared the memory performance (recognition scores) of the four groups, using analysis of variance (see [32] for a nice introduction) followed by pairwise comparisons of means (see [37] for a nice introduction). Table 2.1 describes the recognition scores in each group.

Analysis of variance can be used to test the hypothesis

$$H_0 : \mu_{con} = \mu_{amn} = \mu_{pat} = \mu_{sim} \tag{2.3}$$

versus

$$H_2 : \mu_{con}, \mu_{amn}, \mu_{pat}, \mu_{sim}, \tag{2.4}$$

that is, testing "nothing is going on" versus "something is going on but I don't know what." Note that this is not what the researchers wanted to know. They wanted to know which of the hypotheses H_{1a} and H_{1b} was the best. As can be seen in Table 2.2, the significance of H_0 is .00, implying that "something is going" on. To further clarify this, Scheffe's posthoc tests were computed. All pairwise tests had significance smaller than .05, except the comparison of DID-patients with Simulators (a significance of .11). We additionally conducted other pairwise comparison procedures (Tukey HSD, Sidak, Gabriel, Hochberg), which all rendered the same result. These results are in accordance with H_{1b} but not with H_{1a}. Further support for H_{1b} is obtained from a visual inspection of the means in the four groups in Table 2.1: Controls indeed scored higher than True amnesiacs and both groups scored higher than DID-patients and Simulators.

All in all, outcomes indicated that, in terms of memory performance, the performance of DID-patients was worse than the performance of True amnesiacs and close to those who simulated DID. These findings led Huntjens et

Table 2.2. Significances for the analysis of variance and Scheffe's post-hoc tests

Hypothesis	Significance
$H_0 : \mu_{con} = \mu_{amn} = \mu_{pat} = \mu_{sim}$	.00
$H_0 : \mu_{con} = \mu_{pat}$	.00
$H_0 : \mu_{pat} = \mu_{sim}$	.11
$H_0 : \mu_{amn} = \mu_{pat}$	.04
$H_0 : \mu_{con} = \mu_{sim}$	.00
$H_0 : \mu_{con} = \mu_{amn}$	.00
$H_0 : \mu_{amn} = \mu_{sim}$	.00

al. [17] to conclude that DID-patients are similar to Simulators in providing incorrect answers to questions about the recognition of information to which they were previously subjected.

The study represents an elegant example of testing amnesia in DID-patients, without them being made aware that amnesia is tested. As such, it adds to our knowledge of DID in showing that it is an oversimplification to say that DID-patients indeed suffer from true amnesia. However, it is questionable whether hypothesis testing using analysis of variance and Scheffe's post hoc tests is the most elegant way to evaluate H_{1a} and H_{1b} (note that the authors did not have access to the Bayesian procedures that will be introduced in this book). First of all, the authors were not particularly interested in H_0 and H_2. Hence, testing these hypotheses is a rather indirect manner to evaluate H_{1a} and H_{1b}. Second, testing H_0 versus H_2 did not provide all answers: post hoc tests needed to be executed, followed by a visual inspection of the sample means, in order to be able to reach a conclusion. A number of issues threaten the viability of this procedure:

- The results might not be in agreement with either H_{1a} or H_{1b}, in which case no conclusion can be obtained.
- For the example at hand, the procedure consists of testing seven hypotheses. Of course Scheffe can be used to control the probability of type I errors, also known as errors of the first kind (i.e., the probability of incorrectly rejecting null hypotheses), but also leads to a reduction of power, which will increase the probability of type II errors, also known as an errors of the second kind (i.e., incorrectly accepting one or more null hypotheses). Since power issues are important in psychological research because sample sizes are often limited, this is an undesirable feature of procedures that are used to control the amount of type I errors.
- A visual inspection of means is not a well-established formalized procedure that can be used to evaluate informative hypotheses (i.e., hypotheses formulated using inequality constraints). If four means are expected to be ordered from small to large, and the sample means are 0.2, 0.1, 4, and 8, a formal procedure might show substantial support for the expectation even though the order of the first two means is reversed.
- Finally, the results of pairwise comparisons of means may be inconsistent. Results like "accept H_0: $\mu_A = \mu_B$," "accept H_0: $\mu_B = \mu_C$," and "reject H_0: $\mu_A = \mu_c$" cannot be straightforwardly interpreted because they are logically inconsistent: The three conclusions cannot be simultaneously true (see also [12]).

In the next chapters, Bayesian inequality constrained analysis of variance will be introduced. There are a number of differences between this Bayesian and traditional analysis of variance:

- The Bayesian approach requires a researcher to translate a number of competing theories into inequality constrained hypotheses before looking at the data (see (2.1) and (2.2) and the examples that will be given in the next sections).

- Subsequently the data will be used to quantify the support for each hypothesis under investigation. This quantification is called the posterior probability and will be introduced and discussed in Chapter 4 and later chapters, the interested reader is also referred to Chapter 7 of [14]. For the example at hand it might turn out that the posterior probabilities of H_{1a} and H_{1b} are .2 and .8, respectively. Such a result would imply that after observing the data, H_{1b} is four times as likely as H_{1a}.
- With the Bayesian approach, a conclusion with respect to H_{1a} and H_{1b} will always be obtained (a possible conclusion is that both are equally likely). There is no multiple testing problem and a visual inspection of means is not necessary in order to reach a conclusion.

However, Bayesian analysis can benefit from an evaluation of H_0 and H_2 in addition to H_{1a} and H_{1b}: If neither H_{1a} nor H_{1b} is a better model than H_2, it can be concluded that both sets of constraints are not in accordance with the data. Furthermore, if H_0 is a better model than H_{1a} and H_{1b}, the conclusion for the present DID example could be that the experimental manipulation has completely failed.

Note, finally, that there is an alternative to the formulation of H_{1a} and H_{1b}:

$$H_{1c} : \mu_{con} > \{\mu_{amn}, \mu_{pat}\} > \mu_{sim} \tag{2.5}$$

and

$$H_{1d} : \mu_{con} > \mu_{amn} > \{\mu_{pat}, \mu_{sim}\}. \tag{2.6}$$

Here it is no longer required that $\mu_{amn} = \mu_{pat}$ but only that they are located between μ_{con} and μ_{sim}. Something similar can be observed for H_{1b}. The alternative formulation is less restrictive than the original, without altering the core message of each hypothesis. Which formulation should be preferred, if any and whether or not it is wise to include H_0 and H_2 in the set of hypotheses under investigation will be further discussed in Chapter 4.

2.3 The Effect of Peer Evaluation on Mood in Children High and Low in Depression

A second study that will be used to illustrate the subsequent chapters addressed the influence of depression severity on emotional reactivity after different types of peer evaluation feedback in preadolescent children [26]. The interplay of moods (diffuse, slow moving feeling states) and emotions (quick moving reactions) is an intriguing study topic. Intuitively, it makes sense to think that moods potentiate like-valenced or matching emotions in a way that irritable mood strengthens angry reactions, anxious mood facilitates the experience of panic, and depressed mood facilitates reactions of sadness [29].

Yet, studies on the interplay of moods and emotions do not unambiguously support this so-called mood-facilitation hypothesis; there is evidence for this mood facilitation effect in anxiety [2], but little evidence that depressed mood inflates depressed emotional reactivity [29].

Reijntjes et al. [26, 27] studied the interaction between depressed mood and emotions in preadolescent children as one of the topics in a larger research program designed to enhance knowledge on emotional reactivity, emotional regulation, and depression in children. In this program, social evaluation by peers, manipulated by the authors, was chosen as a means to generate change in emotions. The reason to choose peer evaluation as a way of manipulating affect was chosen, because peer praise and rejection are common emotion-eliciting events in childhood [11] and exert an important influence on the development and maintenance of both externalizing problems (e.g., conduct disorder) [13] and internalizing problems (e.g., depression, cf. [21]). For the purpose of the present book, we will focus on a small aspect of this research program, in which the link between depressed mood and changes in current emotions in response to the manipulated peer evaluation was explored. This issue was addressed in detail in one of the studies of Reijntjes et al. [27].

In this study, 139 children, between the ages of 10 and 13 years, were led to believe that they were participating in an Internet version of a peer evaluation contest that was based on and named after an American television show called *Survivor*. Each participant was seated in front of a laptop computer and told that he/she participated in the online computer contest, together with four same-sex contestants. In actuality, contestants were fictitious. The objective of the game was presented as getting the highest "likeability" score from a jury consisting of 16 members, 8 boys and 8 girls. To this end, children were asked to provide information about themselves. They were asked questions about their favourite musical group, hobbies, and future occupation, and a number of character traits (e.g., sense of humour, agreeableness, intelligence, trustworthiness). Apart from that, their picture was taken with a web-cam.

All participants were led to believe that all the information and the picture would be transmitted to the judges over the Internet, who would then give them a "likeability" score ranging from 0 to 100. A short while after transmitting the information, the outcome of the likeability contest was presented. Specifically, the names of the players with the highest and the lowest score appeared on the screen. This is where the manipulation of the authors came in, that is, all participants were randomly assigned to a success feedback, a failure feedback, and a neutral feedback condition. Participants in the *success feedback* condition were told that they had obtained the highest score, those in the *failure feedback* condition were led to believe that they got lowest score, and those in the *neutral feedback* condition were told that they received neither the highest nor the lowest score. This last condition represented a control condition.

All participants completed a number of questionnaires over the course of the experiment. For the assessment of depressed mood, they completed the

Children's Depression Inventory (CDI) before the contest. The CDI is a 27-item measure for the assessment of social, behavioural, and affective symptoms of depression in children [19]. Each of the 27 items asks children to pick out the statement that applies to them from 3 alternatives (e.g., I like myself, I do not like myself, I hate myself). For each question, the first answer (absence of the symptom) is rated as 0, the second answer (mild symptom) is rated as 1, and the third answer (definite symptom) is rated as 2. Item scores are summed to form an overall depression score that can range from 0 to 54.

To assess changes in affect induced by the peer evaluation, children completed the Positive And Negative Affect Schedule (PANAS; [38]) before and after the contest. In the current illustration, we focus on the Positive Affect scale, the PANAS-P. This scale presents children with 10 mood-related adjectives (e.g., enthusiastic, active, alert) and asks them to rate the extent to which these moods are experienced on 5-point scales ranging from "very slightly or not at all" to "extremely". Scale scores can range from 10 to 50.

The data of Reijntjes et al. [27] allowed us to test several hypotheses about the influence of depressed mood on emotional reactivity induced by peer evaluation. The *improvement in affect/emotion*, defined as post-Survivor positive affect minus pre-Survivor positive affect, represents emotional reactivity and was the dependent variable. Positive values of this emotional reactivity (like in the high depressed success feedback group; see Table 2.3) indicate an improvement in positive mood, and negative values (like in the low depressed failure feedback group) indicate a decrease in positive mood. To compare between children with different levels of depression, the sample was divided in three groups of equal size: the "Low Depressed" group, the "Moderately Depressed" group, and, the "High Depressed" group. The hypotheses for this study addressed differences in emotional reactivity for children in these three groups, who were subjected to success, failure, or neutral peer feedback. This created nine groups of children. Table 2.3 shows the labelling of these groups, as well as the mean Emotional Reactivity Scores across groups.

We were particularly interested in reactivity after success feedback and after failure feedback. Mood improvement after success feedback is defined as emotional reactivity in the success condition minus emotional reactivity in the neutral condition ($\mu_1 - \mu_2$, $\mu_4 - \mu_5$, and $\mu_7 - \mu_8$). Note that the subscripts correspond to the group labelling displayed between [.] in Table 2.3. Positive differences between means indicate that mood improved more in the success condition than in the neutral condition. Negative differences indicate the opposite. Similarly, mood improvement after failure feedback is defined as emotional reactivity in the failure condition minus emotional reactivity in the neutral condition ($\mu_3 - \mu_2$, $\mu_6 - \mu_5$, and $\mu_9 - \mu_8$). Here positive differences between means indicated that mood improved more in the failure than in the neutral condition, and negative differences indicate that the opposite occurred. The neutral condition was chosen as a reference group because it may well be that the emotional reactivity in the neutral condition is different for different levels of depression. Our interest was not primarily in whether

Table 2.3. Emotional reactivity: Mean (M), standard deviation (SD), and sample size (N) per subgroup of depression by feedback condition

	Feedback condition											
	Positive				Neutral				Negative			
Depression		M	SD	N		M	SD	N		M	SD	N
Low	[1]	0.27	4.67	18	[2]	0.29	4.76	17	[3]	−9.33	8.77	12
Moderate	[4]	0.41	4.96	12	[5]	−1.50	5.99	14	[6]	−5.78	5.75	19
High	[7]	5.76	4.39	17	[8]	−0.56	4.25	16	[9]	−3.85	6.49	14

Note: Numbers in square brackets denote the group labelling as used in the hypotheses.

emotional reactivity differs among depression levels in the success an failure conditions, but whether these differences persisted if we controlled for differences in emotional reactivity in the neutral condition.

Considering the issue of emotional reactivity, we could define at least three competing theories on the influence of depressed mood on emotional reactivity after peer praise (i.e., success feedback) and peer rejection (i.e., failure feedback). As noted, the mood-facilitation hypothesis states that mood potentates matching emotions. Based on this hypothesis, we could expect that success feedback would elicit a less pronounced mood improvement in depressed compared to nondepressed children and that failure feedback would generate a more pronounced worsening of mood in depressed compared to nondepressed children; cf. [28]. Translated into an inequality constrained hypothesis this becomes

$$\begin{aligned} H_{1a} : &\{\mu_7 - \mu_8\} < \{\mu_4 - \mu_5\} < \{\mu_1 - \mu_2\}, \\ &\{\mu_9 - \mu_8\} < \{\mu_6 - \mu_5\} < \{\mu_3 - \mu_2\}. \end{aligned} \tag{2.7}$$

One alternative hypothesis can be drawn from recent research on emotional context insensitivity in depression. Emotional context insensitivity comes down to the notion that the presence of depression causes attenuated emotional reactivity to both positive and negative stimuli; that is, depressed mood states are said to prompt withdrawal and to cause reduction in motivated activity and pessimism, such that both positive and negative triggers from the environment lead to little or no emotional reactivity [30]. In a recent study among adults, Rottenberg et al. [31] found evidence that depression coincides with such insensitivity. Based on this viewpoint, it could be expected that success feedback would lead to a weaker increase in mood in depressed compared to nondepressed children and that failure feedback would lead to stronger mood improvement (or, stated otherwise, failure feedback has less impact on depressed children and will thus lead to a smaller reduction in positive mood) in depressed compared to nondepressed children:

$$H_{1b} : \{\mu_7 - \mu_8\} < \{\mu_4 - \mu_5\} < \{\mu_1 - \mu_2\},$$
$$\{\mu_9 - \mu_8\} > \{\mu_6 - \mu_5\} > \{\mu_3 - \mu_2\}. \tag{2.8}$$

We could think of an additional third hypothesis, one that is interesting but not explicitly mentioned and examined in existing literature. It is possible that, since depressed mood coincides with negative cognitions about self-worth [3], children higher in depression expected to get lower scores from their peers in the Survivor contest. From this viewpoint it could be expected that success feedback would actually lead to a stronger increase in mood in depressed compared to nondepressed children, because this success was highly discrepant with their self-view and expectations. In a related vein, it would be possible that failure feedback would lead to a weaker mood reduction (or, taking into account the way that we have defined our dependent variable, a stronger mood improvement) in depressed compared to nondepressed children, because, at some level, the depressed children were already expecting to fail; cf. [15]. Translated into an inequality constrained hypothesis, what could be called discrepancy hypothesis could be formulated as

$$H_{1c} : \{\mu_7 - \mu_8\} > \{\mu_4 - \mu_5\} > \{\mu_1 - \mu_2\},$$
$$\{\mu_9 - \mu_8\} > \{\mu_6 - \mu_5\} > \{\mu_3 - \mu_2\}. \tag{2.9}$$

Evaluation of H_{1a}, H_{1b}, and H_{1c} using a traditional two-way analysis of variance does not render a straightforward evaluation of the hypotheses of interest. Both main effects are significant ($p = .00$ for the feedback condition; and, $p = .01$ for the depression condition) but the interaction effect is not (a significance of .08).

Using Figure 2.1 to interpret the significant main effects, the following conclusions seem valid: The main effect of feedback condition can be described as a decrease of emotional reactivity from the success via the neutral to the failure feedback condition. The main effect of depression level is less clear, but there is a tendency for more emotional reactivity in children with high levels in depression compared to those with moderate and low levels of depression. For a number of reasons, it is questionable whether these results can and should be used to evaluate H_{1a}, H_{1b}, and H_{1c}:

- The significant main effects do not address emotional reactivity in the success and failure condition corrected for emotional reactivity in the neutral condition.
- As mentioned earlier, if the null hypothesis ("nothing is going on") is rejected in favour of the alternative hypothesis ("something is going on but I don't know what"), a further inspection of tabled data summaries (e.g., a table of means) or a visual representation of the data (e.g., using figures like Figure 2.1) may be needed in order to determine what is going on. Evaluation of tabled data summaries and figures is never straightforward.

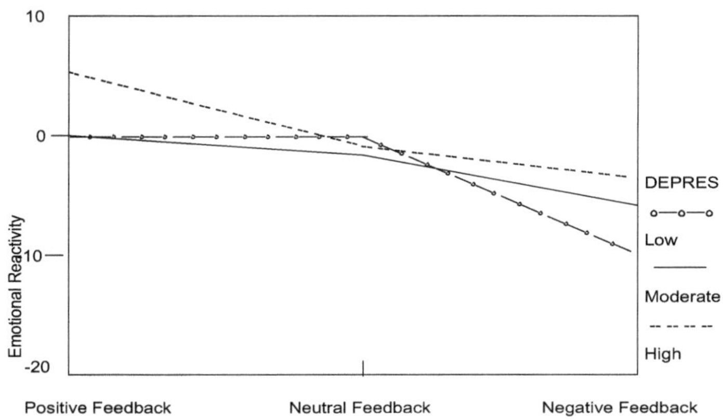

Fig. 2.1. Visual display of average emotional reactivity in the nine experimental groups

Without additional testing it is not clear whether the main effect of depression level was significant because of mean differences in one or more of the success, neutral, and failure feedback conditions. If, for example, standard errors of the means in the success and failure feedback condition would be large compared to the standard errors in the neutral feedback condition, the main effect of depression level would mainly be caused by the neutral feedback condition, which would change the "prima facie" interpretation of Figure 2.1 such that it is not in accordance with either of the three hypotheses.

- With additional testing to support the interpretation of Figure 2.1 (e.g., a pairwise comparison of means within each of the three feedback conditions), there is again a multiple testing problem; that is, due to the fact that more than one hypothesis is tested, the probability of type I errors will increase. This can be remedied using procedures that control the probability of type I errors like Scheffe and Bonferroni. However, this will lead to an increase of type II errors, that is, a reduction in power (the probability to correctly reject the null hypothesis). Furthermore, results may not be consistent with either one of H_{1a}, H_{1b}, or H_{1c}.
- A better approach is to execute four smaller two-way analyses of variance. In each analysis the focus is on the interaction effect in the subtable consisting of the first or last two groups of the factors feedback condition and depression level. Consider, for example, the subtable consisting of low/moderate levels of depression, and success/neutral feedback, that is, groups 1, 2, 4, and 5. If $\{\mu_4 - \mu_5\} < \{\mu_1 - \mu_2\}$ like in H_{1a}, or $\{\mu_4 - \mu_5\} > \{\mu_1 - \mu_2\}$ like in H_{1c}, there should be an interaction among feedback and depression in the subtable at hand. However, the significance

of the interaction was .46. This implies that $\{\mu_4 - \mu_5\} = \{\mu_1 - \mu_2\}$, which is not consistent with either H_{1a}, H_{1b}, or H_{1c}. In a similar manner the interaction in the other three subtables were tested. Neither of these was significant (moderate/high and success neutral, $p = .09$; low/moderate and neutral/failure, $p = .10$; moderate/high and neutral/failure, $p = .12$) and thus not helpful to choose the best of H_{1a}, H_{1b}, or H_{1c}.

Briefly stated, it appears that traditional hypothesis testing using two-way analyses of variance is not very helpful when evaluating H_{1a}, H_{1b}, and H_{1c}. In the following chapters it will be shown that Bayesian inequality constrained analysis of variance renders informative results and is able to select the best of the three hypotheses. Like in the previous example, one can extend H_{1a}, H_{1b}, and H_{1c} with the traditional null hypothesis (i.e., "nothing is going on") and an alternative hypothesis (i.e., "something is going on but I don't know what"). As in the DID example these hypotheses could be used to verify if the experimental manipulation was at all successful and whether at least one of H_{1a}, H_{1b}, and H_{1c} is supported by the data in the sense that it is a better model than the unconstrained alternative hypothesis.

2.4 Coping with Loss: The Influence of Gender, Kinship, and Time from Loss

Gender differences in psychological problems after stressful life events have received considerable attention in the literature. A recent review of the literature on gender differences in posttraumatic stress disorder (PTSD) has shown that in the general population this disorder is more prevalent among women than among men, even when controlling for the fact that men generally have a greater chance of being confronted with events that are potentially traumatizing [36]. Stated otherwise, among people who are confronted with traumatic events, women have a greater chance of developing PTSD than do men. It has been argued that this difference may well be due to the fact that bases rates of psychopathology are higher in women than in men. Indeed, studies have shown that women are generally more prone to develop anxious and depressive symptoms (e.g., [22, 33]). Nevertheless, there is evidence that women have a greater risk of developing problems, even when controlling pretrauma levels of distress [9].

The issue is obviously a complex one. For instance, some have noted that the relative severity of posttrauma psychopathology levels in women compared to men may well be at least partially attributable to the fact that men generally tend to underreport depressive and anxious symptoms [34, 36]. In addition, it is conceivable that gender differences in emotional problems after traumatizing events are moderated by demographic variables such as age and marital status and trauma-related variables such as the nature of the event and the severity of the event. With respect to the severity, it is noteworthy that Kendler et al. [18] found women to have a considerable greater chance of developing depression after exposure to a relatively minor threat, whereas the excess risk in women nearly disappeared in groups exposed to a more severe threat.

So, although female gender seems to be a risk factor for the development of psychological problems after confrontation with adversity, this conclusion is far from definitive to say the least. As a third example of Bayesian inequality constrained analysis of variance, we focussed on gender differences in the consequences of a stress full life-event different from trauma, namely the loss of a loved one. More specifically, we sought to address gender differences in the development of complicated grief. Complicated grief refers to a group of grief-specific symptoms that have been found to be distinct from depression and anxiety and to be predictive of severe mental health impairments [24]. These include intense and persistent yearning, difficulties in accepting the loss, avoidance, and shattered worldview.

Little is known about gender differences in this area. To our knowledge, there are four studies that addressed this issue, using the same conceptualization of complicated grief and the same questionnaire to measure it. Very roughly, these generated different results: In one study there were indications that men were worse off after a loss [4], in another study no differences were found [5], and two studies showed that women suffered more [10, 41]. As an additional illustration of Bayesian approaches, we examined the impact of gender further. In doing so, we focussed on the consequences of losing a partner and the consequences of losing a child, given that these losses are generally regarded as the most devastating (and mostly studied) losses that people may suffer [35]. Our aim was not to get a definitive answer to the question "who suffers more?" but, as with the other studies described in this chapter, to illustrate how the Bayesian inequality constrained analysis of variance can be used to test hypotheses about scores of groups on a single variable.

There were two hypotheses that guided our examination. The first of these is that women have a greater risk of developing complicated grief than do men. This hypothesis was fuelled by the fact that two of four studies that addressed gender differences found women to suffer more and by the fact that research on PTSD has shown that women are more at risk for problems after other types of adversity [36]. The second hypothesis was fuelled by the notion that the loss of partner or child may well work out differently for men and women. It has been claimed that women generally grow more attached to their children than do men [8]. This could make them more prone to develop emotional complications after loss. At the same time, there are reasons to believe that men are more vulnerable to get problems after the loss of a partner for instance, because their well-being is more strongly dependent on their relationship with the partner than that of women (cf. [25]). So, our second key hypothesis was that the influence of gender is moderated by kinship such that women suffer more after losing a child and men suffer more after losing a partner. In keeping with this notion, the study that addressed gender differences in parental bereaved

Table 2.4. Complicated grief: Mean (M), standard deviation (SD) and sample size (N) per subgroup of the Gender by Kinship by Time from loss design

		Time from loss							
		Recent				Remote			
Gender	Kinship		M	SD	N		M	SD	N
Men	Partner	[1]	84.91	21.59	106	[2]	78.60	20.31	131
	Child	[3]	79.77	21.88	26	[4]	77.79	22.37	52
Women	Partner	[5]	86.42	18.56	229	[6]	78.36	19.28	374
	Child	[7]	84.88	17.33	91	[8]	83.02	21.74	165

Note: Numbers in square brackets denote the group labelling as used in the hypotheses.

individuals showed that women suffered more [41], whereas such results were not found in two studies that focussed on conjugal loss [4, 6].

To test these hypotheses, data were obtained from research programs on grief conducted by the first author of this chapter. We selected 760 bereaved partners and parents from a group of 1321 mourners who participated in a study on the distinctiveness of complicated grief, depression, and anxiety [6] and 419 similarly bereaved individuals from a group of 943 mourners included in a study on the dimensionality of complicated grief [7]. The final sample for this illustration thus included $N = 1179$ mourners.

To assess complicated grief symptoms, all participants completed the Dutch version of the Inventory of Complicated Grief-revised (ICG-r). The ICG-r is a self-report measure that taps each of the symptoms of complicated grief as well as other potentially problematic responses to loss [23]. The 29-item Dutch version was developed by Boelen et al. [7]. Examples of items are "I feel that I have trouble accepting the death" and "I feel myself longing and yearning for the lost person." Respondents rate the presence of these symptoms in the preceding month, on 5-point scales ranging from "almost never" to "always." The overall complicated grief severity score is calculated by summing item scores. This score ranged from 29 to 145.

As we wished to include possible effects of time from loss, the group was divided into those who were bereaved less than 1 year (the "recent loss" group) and those whose losses occurred more than 1 year ago (the "remote loss" group). Table 2.4 shows complicated grief severity scores across the groups that were included in the analyses.

As noted, our key hypotheses were that (a) women generally have a higher risk of developing severe complicated grief symptoms than do men and, alternatively, (b) that losing a child leads to higher complicated grief levels in women, whereas the loss of a partner leads to higher complicated grief levels in men. Yet, for illustrative purposes, we included these hypotheses in a sequence of expectations, also including (main) effects of time from loss and kinship to the deceased. The first hypothesis in this sequence was based on the fact that studies have convincingly shown that, in general, complicated grief levels are stronger in the early months of bereavement than later [24]. So, the first hypothesis was formalized as follows:

- The directional (an inequality constraint is used to specify the direction of the effect) simple effect of time in the three-factor design at hand; that is, recent loss leads to more grief that remote loss: H_{1a}: $\mu_1 > \mu_2$, $\mu_3 > \mu_4$, $\mu_5 > \mu_6$, $\mu_7 > \mu_8$. Note that the subscripted numbers refer to the group numbers in Table 2.4 listed between [.].

To investigate our first key hypothesis that women always have higher complicated grief levels than men, H_{1a} could be extended with the following:

- The directional simple main effect of gender in the three-factor design at hand; that is, the grief scores for the women are always higher than the grief scores for the men in the corresponding groups to render H_{1b}: $\mu_5 > \mu_1$, $\mu_6 > \mu_2$, $\mu_7 > \mu_3$, $\mu_8 > \mu_4$, $\mu_1 > \mu_2$, $\mu_3 > \mu_4$, $\mu_5 > \mu_6$, $\mu_7 > \mu_8$. Note that this is a model containing the simple main effects of time and gender.

Although there may be an interaction between gender and kinship, some scholars have claimed that, both for men and for women, losing a child poses a stronger risk for the development of complicated grief than does losing a partner (cf. [41]). This could be investigated via an extension of H_{1b} with the following:

- The directional simple effect of kinship in the three-factor design at hand; that is, losing a child leads to more complicated grief than losing a partner: H_{1c}: $\mu_3 > \mu_1$, $\mu_4 > \mu_2$, $\mu_7 > \mu_5$, $\mu_8 > \mu_6$, $\mu_5 > \mu_1$, $\mu_6 > \mu_2$, $\mu_7 > \mu_3$, $\mu_8 > \mu_4$, $\mu_1 > \mu_2$, $\mu_3 > \mu_4$, $\mu_5 > \mu_6$, $\mu_7 > \mu_8$. Note that this is a model containing the simple main effects of time, gender, and kinship.

Whether losing a child is more devastating for women could be investigated with the following hypothesis, which contains three elements:

- The simple effect of time (i.e., H_{1a}), the notion that losing a child is more severe for women than losing a partner and the notion that losing a child is more severe for women than for men. Together these three components render the following hypothesis: H_{1d}: $\mu_1 > \mu_2$, $\mu_3 > \mu_4$, $\mu_5 > \mu_6$, $\mu_7 > \mu_8$, $\mu_7 > \mu_5$, $\mu_8 > \mu_6$, $\mu_7 > \mu_3$, $\mu_8 > \mu_4$.

Our second key hypothesis that losing a child is more devastating for women, whereas men are worse of after partner loss could be investigated via an extension of the previous hypothesis with two elements:

- Complicated grief levels of men who lose a partner are higher than complicated grief levels of men who lose a child, and complicated grief levels of men who lose a partner are higher than complicated grief levels of women who lose a partner. This renders the following model: H_{1e}: $\mu_1 > \mu_2$, $\mu_3 > \mu_4$, $\mu_5 > \mu_6$, $\mu_7 > \mu_8$, $\mu_7 > \mu_5$, $\mu_8 > \mu_6$, $\mu_7 > \mu_3$, $\mu_8 > \mu_4$, $\mu_1 > \mu_3$, $\mu_2 > \mu_4$, $\mu_1 > \mu_5$, $\mu_2 > \mu_6$.

Table 2.5. Three-way analysis of variance of complicated grief scores

Effect	Significance
Main effect of Gender	.064
Main effect of Time	.004
Main effect of Kinship	.652
Interaction between Gender and Time	.794
Interaction between Gender and Kinship	.148
Interaction between Time and Kinship	.093
Three-way interaction Gender, Time, Kinship	.765

As in the previous two examples, it may be useful to add the traditional null hypothesis and an unconstrained hypothesis to H_{1a} until H_{1e}. The unconstrained hypothesis H_2: μ_1, ..., μ_8 is particularly useful in this example. If none of the models under investigation has a better fit than H_2, none of the sets of constraints is supported by the data.

It is difficult to obtain information with respect to the inequality constrained hypotheses under investigation, using traditional analysis of variance. A straightforward three-way analysis of variance does not test the hypotheses under investigation (combinations of sets of directional simple effects where inequality constraints are used to specify the direction of the effects) since it evaluates only main, two- and three-way interaction effects. For illustrative purposes we executed such a three-way analysis of variance; the results are displayed in Table 2.5. As can be seen, only the main effect of time was found to have a significance smaller than .05. Although this provides some support for H_{1a}, in subsequent chapters it will be shown that using these results to evaluate H_{1a} through H_{1e} will render misleading results, because the hypotheses tested are not directly related to the hypotheses under investigation.

Another approach could be to apply Scheffe's procedure for pairwise comparisons of means to the eight groups that constitute the three-way design at hand (note that other pairwise comparison procedures rendered similar results). The results with a significance smaller than .10 are displayed in Table 2.6. As can be seen, the results are not clearly in agreement with one of the hypotheses under investigation. Furthermore, as will be shown in subsequent chapters, it would be wrong to conclude that none of the hypotheses under investigation is supported by the data. In following chapters it will also be shown how Bayesian inequality constrained analysis of variance can straight-

Table 2.6. Pairwise comparisons with a significance smaller than .10 for the complicated grief scores

Group		Group	Significance
Women, Recent, Partner	vs.	Women, Remote, Partner	.002
Women, Recent, Partner	vs.	Men, Remote, Partner	.077

forwardly be used to determine to which degree each hypothesis is supported by the data.

2.5 Conclusions

In the context of analysis of variance models this chapter has illustrated at least two things. First, theories can often be translated into hypotheses that are formulated using inequality constraints among a set of means. Second, it may be difficult, relatively ad hoc, or even impossible to evaluate these hypotheses using traditional null hypotheses testing. The following features appeared in the three examples that have so far been discussed:

- Traditional null ("nothing is going on") and alternative hypotheses ("something is going on but I don't know what") do not render straightforward evaluations of the informative (specified using inequality constraints) hypotheses under investigation.
- An informal evaluation of the sample means is often needed in addition to traditional hypothesis testing to be able to evaluate informative hypotheses. Since informal analysis, sometimes also referred to as "eyeball testing," can be misleading, this is not a proper mode of inference.
- Results obtained with traditional hypothesis testing may not be in agreement with any of the informative hypotheses under investigation.
- Often multiple smaller null hypotheses and alternative hypotheses have to be tested in order to be able to evaluate informative hypotheses. There are two drawbacks of multiple hypothesis testing: It is hard to control both the type I and type II errors and results obtained using multiple hypothesis testing may be mutually inconsistent.

In the following chapters it will be shown that it is relatively easy to evaluate a set of informative hypotheses using Bayesian inequality constrained analysis of variance. The use of inequality constraints is not limited to analysis of variance type models. In the third part of the book it will be shown how inequality constrained hypotheses can be formulated in the context of analysis of covariance models, latent class models, multilevel models, and models for contingency tables.

The following types of constraints have been used in this chapter:

- Mean restricted to be larger than another mean (e.g., $\mu_1 > \mu_2$).
- Combination of means restricted to be larger than another combination of means (e.g., $\{\mu_1 - \mu_2\} > \{\mu_3 - \mu_4\}$). Note that the restriction in which the minus sign is replaced by a plus sign is also possible.

Restrictions that have not been used but that can also be helpful to translate a theory into an inequality constrained hypothesis are as follows:

- Mean restricted to be larger than a number (e.g., $\mu > 0$). This restriction will be used in Chapter 13 of this book, which deals with inequality constrained multilevel analysis.
- Mean restricted to be larger than another mean plus a number, for instance, $\mu_1 > \{\mu_2 + 2\}$. This restriction can be used if researchers want to include a minimally required effect size in their hypotheses; that is, the first mean should be at least 2 larger than the second mean. The number 2 is not the traditional effect size in terms of Cohen's d, in which case it would be written as $\frac{\mu_1 - \mu_2}{\sigma} > d$. However, rewriting Cohen's d, we obtain $\mu_1 - \mu_2 > d\sigma$; stated otherwise, the number 2 is Cohen's d multiplied with the within-group standard deviation σ.
- Absolute difference between means restricted to be smaller than a number (e.g., $|\mu_1 - \mu_2| < 2$). This restriction can be used to specify that two means are not relevantly different, where, again, researchers should specify the minimally required effect size needed for two means to be required relevantly different.

All these types of restrictions can be combined, thus rendering a flexible tool that can be used to translate one or more theories into a number of competing hypotheses. In this book many examples will be given, and references to software with which these hypotheses can be evaluated will be provided.

References

[1] American Psychiatric Association: Diagnostic and Statistical Manual of Mental Disorders (4th ed. Text Revision). Washington, DC, American Psychiatric Association (2000)

[2] Barlow, D.H.: Anxiety and its Disorders: The Nature and Treatment of Anxiety and Panic (2nd ed). New York, Guilford Press (2002)

[3] Beck, A.T., Rush, A.J., Shaw, B.F., Emery, G.: Cognitive Therapy of Depression. New York, Guilford Press (1979)

[4] Bierhals, A.J., Prigerson, H.G., Fasiczka, A., Frank, E., Miller, M., Reynolds III, C.: Gender differences in complicated grief among the elderly. Omega: Journal of Death and Dying, **32**, 303–317 (1996)

[5] Boelen, P.A., Bout, J. van den: Gender differences in traumatic grief symptom severity after the loss of a spouse. Omega: Journal of Death and Dying, **46**, 183–198 (2003)

[6] Boelen, P.A., Bout, J. van den: Complicated grief, depression, and anxiety as distinct post-loss syndromes: A confirmatory factor analysis study. American Journal of Psychiatry, **162**, 2175–2177 (2005)

[7] Boelen, P.A., Keijser, J. de, Bout, J. van den: Psychometrische eigenschappen van de Rouw VragenLijst (RVL) [Psychometric properties of the Inventory of Traumatic Grief]. Gedrag & Gezondheid, **29**, 172–185 (2001)

[8] Bowlby, J.: Attachment and Loss: Volume 3, Loss: Sadness and Depression. New York, Basic Books (1980)

[9] Breslau, N., Davis, G.C., Andreski, P., Peterson, E.L., Schultz, L.R.: Sex differences in posttraumatic stress disorder. Archives of General Psychiatry, **54**, 1044–1048 (1997)

[10] Chen, J.H., Bierhals, A.J., Prigerson, H.G., Kasl, S.V., Mazure, C.M., Jacobs, S.: Gender differences in the effects of bereavement-related psychological distress in health outcomes. Psychological Medicine, **29**, 367–380 (1999)

[11] Coie, J.D.: Toward a theory of peer rejection. In: Asher, S.R., Coie, J.D. (eds) Peer Rejection in Childhood (pp. 365–401). New York, Cambridge University Press (1990)

[12] Dayton, C.M.: Information criteria for pair wise comparisons. Psychological Methods, **8**, 61–71 (2003)

[13] Dodge, K.A., Lansford, J.E., Salzer Burks, V., Bates, J.E., Pettit, G.S., Fontaine, R., Price J.M.: Peer rejection and social information-processing factors in the development of aggressive behavior problems in children. Child Development, **74**, 374–393 (2003)

[14] Gill, J.: Bayesian Methods. A Social and Behavioral Sciences Approach. London, Chapman & Hall (2002)

[15] Gladstone, T.R.G., Kaslow, N.J.: Depression and attributions in children and adolescents: A meta-analytic review. Journal of Abnormal Child Psychology, **23**, 597–606 (1995)

[16] Huntjens, R.J.C.: Apparent Amnesia. Interidentity Memory Functioning in Dissociative Identity Disorder. Ph.D. thesis, Utrecht University (2003)

[17] Huntjens, R.J.C., Peters, M.L., Woertman, L., Bovenschen, L.M., Martin, R.C., Postma, A.: Inter-identity amnesia in dissociative identity disorder: A simulated memory impairment? Psychological Medicine, **36**, 857–863 (2006)

[18] Kendler K.S., Kuhn J.W., Prescott C.A.: The interrelationship of neuroticism, sex, and stressful life events in the prediction of episodes of major depression. American Journal of Psychiatry, **161**, 631–636 (2004)

[19] Kovacs, M.: Rating scales to assess depression in school-aged children. Acta Paedo Psychiatra, **46**, 305–315 (1981)

[20] Lilienfeld, S.O., Lynn, S.J.: Dissociative identity disorder: Multiple personality, multiple controversies. In: Lilienfeld, S.O., Lohr, J.M., Lynn, S.J. (eds) Science and Pseudoscience in Clinical Psychology. New York, Guilford Press (2003)

[21] Nolan, S.A., Flynn, C., Garber, J.: Prospective relations between rejection and depression in young adolescents. Journal of Personality and Social Psychology, **85**, 745–755 (2003)

[22] Nolen-Hoeksema, S.: Sex Differences in Depression. Stanford, CA, Stanford University Press (1990)

[23] Prigerson, H.G., Jacobs, S.C.: Traumatic grief as a distinct disorder: A rationale, consensus criteria, and a preliminary empirical test. In: Stroebe, M.S., Hansson, R.O., Stroebe, W., Schut, H.A.W. (eds) Handbook of Bereavement Research. Consequences, Coping, and Care (pp. 613–647). Washington, DC, American Psychological Association Press (2001)

[24] Prigerson H.G., Vanderwerker L.C., Maciejewski P.K.: Prolonged grief disorder as a mental disorder: Inclusion in DSM. In Stroebe, M.S., Hansson,

R.O., Stroebe, W., Schut, H.A.W. (eds) Handbook of Bereavement Research and Practice: 21st Century Perspectives. Washington, DC, American Psychological Association Press (in press)
[25] Rando, T.A.: Treatment of Complicated Mourning. Champaign, IL, Research Press (1993)
[26] Reijntjes, A.: Emotion-Regulation and Depression in Pre-adolescent Children. Ph.D. thesis, Free University Amsterdam (2004).
[27] Reijntjes, A., Dekovic, M., Vermande, M., Telch, M.: The role of depressive symptoms in early adolescents online emotional responding to a peer evaluation challenge. Depression and Anxiety (in press)
[28] Rosenberg, E.L.: Levels of analysis and the organization of affect. Review of General Psychology, **2**, 247–270 (1998)
[29] Rottenberg, J.: Mood and emotion in major depression. Current Directions in Psychological Science, **14:3**, 167–170 (2005)
[30] Rottenberg, J., Gotlib, I.H.: Socioemotional functioning in depression. In: Power, M. (ed) Mood Disorders: A Handbook of Science and Practice (pp. 61–77). New York, Wiley (2004)
[31] Rottenberg, J., Gross, J. J., Gotlib, I. H.: Emotional context insensitivity in major depressive disorder. Journal of Abnormal Psychology, **114**, 627–639 (2005)
[32] Rutherford, A.: Introducing ANOVA and ANCOVA, a GLM Approach. London, Sage (2001)
[33] Shear, M. K., Feske, U., Greeno, C.: Gender differences in anxiety disorders: Clinical implications. In: Frank, E. (ed) Gender and its Effects on Psychopathology (pp. 151–165). Washington, DC, American Psychiatric Publishing (2000)
[34] Sigmon, S.T., Pells, J.J., Boulard, N.E., Whitcomb-Smith, S., Edenfield, T.M., Hermann, B.A., LaMattina, S.M., Schartel, J.G., Kubik, E.: Gender differences in self-reports of depression: The response bias hypothesis revisited. Sex Roles, **53**, 401–411 (2005)
[35] Stroebe, W., Schut, H.: Risk factors in bereavement outcome: A methodological and empirical review. In: Stroebe, M.S., Hansson, R.O., Stroebe, W., Schut, H. (eds) Handbook of Bereavement Research. Consequences, Coping, and Care (pp. 349–372). Washington, DC, American Psychological Association Press (2001)
[36] Tolin, D., Foa, E.B.: Sex differences in posttraumatic stress disorder: A quantitative review of 25 years of researcher. Psychological Bulletin, **132**, 959–992 (2006)
[37] Toothaker, L.E.: Multiple Comparison Procedures. London, Sage (1993)
[38] Watson, D., Clark, L.A., Tellegen, A.: Development and validation of brief measures of positive and negative affect: The PANAS scales. Journal of Personality and Social Psychology, **54**, 1063–1070 (1988)
[39] Wechsler, D.: Wechsler Memory Scale-Revised. San Antonio, TX, The Psychological Corporation (1987)
[40] Widiger, T.A., Clark, L.A.: Toward DSM-V and the classification of psychopathology. Psychological Bulletin, **126**, 946–963 (2000)
[41] Wijngaards-de Meij, L., Stroebe, M., Schut, H., Stroebe, W., Bout, J. van den, Heijden, P. van der, Dijkstra, I.C.: Couples at risk following the death of their child: Predictors of grief versus depression. Journal of Consulting and Clinical Psychology, **73**, 617–623 (2005)

3

Bayesian Estimation for Inequality Constrained Analysis of Variance

Irene Klugkist and Joris Mulder

Department of Methodology and Statistics, Utrecht University, P.O. Box 80140, 3508 TC Utrecht, the Netherlands `i.klugkist@uu.nl` and `j.mulder3@uu.nl`

3.1 A Short Introduction to Bayesian Statistics

In Chapter 2, several examples of research questions in the analysis of variance (ANOVA) context were presented. The model parameters of an ANOVA are two or more population means and the common and unknown residual variance. In the examples, the hypotheses or research questions of interest impose inequality constraints on the means. For instance, for the four-group ANOVA in the Dissociative Identity Disorder (DID) data from Huntjens et al. [10], the hypothesis $\mu_{con} > \mu_{amn} > \{\mu_{pat}, \mu_{sim}\}$ represents one of the theories of the researchers. In this chapter, Bayesian estimation of the parameters of inequality constrained ANOVA models is introduced.

In the Bayesian approach, knowledge about model parameters is represented by a probability distribution. Two important ingredients of Bayesian analyses are the *prior* distribution and the *posterior* distribution of the parameters. The prior represents the knowledge (or uncertainty) about model parameters before observing the data. After observing data, the information in the data is combined with the information in the prior, leading to the posterior. Stated otherwise, the posterior distribution represents the knowledge about model parameters after observing the data.

In Bayesian analyses, the role of prior distributions is a point of discussion. Many prior specifications can be made and different prior distributions may lead to different conclusions. Before moving to the inequality constrained ANOVA model, the basic principles of Bayesian analyses and the role of prior distributions herein will be explained based on a simple, one-parameter problem. The example deals with estimation of a population mean μ assuming that the variance is known. Elaborate introductions into Bayesian methodology are provided by, for instance, [2, 6, 8, 14].

Table 3.1. Recognition scores of DID-patients

Patients' scores	0 1 2 2 2 2 3	$M = 3.11$
	3 3 3 3 3 3	$SD = 1.59$
	4 4 4 4 6 7	$N = 19$

3.1.1 The Data

The example will be illustrated by part of the DID data introduced in Chapter 2. The focus is on estimation of the mean recognition score for DID-patients. The outcome variable Recognition is measured on a scale with a range of possible scores between 0 and 15. The observed recognition scores for the N=19 DID-patients can be found in Table 3.1. Also, the sample mean (M) and sample standard deviation (SD) are provided.

It is assumed that the data are normally distributed with a mean μ and variance σ^2. Both in classical and Bayesian approaches, a key element of any analysis is the function that represents the information in the data with respect to the model parameters. This is called the likelihood function. It is a formal representation of the knowledge with respect to μ and σ^2 contained in the data.

Assuming normally distributed data, the likelihood function of $\boldsymbol{y} = \{y_1, \ldots, y_N\}$ given μ and σ^2 is

$$f(\boldsymbol{y}|\mu, \sigma^2) = (2\pi\sigma^2)^{-N/2} \exp\left\{\frac{-1}{2\sigma^2} \sum_{i=1}^{N} (y_i - \mu)^2\right\}. \tag{3.1}$$

The observed recognition scores of N=19 DID-patients are used for further illustration. To simplify the example, the population variance will be assumed known and equal to 3. The model parameter of interest is therefore just μ and the likelihood function is

$$f(\boldsymbol{y}|\mu, \sigma^2 = 3) \propto (2\pi \cdot 3)^{-19/2} \exp\left\{\frac{-1}{2 \cdot 3} \sum_{i=1}^{19} (y_i - \mu)^2\right\}. \tag{3.2}$$

From (3.2) it can be seen that the information about μ follows a normal distribution with the mean equal to $\bar{y} = 3.11$ and variance equal to the squared standard error $\sigma^2/N = 3/19 = 0.16$; that is, $\mathcal{N}(\mu|3.11, 0.16)$. Figure 3.1 provides a graphical representation of (3.2). The likelihood summarizes the information with respect to μ that is available in the data. For example, the value $\mu = 3.11$ has the largest likelihood, because the sample mean is 3.11. Values for μ that are further away from 3.11 have decreasing likelihoods. Stated otherwise, the likelihood tells us which values of μ are more and less supported by the data.

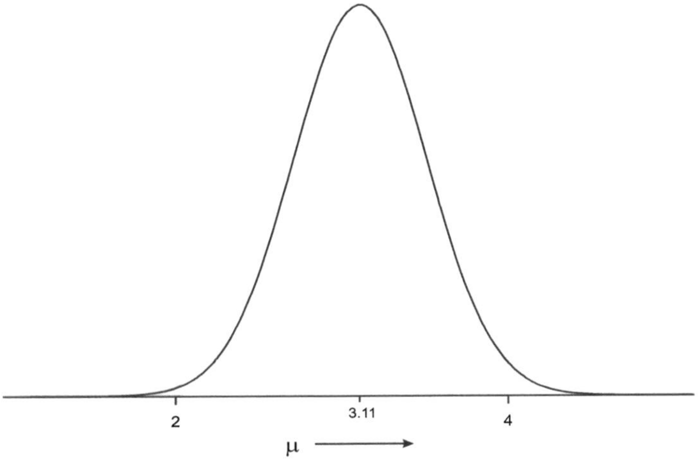

Fig. 3.1. Likelihood function for DID-patients' recognition data

3.1.2 Prior Distributions

Summarizing the information contained in the data with respect to model parameters is something that classical and Bayesian approaches have in common. However, Bayesian estimation also requires the specification of a prior distribution for μ (i.e., the knowledge with respect to μ *before seeing the data*). A simple and not uncommon choice for the prior could be $p(\mu) \propto 1$. Specifying $p(\mu) \propto 1$ means that every value for μ between $-\infty$ and ∞ is equally likely a priori and is known as an unbounded uniform or a constant prior. It is uninformative with respect to the parameter μ, since the prior does not add information. Bayesian estimates of μ based on this prior will lead to results that are comparable with non-Bayesian (e.g., maximum likelihood) estimates.

One could also argue that in this particular example, a prior that allows every value for μ (i.e., $-\infty < \mu < \infty$) is somewhat strange, because the recognition scores and consequently the mean of this variable can only obtain values between 0 and 15. Therefore, a bounded uniform prior is suggested in this example. The prior distribution, denoted $p_1(\mu)$, adds no other information about μ than that it can only have values between 0 and 15. Stated differently, it is flat between 0 and 15 and zero otherwise. In Figure 3.2, $p_1(\mu)$ is represented by the solid line. The function describing this prior is

$$p_1(\mu) = \mathcal{U}(\mu|lb = 0, ub = 15) = \begin{cases} 1/15 & \text{for } \mu \in [0, 15] \\ 0 & \text{otherwise,} \end{cases} \tag{3.3}$$

where lb denotes the lower and ub the upper bound of the uniform distribution. Note that $p_1(\mu)$ is not informative with respect to the parameter of interest and only allows values that can actually be obtained. Since the range of variables is indeed regularly bounded in the social sciences (often scales

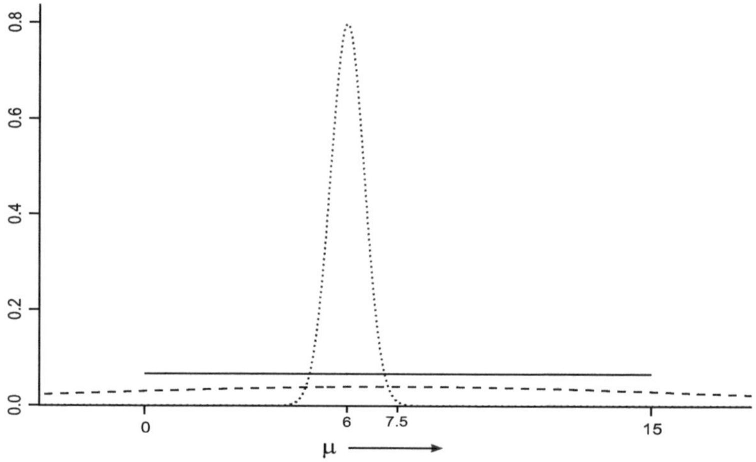

Fig. 3.2. Three suggestions for $p(\mu)$: a bounded uniform distribution $\mathcal{U}(\mu|lb = 0, ub = 15)$ (solid line); a diffuse normal distribution $\mathcal{N}(\mu|\mu_0 = 7.5, \tau_0^2 = 100)$ (dashed line); an informative normal distribution $\mathcal{N}(\mu|\mu_0 = 6, \tau_0^2 = 0.25)$ (dotted line)

with fixed lower and upper bounds are used), this appears to be a useful choice when no (subjective) prior information is available.

Alternatively, another common choice for the prior distribution is a so-called *conjugate* prior [15, 16]. Conjugacy means that the functional form of the prior distribution is such that combined with the likelihood, it leads to a posterior distribution of the same functional form. Note again that the posterior distribution represents the knowledge about model parameters after observing data. The posterior is obtained by combining the information in prior and likelihood (this will be elaborated in Section 3.1.3). For the example at hand, where data are assumed to be normally distributed, a normal prior distribution for μ is conjugate because it leads to a posterior for μ that is again a normal distribution. The prior distribution is further specified by assigning values for the mean and variance of the normal prior, denoted by μ_0 and τ_0^2, respectively. Assume that we still want to specify a (relatively) uninformative prior. This is obtained by specifying a large value for the prior standard deviation. The dashed line in Figure 3.2 represents the second prior, denoted $p_2(\mu)$, and is a normal distribution centered around a mean of 7.5 and with a standard deviation of 10. It can be seen that the distribution is relatively flat and thus providing little or no prior information in the region of interest ($0 \leq \mu \leq 15$). The function representing $p_2(\mu)$ is the formula of a normal distribution with values $\mu_0 = 7.5$ and $\tau_0^2 = 10^2 = 100$, leading to

$$
\begin{aligned}
p_2(\mu) &= \mathcal{N}(\mu|\mu_0 = 7.5, \tau_0^2 = 100) \\
&= (2\pi \cdot 100)^{-1/2} \exp\left\{\frac{(\mu - 7.5)^2}{-2 \cdot 100}\right\}.
\end{aligned} \tag{3.4}
$$

However, it is also possible to model (subjective) prior knowledge about the mean recognition score for DID-patients. Assume it is known from previous research that values below 4.5 and above 7.5 are highly unlikely and that an average of about 6 is usually found. This prior knowledge can be represented by, for instance, an informative normal prior distribution for μ with mean 6 and standard deviation 0.5 (dotted line in Figure 3.2); that is,

$$
\begin{aligned}
p_3(\mu) &= \mathcal{N}(\mu|\mu_0 = 6.0, \tau_0^2 = 0.25) \\
&= (2\pi \cdot 0.25)^{-1/2} \exp\left\{\frac{(\mu - 6.0)^2}{-2 \cdot 0.25}\right\}.
\end{aligned} \tag{3.5}
$$

This prior distribution gives high probabilities to values close to 6, as can be seen by the high peak at 6 in the figure, and, a very small probability (approximately 0.1%) to values larger than 7.5 and smaller than 4.5.

3.1.3 Posterior Distributions

The observed data are used to "update" the knowledge that is represented by the prior distribution. This leads to a new probability distribution, the so-called posterior. The posterior distribution represents the knowledge about model parameters after observing the data taking both the prior knowledge and the observed data into account.

The foundations are found in Bayes' theorem which (loosely formulated) states that posterior knowledge is the product of prior knowledge and knowledge provided by the data. To provide the theorem more formally and in general notations, let $p(\boldsymbol{\theta})$ denote the prior for a parameter vector $\boldsymbol{\theta}$ (for instance, in a four-group ANOVA: $\boldsymbol{\theta} = \{\mu_1, \mu_2, \mu_3, \mu_4, \sigma^2\}$). Let $f(\boldsymbol{y}|\boldsymbol{\theta})$ denote the likelihood of data $\boldsymbol{y}$ given $\boldsymbol{\theta}$ and let $p(\boldsymbol{\theta}|\boldsymbol{y})$ denote the posterior distribution.

$$
\textbf{Bayes' theorem: } p(\boldsymbol{\theta}|\boldsymbol{y}) = \frac{f(\boldsymbol{y}|\boldsymbol{\theta})p(\boldsymbol{\theta})}{m(\boldsymbol{y})} \propto f(\boldsymbol{y}|\boldsymbol{\theta})p(\boldsymbol{\theta}). \tag{3.6}
$$

The numerator $m(\boldsymbol{y})$ is called the marginal density or the normalizing constant. However, often only information about the *shape* of the posterior distribution is required (i.e., the posterior up to proportionality). This is also the case in this chapter: Later on we will *sample* the posterior distribution and for this sampling the normalizing constant is not required. In other situations the marginal density is crucial. This will be elaborated in subsequent chapters about model selection.

Returning to the one-parameter example, we can derive the posterior distribution for each of the priors in Figure 3.2. The posterior following from the first prior is

$$\begin{aligned} p_1(\mu|\boldsymbol{y},\sigma^2=3) &\propto f(\boldsymbol{y}|\mu,\sigma^2=3)p_1(\mu) \\ &= \mathcal{N}(\mu|3.11,0.16)\ \mathcal{U}(\mu|lb=0,ub=15) \\ &= \mathcal{N}(\mu|\mu_N=3.11,\tau_N^2=0.16)\ I_{\mu\in[0,15]}, \end{aligned} \tag{3.7}$$

where μ_N and τ_N^2 denote the mean and variance of the posterior normal distribution, and $I_{\mu\in[0,15]}$ is an indicator function that has the value one if μ is between 0 and 15, and zero otherwise. Stated otherwise, the posterior is a normal distribution proportional to the likelihood, but with the tails truncated at the values 0 and 15. In Figure 3.3, the solid line represents this posterior distribution and it can be seen that $p_1(\mu|\boldsymbol{y},\sigma^2=3)$ is indeed equivalent to the likelihood function presented in Figure 3.1. Because in this example the posterior is centered around 3.11 with a small posterior variance (0.16), the truncation at values that are highly unlikely does not noticeably change the distribution (compared to the same normal distribution without the indicator function). Later in this chapter, however, truncation of normal distributions will play an important role in the specification of and sampling from posterior distributions under inequality constraints.

The two normal priors are conjugate, and derivation of the posterior distribution of a mean (with known variance) using a conjugate prior is standard and can be found in the literature. See, for instance, [6, pp. 46 – 49]. It is also briefly summarized here:

- Given a normal prior $\mathcal{N}(\mu_0,\tau_0^2)$, the mean of the posterior normal distribution is equal to
$$\frac{\mu_0/\tau_0^2+M(N/\sigma^2)}{1/\tau_0^2+N/\sigma^2}.$$
It can be seen that the posterior mean is a combination of the prior mean μ_0 and the sample mean M. The weight given to the prior mean is based on the prior variance τ_0^2. A large value (i.e., a diffuse, vague prior distribution) gives a small weight to the prior mean. The term N/σ^2 is the inverse of the squared standard error (σ^2/N) and gives a weight to the sample mean M. Here it can, for instance, be seen that for larger sample sizes, the sample mean gets a larger weight.
- A similar mix of sources of information is seen in the formula for the posterior variance:
$$(1/\tau_0^2+N/\sigma^2)^{-1}.$$

Application of these formulas to the data and normal priors of our example leads to the following two posteriors:

$$\begin{aligned} p_2(\mu|\boldsymbol{y},\sigma^2=3) &\propto f(\boldsymbol{y}|\mu,\sigma^2=3)p_2(\mu) \\ &= \mathcal{N}(\mu|3.11,0.16)\ \mathcal{N}(\mu|\mu_0=7.5,\tau_0^2=100) \\ &= \mathcal{N}(\mu|\mu_N=3.11,\tau_N^2=0.16) \end{aligned} \tag{3.8}$$

and

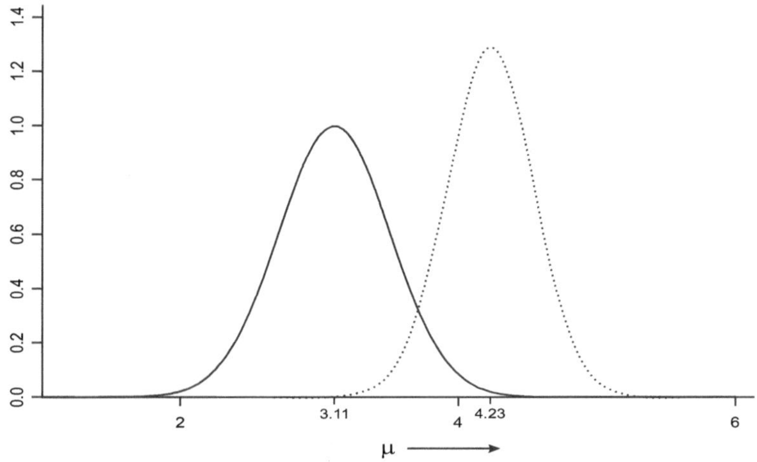

Fig. 3.3. Posterior resulting from uninformative priors $p_1(\mu)$ and $p_2(\mu)$ (solid line) and the posterior resulting from the subjective prior $p_3(\mu)$ (dotted line)

$$
\begin{aligned}
p_3(\mu|\boldsymbol{y}, \sigma^2 = 3) &\propto f(\boldsymbol{y}|\mu, \sigma^2 = 3)p_3(\mu) \\
&= \mathcal{N}(\mu|3.11, 0.16)\ \mathcal{N}(\mu|\mu_0 = 6.0, \tau_0^2 = 0.25) \\
&= \mathcal{N}(\mu|\mu_N = 4.23, \tau_N^2 = 0.10).
\end{aligned}
\tag{3.9}
$$

In (3.8), it can clearly be seen that the data dominate the prior. Compared to the information in the prior (which was indeed chosen to be low– or uninformative) there is far more information in the N=19 observations: The posterior mean is equal to the data mean and the posterior variance is equal to the squared standard error. Stated differently, $p_2(\mu|\boldsymbol{y}, \sigma^2 = 3)$ is proportional to the likelihood function. The only difference between this posterior distribution and $p_1(\mu|\boldsymbol{y}, \sigma^2 = 3)$ is the truncation of the tails for values smaller than 0 and larger than 15 in the latter. However in the range where the posterior has considerable density (i.e., the range visible in Figure 3.3), the first two posteriors overlap. It can be concluded that the two uninformative priors $p_1(\mu) = \mathcal{U}(\mu|lb = 0, ub = 15)$ and $p_2(\mu) = \mathcal{N}(\mu|\mu_0 = 7.5, \tau_0^2 = 100)$ lead to virtually the same posterior distribution.

The results for the last prior are different. The corresponding posterior is represented by the dotted line in Figure 3.3. It can be seen that the posterior mean has a value of 4.23, which shows indeed that the posterior is a compromise between the information contained in the data (M=3.11) and the information contained in the prior distribution ($\mu_0 = 6.0$). The difference between the two posteriors is not only seen in the posterior means but also in the posterior variances (0.16 and 0.10, respectively). This can be explained by the fact that the last posterior is based on more information; that is, the same amount of information is present in the data, but this information is

Table 3.2. Sampled values for μ from the three posterior distributions

Iteration	$p_1(\mu\|\boldsymbol{y}, \sigma^2 = 3)$	$p_2(\mu\|\boldsymbol{y}, \sigma^2 = 3)$	$p_3(\mu\|\boldsymbol{y}, \sigma^2 = 3)$
1	2.75	3.31	4.61
2	3.10	3.57	4.16
3	3.00	3.85	4.19
4	2.97	3.84	4.02
5	2.37	2.94	4.66
6	2.40	3.72	4.37
7	2.98	3.50	3.54
8	3.76	4.09	4.04
9	3.04	3.48	4.19
⋮	⋮	⋮	⋮
997	2.70	3.23	4.53
998	2.90	2.91	4.34
999	2.54	2.90	3.69
1000	3.34	3.34	4.16

combined with the information in the prior distribution. In Section 3.1.5 the topic of uninformative versus informative priors is further elaborated.

3.1.4 Bayesian Estimation

In simple examples, the posterior distribution and posterior estimates (like the mean and standard deviation) can be derived analytically. This is what we have just done for the one-mean example. However, in models with more than one parameter, this is often not possible. For instance, for inequality constrained analysis of variance models with several means and a variance parameter, analytic derivation of the posterior distribution and its summary statistics is not at all straightforward. Therefore, one often uses sampling methods to obtain the posterior distribution and subsequent posterior estimates.

To illustrate the use of sampling methods, samples from the three posterior distributions of the one-mean example are drawn. Note that the sampling for this simple example is straightforward and basically redundant, because we already derived exactly what the form of the posterior is. Here we just want to illustrate that sampling methods provide the same answers as analytic derivation and thus can provide a good alternative when models are more complex.

Sampling from the second and third posterior is very straightforward since both are normal distributions with known mean and variance. Most statistical programs (e.g., SPSS) provide the option to draw random numbers from a specified distribution. A sample from the posterior consists of several of these random draws, also called iterations. In Table 3.2, it can, for instance, be seen that the first randomly sampled number from $p_2(\mu|\boldsymbol{y}, \sigma^2 = 3)$ (i.e., from

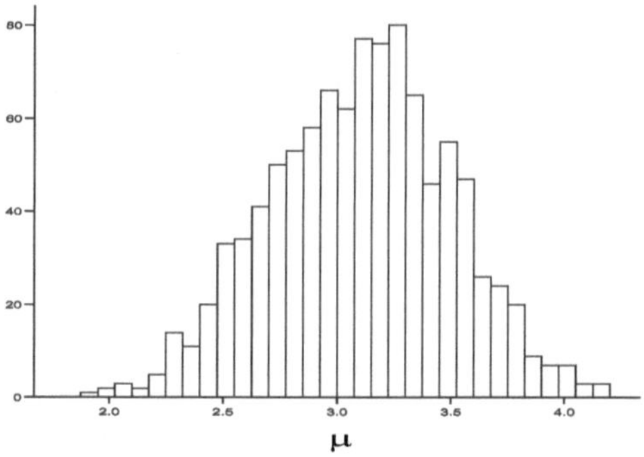

Fig. 3.4. A histogram of 1000 iterations from $p_1(\mu|\boldsymbol{y}, \sigma^2 = 3)$

$\mathcal{N}(\mu_N = 3.11, \tau_N^2 = 0.16))$ delivered the value 3.31. In subsequent iterations the values 3.57, 3.85, 3.84, and so on are obtained. The last column of Table 3.2 shows the sample (consisting of 1000 iterations) from the third posterior (i.e., from $\mathcal{N}(\mu_N = 4.23, \tau_N^2 = 0.10)$ providing values like 4.61, 4.16, 4.19 and so on).

Sampling from $p_1(\mu|\boldsymbol{y}, \sigma^2 = 3)$ seems somewhat more complicated because the posterior is a *truncated* normal distribution. A sample from a truncated distribution is, however, easily obtained by sampling from the corresponding not-truncated distribution and subsequently discarding all values that are not allowed. In general, this is not the most efficient method (this will be elaborated in Section 3.2), but for the example at hand it works very well since values outside the 0–15 range are highly unlikely and will therefore rarely or never be sampled. The first column of Table 3.2 provides 1000 randomly sampled values from $p_1(\mu|\boldsymbol{y}, \sigma^2 = 3)$ and indeed all 1000 values were within the 0–15 range, so no iterations needed to be discarded.

The 1000 draws provide a good approximation of the posterior distribution and this can best be seen by plotting all iterations in a histogram. To provide an illustration, the sample from $p_1(\mu|\boldsymbol{y}, \sigma^2 = 3)$ is plotted in Figure 3.4. Any level of precision of the approximation of the normal posterior can be obtained by increasing the number of iterations.

With a sample from the posterior, the computation of posterior estimates is straightforward. For instance, the posterior mean is the average of all the sampled values. Likewise, the posterior standard deviation is the standard deviation of the 1000 sampled values. By ordering the values from low to high and taking the 2.5th and 97.5th percentile, a Bayesian 95% central credibility interval (CCI) is obtained (the Bayesian counterpart of confidence intervals). The summary statistics for the three posteriors are presented in Table 3.3. Note, again, that the resulting posterior mean and standard deviation are (not surprisingly) equal to the mean and standard deviation in $p_1(\mu|\boldsymbol{y}, \sigma^2 = 3)$, $p_2(\mu|\boldsymbol{y}, \sigma^2 = 3)$, and $p_3(\mu|\boldsymbol{y}, \sigma^2 = 3)$. The fact that these values are equal

Table 3.3. Posterior estimates based on a sample of 1000 iterations

	$p_1(\mu \vert \boldsymbol{y}, \sigma^2 = 3)$	$p_2(\mu \vert \boldsymbol{y}, \sigma^2 = 3)$	$p_3(\mu \vert \boldsymbol{y}, \sigma^2 = 3)$
Posterior mean:	3.11	3.11	4.23
Posterior standard deviation:	0.41	0.40	0.31
Lower bound 95% CCI:	2.31	2.31	3.62
Upper bound 95% CCI:	3.91	3.92	4.87

shows that the sample of 1000 iterations is large enough to provide a good approximation of the posterior. An elaboration on sampling methods and the required number of iterations is provided in Sections 3.2 and 3.3, where models with several (constrained) parameters are discussed.

3.1.5 Objective Versus Subjective

Prior distributions are often seen as the bottleneck of Bayesian analyses because conclusions (e.g., estimates) depend on the specification of the prior. Stated differently, critical remarks are often made about Bayesian methods being subjective. In many applications however, uninformative priors can be used leading to results that are determined just by the observed data. In the example above, both the uniform and the normal prior with large standard deviation can be considered uninformative for Bayesian estimation because a priori every value for μ is (approximately) equally likely, and the resulting estimates do not depend on the prior. Bayesian analyses with uninformative priors are also referred to as "objective" Bayesian methods. Both terms – uninformative and objective – should, however, be used with care. One reason is that an uninformative prior for a certain parameter (e.g., a uniform prior for a mean) is not necessarily uniform for a transformation of this parameter (e.g., the logarithm of the mean). For an elaboration on this issue, see, for instance, [2, 6, 11, 12]. A second reason to be careful with the labels objective and uninformative is that priors that are uninformative with respect to the estimation of model parameters can be extremely informative in model selection. Model selection will be elaborated in later chapters.

Besides problematic, prior distributions can also be seen as an advantage of Bayesian methods. Often, prior knowledge is available before collecting the data (for instance, from previous research) and the Bayesian approach allows formal inclusion of this knowledge in the analysis. The third prior applied to the recognition data is subjective and informative with respect to the unknown mean μ. We have seen that the prior distribution used in the Bayesian analysis can affect the resulting estimates and this is usually the case if the prior specified is informative. The resulting estimate for μ under the third prior was clearly different from the estimate based on the first or second prior (the latter two being uninformative). An interesting question is: Should we conclude that the prior must have been "incorrect" because it is not clearly supported by the observed data (recall that the prior mean was 6 and the data mean was 3.1)? Or should we trust our prior knowledge, which may be the result of many research years in the field and several previous observed samples with a mean around 6? Stated differently: Should we throw our beliefs overboard based on the observation of these 19 patients, or should we update our prior beliefs with this newly collected information? The latter is exactly what is done in a Bayesian analysis, represented in the posterior distribution. We refer to [9] for an elaboration on subjective Bayesian methods.

This book is centered around inequality constrained models. Inequality constraints among parameters are also an example of the inclusion of prior knowledge. A priori, certain combinations or orderings of parameters are not allowed based on a theory or hypothesis stated before observing any data. Examples of these hypotheses were presented in Chapter 2 – for instance, in the DID data, certain expectations existed a priori about the ordering of the four groups (DID-patients, Controls, Simulators, and True amnesiacs) in their ability ro retrieve previously presented information.

In these inequality constrained applications, one approach that can be taken in the specification of the prior distribution is a two-step approach: An initial prior is specified for the unconstrained counterpart of the model of interest, and, subsequently, the inequality constraints are incorporated by truncating this distribution according to the constraints at hand. This is the approach taken in this and the next chapter. Because the main interest lies in the role of the inequality constraints in the model, in most examples the unconstrained prior is specified to be low or uninformative. In this way, the only subjective and informative ingredient of the model is the ordering imposed on certain model parameters. In the remainder of this chapter, the analysis of variance (ANOVA) model with ordered mean parameters and unknown variance is discussed. Bayesian estimation for inequality constrained analysis of variance models was also discussed in [13].

In Section 3.2, likelihood, prior, and posterior of the general ANOVA model are presented, as well as a sampling method that is often used to obtain a sample from a multiparameter posterior. Furthermore, the inclusion of inequality constraints in the sampling method is discussed. Note that Section 3.2 is relatively technical. Readers that prefer to focus on the applications can skip this section and move on to the illustration in Section 3.3. The illustration concerns the DID data including all four groups. A specific theory about the order of the mean recognition scores in these groups is incorporated in the estimation. It will be shown how a sample from the truncated posterior distribution provides estimates like the posterior mean and posterior standard deviation as well as central credibility intervals.

3.2 The Analysis of Variance Model

In Section 3.2.1, the general ANOVA model as well as an uninformative prior and resulting posterior for this model are presented without the inclusion of inequality constraints. Subsequently, in Section 3.2.2 it is shown how the Gibbs sampler provides a sample from the (unconstrained) multidimensional posterior distribution. Section 3.2.3 presents a diagnostic to monitor the adequacy of the size of this sample. The inclusion of inequality constraints in the prior distribution and consequently in the posterior and Gibbs sampler is discussed in Section 3.2.4.

3.2.1 Likelihood, Prior, and Posterior

The general ANOVA model [4, 17] with J groups, N respondents, and criterion variable y_i, for $i = 1, \ldots, N$, is given by

$$y_i = \sum_{j=1}^{J} \mu_j d_{ji} + \varepsilon_i, \tag{3.10}$$

where $\boldsymbol{\mu}$ is a vector of length J with the group means, d_{ji} is the group indicator,

$$d_{ji} = \begin{cases} 1 \text{ if the } i\text{th observation belongs to group } j \\ 0 \text{ otherwise,} \end{cases}$$

and ε_i is assumed to be i.i.d. normally distributed with common and unknown variance σ^2. Note that a nonstandard dummy coding is used with J dummies for J groups, resulting in a model without intercept and parameters μ_j representing group means.

Because the N observations y_i are independent, the likelihood of $\boldsymbol{y}$ is equal to the product of the separate likelihood functions of y_i. Hence,

$$f(\boldsymbol{y}|\boldsymbol{\mu}, \sigma^2, D) = \prod_{i=1}^{N} \frac{1}{\sqrt{2\pi\sigma^2}} \exp\left\{ -\frac{(y_i - \sum_{j=1}^{J} \mu_j d_{ji})^2}{2\sigma^2} \right\}, \tag{3.11}$$

where D is a $J \times N$ matrix with the (j, i)th element equal to d_{ji}.

For the model parameters $\{\boldsymbol{\mu}, \sigma^2\}$, a prior distribution must be specified. For estimation of model parameters without inequality constraints, the reference prior $p(\boldsymbol{\mu}, \sigma^2) = \sigma^{-2}$ will be used. This is a standard uninformative prior for models with mean and variance parameters and implies a constant prior for each mean μ_j and a constant prior for $\log \sigma^2$. The motivation for the latter or reference priors in general will not be elaborated here. The interested reader is referred to [1].

The $J + 1$-dimensional joint posterior $p(\boldsymbol{\mu}, \sigma^2|\boldsymbol{y}, D)$ is proportional to the product of the likelihood and the prior:

$$p(\boldsymbol{\mu}, \sigma^2|\boldsymbol{y}, D) \propto \sigma^{-2} f(\boldsymbol{y}|\boldsymbol{\mu}, \sigma^2, D). \tag{3.12}$$

Although posterior estimates for the general ANOVA model are still relatively easy to obtain, this is not the case when inequality constraints are additionally imposed on model parameters. The Gibbs sampler is a method that simplifies the sampling of multidimensional distributions and can also straightforwardly be applied in the case of inequality constrained parameters. For an introduction of the general Gibbs sampler, see [18]; for the introduction of the constrained parameter Gibbs sampler, see [5]. In the next section, the Gibbs sampler is explained and applied to the unconstrained analysis of variance model. In Section 3.2.4 and [13], the inclusion of inequality constraints is presented.

3.2.2 The Gibbs Sampler

In Section 3.1 it was shown how sampling random numbers from the posterior distribution of a mean μ provides posterior estimates for μ. In the analysis of variance model, the (joint) posterior contains $J + 1$ parameters. The basic idea of the Gibbs sampler is that sampling from a multidimensional posterior distribution can be done via repeatedly sampling from the univariate distribution of each parameter, conditional upon the current values of the other parameters [18]. The conditional posterior distributions generally have simpler structures than the joint posterior, so that they are easier to sample from.

In the Gibbs sampler, parameters are sampled iteratively and in a fixed sequence. Let $s = 1, \ldots, S$ denote the iteration number and $\{\boldsymbol{\mu}^{(s)}, \sigma^{2^{(s)}}\}$ the parameter values sampled in the sth iteration. For the posterior in (3.12), the Gibbs sampler consists of four steps:

1. Specify initial values for the model parameters, $\{\boldsymbol{\mu}^{(0)}, \sigma^{2^{(0)}}\}$.
2. For $j = 1, \ldots, J$, sample $\mu_j^{(s)}$ from $p(\mu_j|\mu_1^{(s)}, \ldots, \mu_{j-1}^{(s)}, \mu_{j+1}^{(s-1)}, \ldots, \mu_J^{(s-1)}, \sigma^{2^{(s-1)}}, \boldsymbol{y}, D)$.
3. Sample $\sigma^{2^{(s)}}$ from $p(\sigma^2|\boldsymbol{\mu}^{(s)}, \boldsymbol{y}, D)$.
4. Repeat steps 2 and 3 until S draws have been obtained for all model parameters.

Each of these four steps will be elaborated below.

Ad 1. To be able to sample parameters from the corresponding conditional distributions in the first iteration of the Gibbs sampler, values must be assigned to each parameter. These so-called initial values can be chosen arbitrarily. This implies, however, that the first iterations will depend on these arbitrary starting values, and, consequently, these iterations are not draws from the joint posterior distribution. Therefore, a first set of iterations must be discarded. This part is referred to as the burn-in period. The length of the required burn-in depends on the starting values, the number of parameters, and the complexity of the model at hand. For the unconstrained ANOVA model, within just a few iterations the effect of the initial values is no longer noticeable. When inequality constraints are included in the ANOVA model, the burn-in period is somewhat longer but is in our experience still within say 100 to 500 iterations (depending mainly on the number of parameters and, moreover, on the number and type of constraints). For each model, a check for the size of burn-in is important and can be done using multiple starting values and examination of the resulting multiple chains, as will be illustrated in the example in Section 3.3. Several formal diagnostics are also available (cf. [3, 7]), and can be used to monitor the chain(s) and provide information whether the size chosen for the burn-in period was sufficient. One such diagnostic, the $\widehat{R}$ presented in [7], will be discussed in Section 3.2.3.

Ad 2. The conditional posterior of the μ_j's can be determined with Bayes' theorem and is proportional to the product of the conditional prior of μ_j and the likelihood. Hence,

$$\begin{aligned} p(\mu_j|\mu_1,\ldots,\mu_{j-1},\mu_{j+1},\ldots,\mu_J,\sigma^2,\boldsymbol{y},D) &\propto f(\boldsymbol{y}|\boldsymbol{\mu},\sigma^2)\pi(\mu_j) \\ &= \mathcal{N}(\mu_j|\mu_{N,j}=M_j,\tau^2_{N,j}=\sigma^2/N_j), \end{aligned} \tag{3.13}$$

where M_j is the sample mean of group j and N_j is the number of observations in group j. The conditional distribution of each μ_j is a normal distribution with posterior mean $\mu_{N,j}$ equal to the jth group sample mean, and, posterior variance $\tau^2_{N,j}$ equal to the current value of σ^2 divided by the jth group sample size. Sampling from a normal distribution with known mean and variance is straightforward.

Ad 3. The conditional posterior of the variance σ^2 can be determined in a similar way:

$$\begin{aligned} p(\sigma^2|\mu_1,\ldots,\mu_J,\boldsymbol{y},D) &\propto f(\boldsymbol{y}|\boldsymbol{\mu},\sigma^2)\pi(\sigma^2) \\ &= \text{Inv-}\chi^2(\sigma^2|\nu_N=N,\sigma^2_N), \end{aligned} \tag{3.14}$$

which is a scaled inverse χ^2-distribution [14] with degrees of freedom ν_N equal to the total sample size N and scale parameter σ^2_N equal to the residual sum of squares:

$$\sigma^2_N = \frac{1}{N}\sum_{i=1}^{N}\left(y_i - \sum_{j=1}^{J}\mu_j d_{ji}\right)^2 .$$

Note that within step 3, each μ_j has a current value making the computation of σ^2_N and, consequently, sampling from the scaled inverse χ^2-distribution straightforward.

Ad 4. As stated in the first step, the initial set of draws depend on the arbitrary starting values. The burn-in period is discarded and the remaining draws provide a sample from the joint posterior distribution. After the burn-in period, the number of draws must be large enough for the sample to be

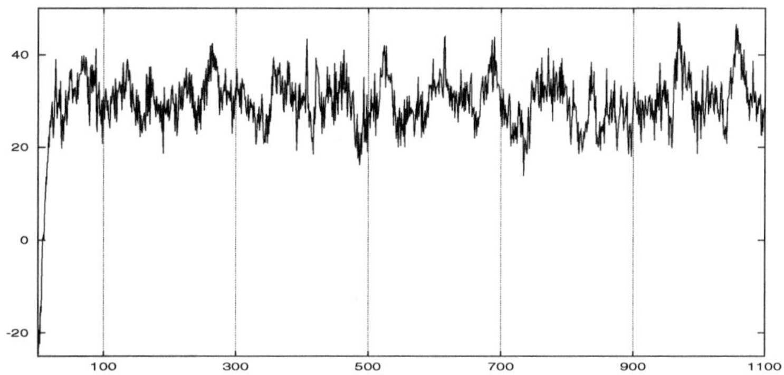

Fig. 3.5. Chain of 1100 sampled values for θ

a good approximation of the posterior distribution. This is known as convergence of the Gibbs sampler. For the model at hand, convergence is reached relatively fast. Like with the burn-in period, it is important to always closely monitor convergence – for instance, by visual inspection of multiple chains. This will also be illustrated in Section 3.3. A formal diagnostic that can be used to monitor burn-in and convergence is $\widehat{R}$ [7] and will be presented in the next section. For a more elaborate presentation and discussion of convergence diagnostics, the interested reader is referred to [3].

3.2.3 The Convergence Diagnostic $\widehat{R}$

Consider the Gibbs output with respect to a parameter θ (θ could, for instance, be one of the means μ_j or σ^2). In Figure 3.5, the results of a sample of 1100 draws from the posterior distribution of θ are presented. On the x-axis the iteration number is plotted; on the y-axis the sampled value for θ. Visual inspection shows that a first set of iterations must be discarded because the first few sampled values are different from the subsequent iterations. Discarding a burn-in period of say 100 iterations leads to a sample that seems to be converged according to an eyeball test. This visual inspection is, for instance, based on comparing several subsets of iterations (e.g., iterations 300–500, 500–700, and 900–1100) and concluding that the subsets look rather similar. The diagnostic $\widehat{R}$ is based on the same principle: The sample after burn-in is divided in K sequences of Q iterations and the between and within sequence variations in the sampled parameter values are evaluated. To compute the diagnostic, let θ_{qk} be the qth iteration in the kth sequence, $\bar{\theta}_k = \frac{1}{Q}\sum_{q=1}^{Q}\theta_{qk}$ and $\bar{\theta} = \frac{1}{K}\sum_{k=1}^{K}\bar{\theta}_k$. The between sequences variance is

$$B = \frac{Q}{K-1}\sum_{k=1}^{K}(\bar{\theta}_k - \bar{\theta})^2, \tag{3.15}$$

and the within-sequences variance is

$$W = \frac{1}{K}\sum_{k=1}^{K}\left(\frac{1}{Q-1}\sum_{q=1}^{Q}(\theta_{qk} - \bar{\theta}_k)^2\right), \tag{3.16}$$

providing the ingredients for

$$\widehat{R} = \sqrt{\frac{\frac{Q-1}{Q}W + \frac{1}{Q}B}{W}}. \tag{3.17}$$

The larger the number of iterations, the closer the value for $\widehat{R}$ gets to one (i.e., $\widehat{R} \rightarrow 1.0$ if $Q \rightarrow \infty$). According to Gelman and Rubin [7], values of $\widehat{R}$ smaller than 1.1 are indicative of convergence of the Gibbs sampler.

For the sampled θ values presented in Figure 3.5, the 1000 iterations remaining after burn-in are divided in $K = 5$ blocks of $Q = 200$ subsequent iterations. Computation of $\widehat{R}$ gives a value of 1.03, confirming the conclusion of our eyeball test. For this example with a sample of 1000 iterations after a burn-in of 100 iterations convergence is reached. Note that in multiple parameter models, $\widehat{R}$ must be smaller than 1.1 for each model parameter.

3.2.4 Constrained Analysis of Variance

In Chapter 2 several illustrations of inequality constrained hypotheses in the analysis of variance model were introduced. An a priori expected ordering of three means can, for instance, be modeled as $\mu_1 < \mu_2 < \mu_3$. Also, an example with constraints on differences between specific means was presented, the corresponding hypothesis being of the form $\{\mu_1 - \mu_2\} < \{\mu_3 - \mu_4\}$. Inequality constraints like these can be included in the Bayesian procedure by incorporation of the constraints in the prior distribution. The estimation of each μ_j is then based on a sample from the constrained posterior; that is, the resulting estimates will be in accordance with the constraints imposed.

Let H_t denote a model that imposes certain inequality constraints on the group means μ_j. The constraints are included in the prior distribution according to

$$p(\boldsymbol{\mu}, \sigma^2 | H_t) \propto \sigma^{-2} I_{\boldsymbol{\mu} \epsilon H_t}, \tag{3.18}$$

where $I_{\boldsymbol{\mu} \epsilon H_t}$ is the indicator function with

$$I_{\boldsymbol{\mu} \epsilon H_t} = \begin{cases} 1 \text{ if } \boldsymbol{\mu} \text{ satisfies the inequality constraints of model } H_t \\ 0 \text{ otherwise.} \end{cases}$$

This means that the prior proposed for the corresponding unconstrained model is truncated on the region that satisfies the inequality constraints on $\boldsymbol{\mu}$ under model H_t.

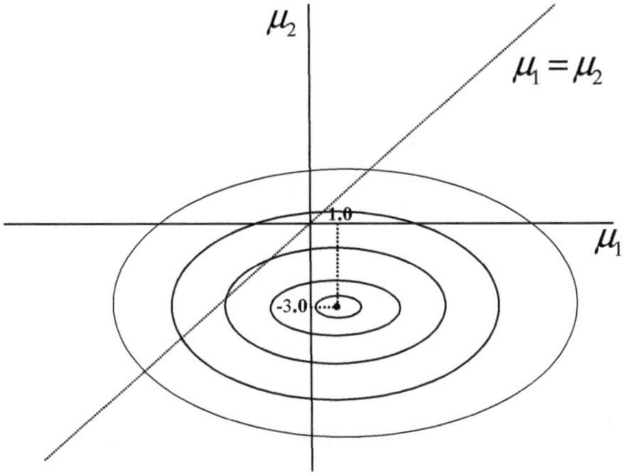

Fig. 3.6. Illustration of truncation of parameter spaces for the model $\mu_1 > \mu_2$

The posterior distribution under model H_t changes accordingly since it is proportional to the product of the truncated prior (3.18) and the likelihood (3.11). Hence,

$$p(\boldsymbol{\mu}, \sigma^2 | \boldsymbol{y}, H_t) \propto \sigma^{-2} f(\boldsymbol{y}|\boldsymbol{\mu}, \sigma^2) I_{\boldsymbol{\mu} \epsilon H_t}. \tag{3.19}$$

This is the joint posterior (3.12) truncated to the region that satisfies $\boldsymbol{\mu} \epsilon H_t$.

In Figure 3.6, the formulation of a prior distribution for an inequality constrained model is illustrated for $J = 2$ and (to simplify the graphical representation) assuming that the variance is known. Assume a uniform prior distribution for both μ_1 and μ_2. The two-dimensional unconstrained prior is plotted in Figure 3.6 with μ_1 and μ_2 on the x- and y-axis, respectively. The prior distribution for the constrained hypothesis $\mu_1 > \mu_2$ is

$$p(\mu_1, \mu_2 | \mu_1 > \mu_2) = p(\mu_1, \mu_2) I_{\mu_1 > \mu_2}.$$

The indicator function denotes that the density is zero in the area that is not in agreement with the constraint (i.e., the area above the line $\mu_1 = \mu_2$). The prior distribution for the part of the parameter space in agreement with the constraint $\mu_1 > \mu_2$ (the area below the line $\mu_1 = \mu_2$) is proportional to the prior in the unconstrained model.

Furthermore, let the ellipses in Figure 3.6 represent the posterior distribution under the unconstrained hypothesis. The center of the ellipses is the point $(1, -3)$ and represents the values μ_1 and μ_2 with the highest posterior density. Larger ellipses that are further away from the center represent values for μ_1 and μ_2 with lower posterior densities. The posterior distribution for the model $\mu_1 > \mu_2$ is proportional to the unconstrained posterior for the area below the line $\mu_1 = \mu_2$ and is zero above this line.

Let us see how the Gibbs sampler is adjusted to incorporate the inequality constraints. Compared to the sampling scheme presented in Section 3.2.2, the Gibbs sampler consists of the same four steps. In step 2, however, the conditional distributions of the μ_j subjected to inequality constraints are now *truncated* normal distributions. In each iteration s and for each j, $\mu_j^{(s)}$ is sampled from

$$p(\mu_j|\mu_1^{(s)},\ldots,\mu_{j-1}^{(s)},\mu_{j+1}^{(s-1)},\ldots,\mu_J^{(s-1)},\sigma^{2(s-1)},\boldsymbol{y},D,H_t)$$
$$= \mathcal{N}(\mu_j|\mu_{N,j} = M_j, \tau_{N,j}^2 = \sigma^2/N_j) I_{\boldsymbol{\mu} \epsilon H_t}. \tag{3.20}$$

The function $I_{\boldsymbol{\mu} \epsilon H_t}$ indicates that each value $\mu_j^{(s)}$ sampled from (3.20) has to satisfy the constraints under model H_t. Stated otherwise, a sampled value from the conditional normal distribution is only accepted if it is allowed according to the constraints of H_t and the current values of the other μ_j's.

Returning to Figure 3.6, consider that within the current iteration of the Gibbs sampler a value of zero was obtained for μ_1. This implies that the (unconstrained) conditional posterior distribution of μ_2 is proportional to the cross section at $\mu_1 = 0$ (the y-axis) of the bivariate normal distribution that is represented by the ellipses. This cross section is a normal distribution with a mean of -3.0 (the μ_2 value with the highest posterior density). For the constrained hypothesis $\mu_1 > \mu_2$, a draw from this normal distribution must be obtained under the condition that only values smaller than 0 are allowed.

A correct method to obtain a sample from the constrained model is sampling from the unconstrained conditionals until a value is sampled that is allowed by the constraints at hand. However, it can be very inefficient, especially if the constraints are such that only a small part of the normal distribution is allowed. A more efficient method is sampling directly from the truncated distribution using inverse probability sampling [5]. Let the truncation be denoted by a and b, where a is the largest current value of all μ_j's that are restricted to be smaller than $\mu_j^{(s)}$. Likewise, b is the smallest current value of all μ_j's that are restricted to be greater than $\mu_j^{(s)}$. Note that a and b can also be $-\infty$ and ∞, respectively, denoting that the conditional distribution is not truncated in one or both of the tails. For instance, in the previous example (Figure 3.6) where μ_2 must be sampled according to $\mu_1 > \mu_2$ and a current value for μ_1 of zero, $a = -\infty$ and $b = 0$.

Inverse probability sampling is illustrated in Figure 3.7. The bottom panel presents a normal distribution with the values a and b to denote the area from which the new value $\mu_j^{(s)}$ must be drawn (i.e., $a < \mu_j^{(s)} < b$). In the top panel, the corresponding cumulative normal distribution is plotted. The values $\Phi(a)$ and $\Phi(b)$ on the y-axis are the cumulative probabilities of a and b, respectively.

To obtain the sth draw of μ_j from its conditional posterior (3.20) a deviate $U^{(s)}$ is randomly drawn from the uniform distribution $\mathcal{U}(\Phi(a), \Phi(b))$ and

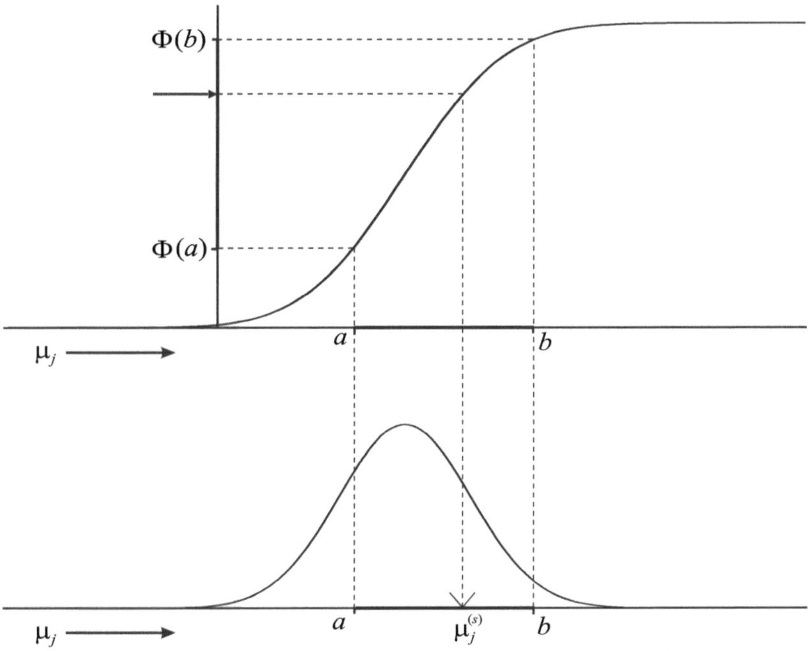

Fig. 3.7. A normal density (bottom) and corresponding cumulative normal distribution (top) to illustrate inverse probability sampling in order to obtain a draw from a truncated normal distribution

subsequently the corresponding value in the truncated normal is computed, by application of

$$\mu_j^{(s)} = \Phi^{-1}(U^{(s)}).$$

In Figure 3.7, the arrow plotted on the y-axis represents the random deviate $U^{(s)}$ (i.e., a randomly drawn value between $\Phi(a)$ and $\Phi(b)$). The corresponding value on the horizontal axis is the new value sampled for μ_j satisfying the constraint $a < \mu_j^{(s)} < b$. This value is a random draw from (3.20).

Inverse probability sampling adjusts Step 2 of the unconstrained Gibbs sampler that was presented in the previous section such that μ_j's are sampled directly from the conditional truncated normal distributions. Step 3 of the unconstrained Gibbs sampler – that is, sampling the variance parameter σ^2 – does not change since constraints are only imposed on group means. With respect to Steps 1 and 4 (i.e., burn-in and convergence), carefully monitoring the chain(s) is important as usual, and more, because inequality constraints in the model can slow down the convergence rate.

In the next section, an illustration for a four-group inequality constrained ANOVA is provided. Gibbs output and estimates are shown, as well as convergence and burn-in diagnostics.

Table 3.4. Recognition scores in the DID data

	M	SD	N
DID-patients	3.11	1.59	19
Controls	13.28	1.46	25
Simulators	1.88	1.59	25
True amnesiacs	4.56	1.83	25

3.3 Illustration

3.3.1 The Data

The illustration is based on the DID data introduced in Chapter 2. Recall that DID-patients were compared with three different types of control groups on their ability to retrieve information that they obtained in a prior phase of the experiment. The groups are DID-patients, Controls, Simulators, and True amnesiacs. The criterion variable is the score on a recognition test and the obtained scores per group are summarized in Table 3.4.

The recognition scores (denoted by $\boldsymbol{y}$) are assumed to be i.i.d. normally distributed. Let μ_j ($j = pat, con, sim, amn$) denote the parameters for the group means and σ^2 denote the common but unknown residual variance. A (nonstandard) dummy coding with four dummy variables, denoted by d_{ji} (with $d_{ji} = 1$ if the ith respondent belongs to subgroup j and zero otherwise), leads to the following model without intercept:

$$y_i = \mu_{pat} d_{pat,i} + \mu_{con} d_{con,i} + \mu_{sim} d_{sim,i} + \mu_{amn} d_{amn,i} + \varepsilon_i, \qquad (3.21)$$

with $i = 1, \ldots, 94$ respondents and $\varepsilon_i \sim \mathcal{N}(0, \sigma^2)$.

A theory or expectation for this experiment is that both DID-patients and Simulators will score lower than respondents who just guess (i.e., the True amnesiacs). Furthermore, it is expected that Controls score higher than each of these three groups. This leads to the following inequality constrained hypothesis: $\mu_{con} > \mu_{amn} > \{\mu_{pat}, \mu_{sim}\}$. In this and the next section it is shown how parameter estimates are obtained by a sample from the posterior conditional on the constraints – that is, after including the inequality constraints as prior knowledge. We refer the interested reader to `http://www.fss.uu.nl/ms/informativehypotheses` for software for the estimation of inequality constrained analysis of variance models.

3.3.2 Prior, Posterior, and Gibbs Sampling (Revisited)

In Section 3.1 we have seen that Bayesian estimation requires the specification of a prior distribution for all model parameters. In this example the parameters are $\mu_{pat}, \mu_{con}, \mu_{sim}, \mu_{amn}$, and σ^2. For unconstrained estimation we will use the prior $\pi(\mu_{pat}, \mu_{con}, \mu_{sim}, \mu_{amn}, \sigma^2) = \sigma^{-2}$. This is a standard uninformative prior for models with mean and variance parameters and implies a constant prior for each mean μ_j and a constant prior for log σ^2. The resulting (unconstrained) posterior $p(\mu_{pat}, \mu_{con}, \mu_{sim}, \mu_{amn}, \sigma^2 | \boldsymbol{y}, \boldsymbol{d}_{pat}, \boldsymbol{d}_{con}, \boldsymbol{d}_{sim}, \boldsymbol{d}_{amn})$ is a five-dimensional distribution and equal to (up to proportionality) the product of prior and likelihood (data).

At this point it is useful to repeat once more that Bayesian *estimation* is not sensitive to the exact specification of the prior as long as it is relatively uninformative. Both constant and conjugate priors with large variances will lead to similar results. This is usually not the case when the goal is comparing different models using a Bayesian approach. Bayesian model selection is much more sensitive to the prior distribution and will be thoroughly discussed and examined for inequality constrained models throughout the remainder of this book.

In the application at hand (i.e., estimating inequality constrained parameters) the initial uninformative prior for the unconstrained model becomes informative with the inclusion of the constraints. The constraints are incorporated in the formulation of the prior distribution, which boils down to truncating the prior parameter space according to the constraints at hand. Truncation of unconstrained prior distributions was explained in Section 3.2 based on the illustration provided in Figure 3.6. This figure represents the prior distribution for an unconstrained model μ_1, μ_2. The area below the line $\mu_1 = \mu_2$ represents the prior (up to proportionality) for the constrained model $\mu_1 > \mu_2$; the area above the line $\mu_1 = \mu_2$ represents the part of the unconstrained prior that contains (μ_1, μ_2) combinations that are not in agreement with the constraint and therefore have zero prior density. This results in a posterior distribution of the parameters for the constrained model that is still equal (up to proportionality) to the product of prior and likelihood for parameter values in agreement with the constraint. The posterior density for the parts of the parameters space that are not allowed by the hypothesis is zero.

To get all the desired estimates for each of the five parameters, a sample from the posterior is drawn. This is done using the so-called Gibbs sampler, described extensively for the general ANOVA model (both unconstrained and including inequality constraints) in the previous section. Here a short summary with the key ideas required for those who skipped Section 3.2 is provided.

In the Gibbs sampler, parameters are sampled iteratively from their conditional posterior distributions; that is, μ_1 is sampled from the distribution of μ_1 given that the other parameters are known (i.e., have a value assigned to them). The reason to apply the Gibbs sampler is that sampling from the joint posterior for the constrained model at hand is not straightforward, whereas all conditional posterior distributions are known and easy to sample.

To be able to sample the first values for each parameter, initial or starting values must be assigned to all parameters. These starting values are arbitrary and thus are not (yet) draws from the posterior at hand. Therefore, a first set of draws of the Gibbs sampler is discarded, the so-called burn-in period. The

Table 3.5. Gibbs output and posterior estimates based on last 1000 iterations

Iteration	μ_{pat}	μ_{con}	μ_{sim}	μ_{amn}	σ^2
1	−4.49	100.87	2.29	2.73	2769.65
2	1.53	22.56	−18.20	2.74	145.69
3	1.49	14.18	−3.22	1.93	13.01
4	1.40	13.46	1.37	4.78	2.69
5	3.18	13.32	1.77	4.74	2.00
⋮	⋮	⋮	⋮	⋮	⋮
101	3.24	13.40	2.18	3.81	3.13
102	2.82	13.08	2.05	4.24	3.31
103	3.96	13.11	2.24	4.79	3.53
104	2.22	13.77	2.30	4.50	2.38
105	3.28	13.40	1.81	4.83	2.58
⋮	⋮	⋮	⋮	⋮	⋮
1096	3.14	13.52	2.08	4.60	2.54
1097	3.59	12.24	2.01	4.26	2.72
1098	3.25	13.65	2.15	4.89	2.38
1099	3.30	12.91	1.74	4.34	1.93
1100	2.89	13.65	1.28	4.75	2.69
Convergence diagnostic $\widehat{R}$:	1.00	1.00	1.00	1.00	1.00
Posterior mean:	3.09	13.28	1.88	4.54	2.78
Posterior SD:	0.38	0.34	0.34	0.32	0.43
Lower bound 95% CCI:	2.39	12.63	1.24	3.89	2.06
Upper bound 95% CCI:	3.86	13.97	2.52	5.19	3.69

remaining set of iterations must be large enough to provide a good approximation of the joint posterior distribution. Convergence must be monitored to examine the size of burn-in and the total number of iterations required.

In the next section, it will be shown how the output of the Gibss sampler provides all posterior parameter estimates, as well as diagnostics for monitoring convergence of the Gibbs sampler.

3.3.3 Posterior Parameter Estimates

In Table 3.5, parts of the output of the Gibbs sampler for the model $\mu_{con} > \mu_{amn} > \{\mu_{pat}, \mu_{sim}\}$ are presented.

The first thing that can be noted is that in each row of the table, the sampled values for the four mean parameters are indeed in accordance with the constraint $\mu_{con} > \mu_{amn} > \{\mu_{pat}, \mu_{sim}\}$. This is how the Gibbs sampler is constructed: In each iteration (that is a step providing one sampled value for each parameter) the constraints are incorporated in the sampling scheme.

Furthermore, looking at the first couple of rows in Table 3.5, the influence of the (randomly chosen) starting values can be seen. However, it can also be seen that for this relatively simple model, the burn-in period is very short: The

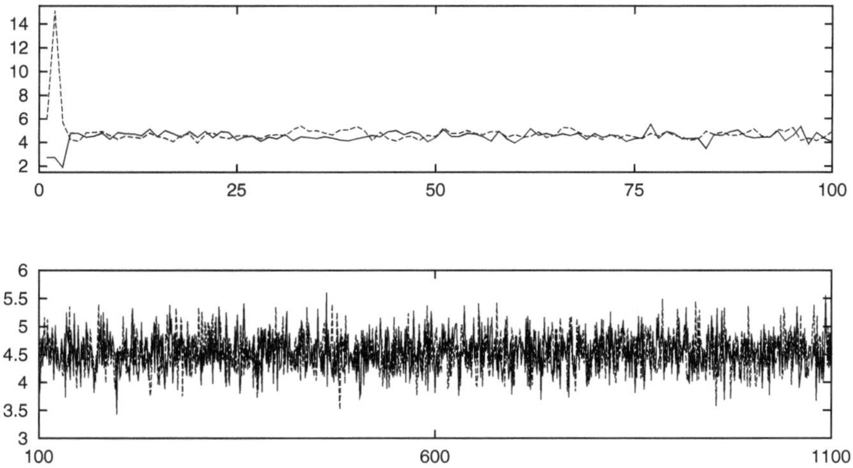

Fig. 3.8. Two chains with sampled values for μ_{amn} in the first 100 (top panel) and subsequent 1000 iterations (bottom panel)

sampled parameter values in the fourth and fifth iterations are already very similar to the draws with an iteration number larger than 100. Furthermore, comparison of subsets of iterations (e.g., comparing iterations 100–200 with 1000–1100; sets that are partly printed in Table 3.5) show similar values based on an eyeball inspection. Burn-in and convergence can also be inspected by drawing plots of sampled values in subsequent iterations. Figure 3.8 shows the results of the first 100 (top panel) and subsequent 1000 (bottom) iterations for one of the parameters (μ_{amn}). Note that two chains are sampled – that is, the Gibss sampler is run twice, each time with a different starting value. It can be seen that after less than 50 iterations the results are very similar, irrespective of the starting values used. This again indicates a short burn-in period and fast convergence. Based on Figure 3.8 and similar plots for the other four parameters, the first 100 iterations are considered the burn-in period, and 1000 iterations after burn-in are considered enough to derive the posterior estimates.

Also according to $\widehat{R}$, a formal diagnostic for convergence [7], this conclusion is supported. Values for $\widehat{R}$ smaller than 1.1 are considered indicative for convergence, and in Table 3.5 it can be seen that all values are (rounded to two decimals) 1.00.

The iterations remaining after discarding the burn-in form an approximation of the posterior distribution. For each parameter, the iterations after burn-in can be plotted in a histogram. This is done for one of the mean parameters (μ_{pat}) and for the residual variance (σ^2) in Figure 3.9.

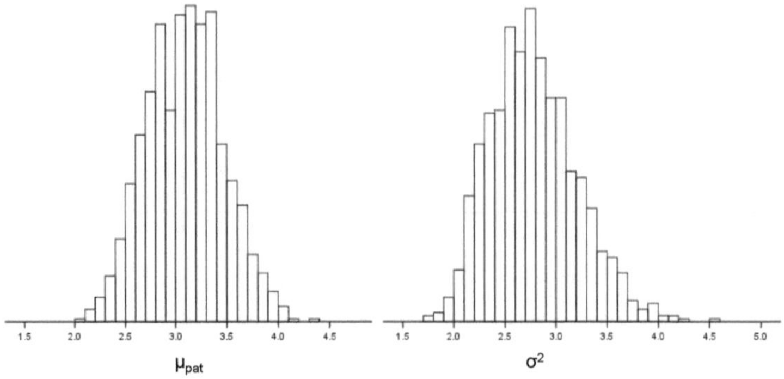

Fig. 3.9. Posterior distributions of μ_{pat} and σ^2 based on a Gibbs sample of 1000 iterations after burn-in

Finally, posterior estimates can be derived from the posterior sample. In the last four rows of Table 3.5, the most common summary measures are provided for each parameter. The posterior mean is computed by taking the average of the 1000 draws from the posterior of the parameter at hand. So, the value 3.09 for μ_{pat} is computed by summing all values (discarding the first 100 iterations) in the corresponding column and dividing the sum by 1000. Likewise, computing the standard deviation of the same column of numbers provides the posterior standard deviation of μ_{pat} (0.38). Central credibility intervals (CCI; the Bayesian alternative for confidence intervals) are obtained by sorting the 1000 draws from smallest to largest and letting the 25th draw be the lower bound and the 975th draw the upper bound of a 95% CCI. For μ_{pat} this provided the values 2.39 and 3.86, respectively. These results correspond to what can be seen in Figure 3.9, in which the left histogram is indeed centered around 3, with approximately lower and upper bound values of 2.4 and 3.9 for a 95% interval.

Taking the a priori assumed ordering $\mu_{con} > \mu_{amn} > \{\mu_{pat}, \mu_{sim}\}$ into account, the estimates for the mean recognition scores for the Controls, True amnesiacs, DID-patients, and Simulators are 13.3, 4.5, 3.1, and 1.9, respectively. The control group clearly shows considerably higher recognition skills than the other three subgroups. This is confirmed by the 95% credibility intervals: the interval for the Controls (12.6; 14.0) has no overlap with any of the other intervals. The posterior mean of 13.3 tells us that "healthy" respondents who are motivated to retrieve as much information as possible score on average more than 13 out of 15 points (the latter being the maximum score on the recognition test).

The True amnesiacs were asked to "retrieve" information that they never received (they skipped the learning phase of the experiment). Since the multiple-choice items had either three (10 items) or five (5 items) answer categories, the expected score based on random guessing is 4.33. The posterior mean of the True amnesiacs (4.5) is therefore almost exactly as expected.

Furthermore, an interesting observation is that the average recognition score for the DID-patients (posterior mean 3.1) is larger than the posterior mean recognition of the Simulators (1.9) but smaller than the True amnesiacs (posterior mean is 4.5). Examination of the 95% credibility intervals shows that the intervals for DID-patients and True amnesiacs are not overlapping, but for DID-patients and Simulators, they do overlap. According to these estimates, DID-patients are more similar to Simulators than to True amnesiacs in terms of their recognition skills. Note, however, that these estimates are based on the information contained in the observed data as well as on information provided by the a priori ordering imposed on the means.

One of the questions of interest in this research (see [10] or Chapter 2) was whether DID-patients suffer from real amnesia or feign their amnesia. Stated differently, we are not in the situation that a priori knowledge about the ordering is available, because there is uncertainty about the order of the means. In this chapter we chose one of the hypotheses as prior knowledge just to serve as an illustration of inequality constrained Bayesian estimation. The real issue in the DID-example is, however, that we have two (or more) conflicting hypotheses, represented by different inequality constrained hypotheses. In the next chapters, choosing the best hypothesis from a set of inequality constrained hypotheses using Bayesian model selection will be discussed and applied to (among others) the DID data.

References

[1] Bernardo, J.M., Smith, A.F.M.: Bayesian Theory. Chichester, Wiley (1994)

[2] Box, G.E.P., Tiao, G.C.: Bayesian Inference in Statistical Analysis. Reading, MA, Addison-Wesley (1973)

[3] Cowles, M.K., Carlin, B.P.: Markov chain Monte Carlo convergence diagnostics: A comparative review. Journal of the American Statistical Association, **91**, 883–904 (1996)

[4] Everitt, B.: Statistics for Psychologists: An Intermediate Course. Mahwah, NJ, Lawrence Erlbaum Associates (2001)

[5] Gelfand, A.E., Smith, A.F.M., Lee, T.: Bayesian analysis of constrained parameter and truncated data problems using Gibbs sampling. Journal of the American Statistical Association, **87**, 523–532 (1992)

[6] Gelman, A., Carlin, J.B., Stern, H.S., Rubin, D.B.: Bayesian Data Analysis (2nd ed.). London, Chapman & Hall (2004)

[7] Gelman, A., Rubin, D.B.: Inference from iterative simulation using multiple sequences. Statistical Science, **7**, 457–511 (1992)

[8] Gill, J.: Bayesian Methods. A Social and Behavioral Sciences Approach. London, Chapman & Hall (2002)

[9] Howson, C., Urbach, P.: Scientific Reasoning: The Bayesian Approach (2nd ed.). Chicago, Open Court Publishing Company (1993)

[10] Huntjens, R.J.C., Peters, M.L., Woertman, L., Bovenschen, L.M., Martin, R.C., Postma, A.: Inter-identity amnesia in dissociative identity disorder: A simulated memory impairment? Psychological Medicine, **36**, 857–863 (2006)
[11] Jeffreys, H.: Theory of Probability (3rd ed.). Oxford, Oxford University Press (1961)
[12] Kass, R.E., Wasserman, L.: The selection of prior distributions by formal rules. Journal of the American Statistical Association, **91**, 1343–1370 (1996)
[13] Klugkist, I., Laudy, O., Hoijtink, H.: Inequality constrained analysis of variance: A Bayesian approach. Psychological Methods, **10**, 477–493 (2005)
[14] Lee, P.M.: Bayesian Statistics: An Introduction. London, Arnold (1997)
[15] Lindley, D.V.: Bayesian Statistics, A Review. Philadelphia, SIAM (1972)
[16] Raiffa, H., Schlaifer, R.: Applied Statistical Decision Theory. Boston, Graduate School of Business Administration, Harvard University (1961)
[17] Rutherford, A.: Introducing ANOVA and ANCOVA; a GLM approach. London, Sage (2001)
[18] Smith, A.F.M., Roberts, G.O.: Bayesian computation via the Gibbs sampler and related Markov chain Monte Carlo methods. Journal of the Royal Statistical Society, B, **55**, 3–23 (1993)

4

Encompassing Prior Based Model Selection for Inequality Constrained Analysis of Variance

Irene Klugkist

Department of Methodology and Statistics, Utrecht University, P.O. Box 80140, 3508 TC Utrecht, the Netherlands i.klugkist@uu.nl

4.1 Competing Theories Based on (In)Equality Constraints

In Chapter 2, three psychological datasets with competing, informative hypotheses were introduced. For instance, with respect to the Dissociative Identity Disorder (DID) data of Huntjens, two competing theories about interidentity amnesia were presented [11]. Some believe that information provided to one identity cannot be retrieved by another identity of the DID-patient; that is, there is no transfer of information between identities. Others concluded that the interidentity amnesia is not "real"; that is, patients simulate their inability to retrieve information provided to an alternative identity. To investigate these conflicting theories, an experiment was performed where the ability to retrieve information of DID-patients (the information was provided to one alter and retrieved from another) was compared with three different control groups. The Controls received information and were asked to retrieve as much as possible in a later phase in the experiment. Simulators were asked to simulate switching to another alter after the learning phase and retrieve the information in the role of the simulated alter. True amnesiacs skipped the learning phase and were therefore asked to retrieve information that they never received.

In both competing theories, the Controls were expected to retrieve the most information. Furthermore, Simulators were expected to perform worse than random guessing (i.e., True amnesiacs), because they will deliberately choose a wrong answer. The theories differ with respect to the position of the DID-patients. Is their score comparable to the mean of the True amnesiacs or do they show more resemblance with the Simulators? In Chapter 2, two different translations of the competing theories were suggested. In the first set of hypotheses, the equivalence of DID-patients with either True amnesiacs or Simulators was formulated using (about) equality constraints. This leads to the following models:

$$H_{1a}: \mu_{con} > \{\mu_{amn} \approx \mu_{pat}\} > \mu_{sim},$$
$$H_{1b}: \mu_{con} > \mu_{amn} > \{\mu_{pat} \approx \mu_{sim}\}.$$

In the Bayesian approach introduced in this chapter, an equality restriction will always be evaluated using about equality constraints (i.e., $\approx$). This implies that, for instance, $\mu_1 = \mu_2$ is evaluated using the constraint $|\mu_1 - \mu_2| < \delta$, with a small δ. The value for δ can be specified by the researcher and reflects *relevant* differences (subjective judgement) or a procedure where δ approaches zero can be applied. The latter will be elaborated in Section 4.3.3.

A less restrictive translation of the competing expectations is using just an imposed ordering on the group means without restricting some means to be (about) equal:

$$H_{1c}: \mu_{con} > \{\mu_{amn}, \mu_{pat}\} > \mu_{sim},$$
$$H_{1d}: \mu_{con} > \mu_{amn} > \{\mu_{pat}, \mu_{sim}\}.$$

The results for both sets of hypotheses may be different and it is up to the researcher to carefully consider which set he or she prefers; that is, which formulation reflects best what the researcher wants to be able to conclude from the analysis. In Section 4.4, for illustrative purposes both sets of hypotheses in the DID example will be analyzed so that results can be compared.

A special feature of the hypotheses considered is that they all consist of parameters that are constrained to be larger, smaller, or about equal to other parameters. As a consequence, each hypothesis under consideration is nested in an unconstrained version of the hypothesis. The unconstrained hypothesis is also called the encompassing model, because it encompasses all the constrained hypotheses. The encompassing model for the DID data is

$$H_2: \mu_{con}, \mu_{amn}, \mu_{pat}, \mu_{sim}, \sigma^2,$$

where a comma is used to denote that no constraints are imposed on any of the parameters. Note that σ^2 is also a parameter of the ANOVA model and for completeness here it is included in the notation. In most cases, the variance term is not included in the presentation of hypotheses, since the focus in all models is on the constraints that are imposed on the μ's. The Bayesian model selection approach presented in this chapter explicitly uses the nesting and the encompassing model and is called the encompassing prior approach.

In Section 4.3, Bayesian model selection using the encompassing prior approach is presented for the general ANOVA model. In Section 4.4, the three illustrative datasets introduced in Chapter 2 (Huntjens' DID data, Reijntjes' emotional reactivity data, and Boelen's complicated grief data) are analyzed and the results discussed. However, first, in Section 4.2 we will present the encompassing prior approach based on a simplified example using just a part of the DID data.

Table 4.1. Recognition scores of True amnesiacs and DID-patients

True amnesiacs	1	1	2	3	3	4	4	4	4	$M_{amn} = 4.56$
	4	4	4	5	5	5	5	5	5	$SD_{amn} = 1.83$
	5	6	6	6	6	8	9			$N_{amn} = 25$
DID-patients	0	1	2	2	2	2	3	3	3	$M_{pat} = 3.11$
	3	3	3	3	4	4	4	4	6	$SD_{pat} = 1.59$
	7									$N_{pat} = 19$

4.2 The Encompassing Prior Approach

In this section the encompassing prior approach to Bayesian model selection will be introduced based on a simple example dealing with two (constrained) means and assuming that the variance is known. Using this example, the concepts marginal likelihoods, Bayes factors, and prior and posterior model probabilities will be explained at an intuitive, nontechnical level. A more elaborate explanation for the general ANOVA model is provided in Section 4.3 but can be skipped by readers who are more interested in practical implications than in the technical details.

To illustrate the two-means example, again part of the DID data will be used. Consider just the data from the True amnesiacs and the DID-patients. The data as well as sample means (M_{amn} and M_{pat}), standard deviations (SD_{amn}, SD_{pat}), and sample sizes (N_{amn}, N_{pat}) are presented in Table 4.1. Three theories are formulated in the following hypotheses: $H_0 : \mu_{amn} \approx \mu_{pat}$, $H_1 : \mu_{amn} > \mu_{pat}$, and $H_2 : \mu_{amn}, \mu_{pat}$, where μ_{amn} and μ_{pat} denote the mean recognition score of the True amnesiacs and DID-patients, respectively. The variance σ^2 is assumed known and equal to 3. Note that with H_0 and H_1 the two types of constraints (inequality and about equality) are represented. Other constrained hypotheses considered in the illustrations are based on combinations and variations of each of these constraints.

To determine which theory or hypothesis best fits the data, a confrontation of each hypothesis with the data is needed. Like in the previous chapter about Bayesian estimation, the two basic ingredients of the analysis are the prior distribution of the model parameters and the information contained in the data as represented by the likelihood function. In the subsequent sections it will be shown how the different hypotheses, stated in terms of different constraints imposed on the model parameters, can be captured in the specification of different prior distributions (Section 4.2.1). As we have also seen in Chapter 3, different priors lead to different posteriors. Likewise, different priors lead to different so-called marginal likelihoods, the latter being the key-ingredients of Bayesian model selection. The marginal likelihood of a model will be explained in Section 4.2.2. Comparison of marginal likelihoods of different models or hypotheses is done using Bayes factors (Section 4.2.3) or posterior model probabilities (Section 4.2.4). This section is concluded with a

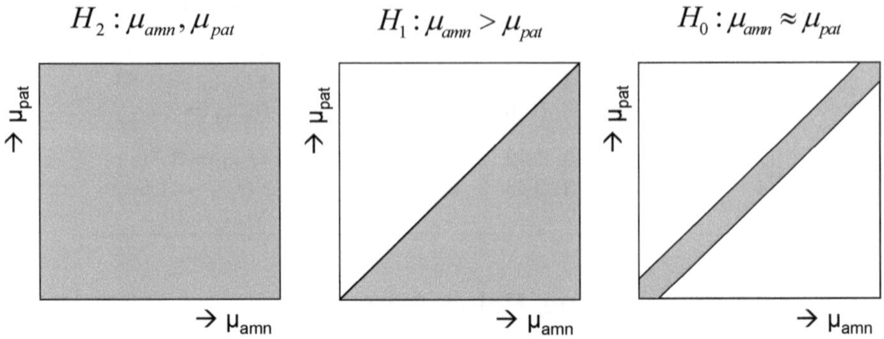

Fig. 4.1. Prior distributions for the three hypotheses H_2, H_1, and H_0

short discussion on sensitivity of results to the prior specification in Section 4.2.5.

4.2.1 Specification of Prior Distributions

In general, a prior distribution for the parameters must be specified for each model or hypothesis under consideration. However, we are dealing with hypotheses that are all nested in the unconstrained, encompassing model; that is, both $H_0 : \mu_{amn} \approx \mu_{pat}$ and $H_1 : \mu_{amn} > \mu_{pat}$ are nested in $H_2 : \mu_{amn}, \mu_{pat}$. The nesting makes the specification of prior distributions considerably easier. In the encompassing prior approach, just one prior is specified, namely the prior for the unconstrained model. The prior distributions for the nested (constrained) hypotheses follow directly from this so-called encompassing prior. The constraints restrict the parameter space according to the hypothesis at hand, as was previously seen in Chapter 3 and is illustrated for the two-means example in Figure 4.1.

In the left graph, the square denotes the prior distribution for the unconstrained model H_2. It can, for instance, represent a flat plane with specific upper and lower bounds for both μ_{amn} and μ_{pat}. But it could also represent a bivariate normal or any other prior distribution $p(\mu_{amn}, \mu_{pat}|H_2)$. The specification of the prior for the unconstrained model will be discussed later in this section. Constraints among model parameters are included in the prior by setting the prior density at zero in the area that is not allowed according to the constraints at hand. In Figure 4.1, the area with nonzero prior density is shaded. In the central graph, the line $\mu_{amn} = \mu_{pat}$ splits the unconstrained prior in two parts: The nonshaded upper triangle is the area that is excluded by the constraint $\mu_{amn} > \mu_{pat}$ and therefore has a prior density of zero. The shaded lower triangle is the part of the prior that has nonzero density because it is in agreement with $H_1 : \mu_{amn} > \mu_{pat}$. In that area, the prior $p(\mu_{amn}, \mu_{pat}|H_1)$ has the same shape as the unconstrained prior (it is proportional to the unconstrained prior), but the density is multiplied by a constant to get the same overall density. For instance, if exactly half of the unconstrained prior density is set to zero in the constrained prior, the density of the latter is equal to the density in the unconstrained prior *times 2* for parameter values that are in agreement with the constraints. The same idea is represented in the right-hand graph for the constraint $\mu_{amn} \approx \mu_{pat}$. The shaded area is the only area that has nonzero prior density and in that area the prior distribution is proportional to the unconstrained prior.

The only prior distribution that needs to be specified is the prior of the unconstrained model (H_2). In the encompassing prior approach, the specification is based on some general guidelines. The first is that all model parameters are a priori considered to be independent. This implies that the joint prior $p(\mu_{amn}, \mu_{pat}|H_2)$ is specified as the product of a prior for μ_{amn} and a prior for μ_{pat}. This is not strictly required for the approach to be applicable but simplifies the prior specification and the resulting computations. The second specification rule is that the prior distributions for all parameters that are constrained in one or more of the hypotheses are equal. For the example, this implies $p(\mu_{amn}|H_2) = p(\mu_{pat}|H_2)$. There are two motivations for this guideline. Since the interest is in the ordering and/or equality of parameters, it is reasonable *not* to benefit any of the orderings a priori. Specifying the prior for μ_{amn} with the mode at say 4 and μ_{pat} with a mode at say 2, a priori supports the ordering $\mu_{amn} > \mu_{pat}$. Since our interest lies in finding out to what extent $H_1 : \mu_{amn} > \mu_{pat}$ is supported, this is considered undesirable. The second motivation for specifying equal prior distributions for parameters that are constrained in one or more of the hypotheses is that it leads to nice properties in terms of (in)sensitivity to the prior specification. This will be elaborated later. The last specification guideline is that the encompassing prior is a relatively uninformative, conjugate distribution for each parameter. The conjugate prior for a mean parameter is a normal distribution, for the example leading to $p(\mu_{amn}|H_2) = p(\mu_{pat}|H_2) = \mathcal{N}(\mu_0, \tau_0^2)$ and this prior is low informative as long as τ_0^2 is large. The final step is to specify μ_0 and τ_0^2, where the value for μ_0 hardly affects the results since the value for τ_0^2 is specified to be relatively large.

The three criteria ensure that the model comparison of H_1 with H_2 is objective (i.e., not sensitive to the exact specification of μ_0 and τ_0^2). This will be shown in Section 4.2.5. However, as will also be illustrated, the results for comparison of H_0 with H_2 are sensitive for the exact value that is specified for τ_0^2. In the encompassing prior approach, we use the range in the observed data to find values for μ_0 and τ_0^2 that provide a low informative but reasonable (not too diffuse) prior distribution. The procedure used to derive the encompassing prior from the range of the observed data is elaborated in Section 4.3.1 and in the illustration discussed in Section 4.4.1. The motivation and the consequences of the criteria for the encompassing prior will be further elaborated in the following sections. The encompassing prior approach was previously introduced and evaluated in [15, 16, 17].

Summarizing, in this section we have seen that theories translated into constrained hypotheses can be represented by different prior distributions. Since the constrained hypotheses are all nested in the unconstrained version of the model, initially just one prior needs to be specified, the encompassing prior. The priors of the constrained models follow from the encompassing prior by truncation according to the constraints. Differences in subsequent marginal likelihoods (discussed in the next section) represent the amount of support the data provide for each of the constrained hypotheses.

4.2.2 The Marginal Likelihood

To determine which theory is mostly supported, a confrontation of each hypothesis with the data is needed. The support in the data is measured by the so-called marginal likelihood. In Figure 4.2, on each row the three hypotheses are presented by three plots (from left to right: H_2, H_1, and H_0). As was introduced in the previous section, the square denotes the unconstrained prior distribution and the shaded area denotes the part of the prior with nonzero density. According to the specification guidelines, the unconstrained prior is the product of two equal and low informative (diffuse, that is almost flat) normal distributions. Therefore, H_2 implies that a priori each combination of values for μ_{amn} and μ_{pat} is about equally likely, H_1 implies that a priori each combination of values for which $\mu_{amn} > \mu_{pat}$ is about equally likely, and H_0 implies that each combination for which $\mu_{amn} \approx \mu_{pat}$ is about equally likely. The nonshaded areas have zero prior density.

Furthermore, in the figure isodensity ellipses denote the density of the data. They contain the information about μ_{amn} and μ_{pat} in the observed data. Loosely formulated, the small ellipse implies that the data are quite likely given these parameter values. For other parameter values, the density of the data is smaller and in Figure 4.2 this is denoted by ellipses on increasing distances from the central ellipse. Each row of three plots represents three hypothetical observed datasets (from top to bottom: $M_{amn} > M_{pat}$, $M_{amn} < M_{pat}$, $M_{amn} = M_{pat}$). In the first row, the data therefore support H_1, in the second row the data are supporting neither H_1 nor H_0, and in the third row the data support H_0.

The support for a model provided by the data is measured by the so-called marginal likelihood. For a model H_t (here $t = 0, 1, 2$), the marginal likelihood is the average density of the data over the prior for the hypothesis at hand (i.e., over all combinations of parameter values that are admitted by the prior). Consider the results on the first row with sample means $M_{amn} > M_{pat}$. H_2 admits all values of μ_{amn} and μ_{pat} thus also including a lot of values for which the density of the data is rather small (the region outside the largest ellipse). H_1 excludes many values of μ_{amn} and μ_{pat} for which the density of the data is rather small. The marginal likelihood of H_1 will therefore be substantially larger than the marginal likelihood of H_2. H_0 excludes the values of μ_{amn} and μ_{pat} for which the density of the data is relatively large (the

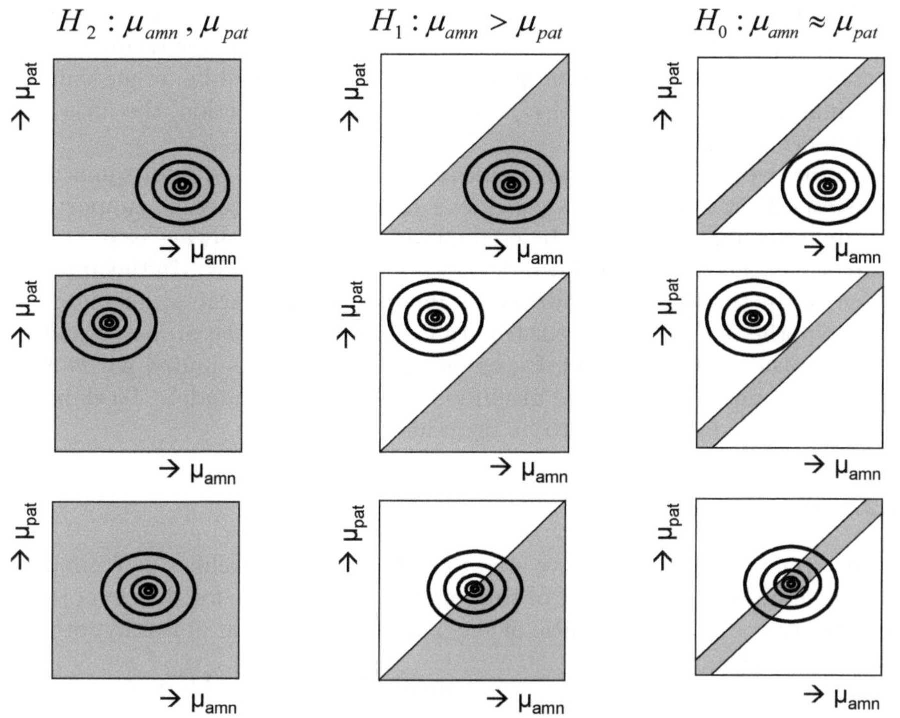

Fig. 4.2. Interpretation of the marginal likelihood for three hypothetical data outcomes: $M_{amn} > M_{pat}$ (first row), $M_{amn} < M_{pat}$ (second row), $M_{amn} = M_{pat}$ (third row)

smallest ellipses), so the marginal likelihood of H_0 will be relatively small. In conclusion, comparing H_0, H_1, and H_2 for the data on the first row, H_1 seems to be the best model; that is, based on visual inspection it has the largest marginal likelihood.

In the three plots on the second row, both H_1 and H_0 exclude those values of μ_{amn} and μ_{pat} that have large densities: The small ellipses are located in the part of the prior with zero density. Therefore, the marginal likelihood of the constrained models will be smaller than the marginal likelihood of the unconstrained model. Neither of the constrained models is supported by these data. Finally, in the bottom row the smallest ellipse falls exactly in the shaded area of H_0. Many parameter values with low densities are excluded; parameter values with high densities remain. Therefore, the marginal likelihood of H_0 will be large compared to the marginal likelihood of H_1 and H_2, where both large and very small density values are included.

Returning to the DID example, the data are in line with the results of the first row. As presented in Table 4.1, $M_{amn} = 4.56$ and $M_{pat} = 3.11$; that is, the likelihood has its top below the line $\mu_{amn} = \mu_{pat}$. This is represented by the smallest ellipse, which therefore falls in the shaded lower triangle of the plot. As a consequence, the marginal likelihood of H_1 will be larger than the marginal likelihood of H_0 and H_2. Based on visual inspection, the data seem to support $H_1 : \mu_{amn} > \mu_{pat}$.

So far, we examined marginal likelihoods intuitively without quantifying the support in numbers. To provide a numerical measure of support, the marginal likelihoods need to be estimated, which is not always easy. Several estimation procedures have been developed (cf. [8, 9, 13, 21]). In the approach presented in this chapter, however, estimation of the marginal likelihood for each hypothesis is not required. Due to the nesting and the proposed method for the specification of priors, Bayes factors are easily computed without the need to estimate the marginal likelihood of the separate models. To elaborate this, the Bayes factor needs to be introduced first.

4.2.3 The Bayes Factor

In the previous section, we have seen that the marginal likelihood of a hypothesis H_t represents the amount of support the data provide for that hypothesis. A Bayes factor (BF) is the ratio of the marginal likelihoods of two hypotheses, say H_t and $H_{t'}$:

$$BF_{tt'} = \frac{m(\boldsymbol{y}|H_t)}{m(\boldsymbol{y}|H_{t'})}. \tag{4.1}$$

The interpretation is straightforward: A Bayes factor of, for instance, 4 implies that the support for H_t is four times larger than the support for $H_{t'}$. Likewise, if $BF_{tt'} = 0.5$, the support for $H_{t'}$ is two times larger than for H_t.

For readers familiar with classical model selection criteria like Akaike's information criterion (AIC [1]) and corrected AIC (CAIC [7]), it will be known that both the fit of a model and the model complexity are taken into account when models are compared. Each of these criteria contains separate terms for model fit and complexity, where the latter is some function of the number of model parameters. Problematic in inequality constrained hypotheses is that the number of parameters of both $H_2 : \mu_1, \mu_2$ and $H_1 : \mu_1 > \mu_2$ is two; however, the complexity of the model, or stated differently, the size of the parameter space, is different.

The Bayes factor is a model selection criterion with an implicit fully automatic Occam's razor [12, 13, 23]; that is, the complexity of the model is accounted for automatically, without needing an explicit penalty term based on the number of parameters. The Bayes factor is therefore useful for model selection in the context of inequality constrained hypotheses. See for a general introduction to the Bayes factor, for instance, [13, 21]; for applications in inequality constrained model comparison, see [15, 16, 17].

In the encompassing prior approach, Bayes factors for constrained models versus the encompassing model are computed; that is, BF_{t2}. It can be shown that as a consequence of the nesting of H_t in H_2, a simple algorithm that does not require the computation of each of the marginal likelihoods provides BF_{t2}. The derivation of this algorithm will not be presented here. The interested reader is referred to Section 4.3. The result of the derivation is that BF_{t2} is equal to the proportion of the encompassing *posterior* in agreement with the constraints of the nested model H_t, divided by the proportion of the encompassing *prior* in agreement with these constraints. In the sequel the two proportions are denoted by $1/d_t$ and $1/c_t$ for posterior and prior, respectively, giving

$$BF_{t2} = \frac{1/d_t}{1/c_t} = \frac{c_t}{d_t}. \tag{4.2}$$

The two proportions represent the fit ($1/d_t$) and the complexity ($1/c_t$) of hypothesis H_t compared to H_2. This will be illustrated using the three plots on the *bottom row* of Figure 4.2. Recall that the square denotes the encompassing prior and the ellipses denote the density of the data. Since the unconstrained prior is relatively uninformative (i.e., almost flat), the unconstrained posterior will be virtually proportional to the density of the data. For the first plot, the posterior has the same shape as the density of the data and is therefore represented by the same ellipses. In the second plot it can be seen that half of the unconstrained posterior mass lies in the area $\mu_{amn} > \mu_{pat}$. Compared to the unconstrained hypothesis, the fit of H_1 is half as good. In the third plot, quite a large part of the unconstrained posterior lies in the area $\mu_{amn} \approx \mu_{pat}$ (the mass is largest within the smallest ellipse); let us say it is 40% of the total mass. Compared to the unconstrained hypothesis, H_0 also has a worse fit. The values for $1/d_t$ are .5 for H_1, .4 for H_0, and (always) 1 for H_2. The best fit is obtained if no constraints are imposed on the means. This is, however, not very informative since the model complexity is not yet taken into account. The proportions of the unconstrained prior in agreement with the constraints of H_1 and H_0, respectively, provide information about the complexity of the models (i.e., the sizes of the parameter spaces). For H_1 it is easily seen that the proportion $1/c_1$ is .5. For H_0 it is a small value, say $1/c_0 = .1$ (estimating that the shaded area is 10% of the total square).

The resulting $BF_{12} = .5/.5 = 1.0$ shows that compared to the unconstrained model, H_1 is neither better nor worse. For H_0, $BF_{02} = .4/.1 = 4$ is obtained. Since the size of the constrained parameter space was .1 of the encompassing model a priori, the posterior proportion of .4 shows an increased amount of support for $H_0 : \mu_{amn} \approx \mu_{pat}$ after inclusion of the information in the observed data. The resulting Bayes factor represents this support.

As a final remark, note that although the algorithm 4.2 is restricted to the estimation of Bayes factors of constrained with the encompassing model, the Bayes factor of two constrained hypotheses easily follows from the BF_{t2} values through

$$BF_{tt'} = \frac{BF_{t2}}{BF_{t'2}}. \tag{4.3}$$

Estimation of BF_{t2}

The BF_{t2} values are computed by estimation of the two proportions $1/d_t$ and $1/c_t$. Estimation of these proportions will be discussed for prior and posterior, respectively.

In Section 4.2.1 we have seen that for $H_1 : \mu_{amn} > \mu_{pat}$, part of the prior parameter space is in agreement with the constraint (the shaded area in the central plot in Figure 4.1). The specification guidelines of the encompassing prior ensured symmetry around the line $\mu_{amn} = \mu_{pat}$ and this leads to a value of .5 for the part of the encompassing prior in agreement with the constraint. So, for some hypotheses the value for $1/c_t$ needs not be estimated: It is known as a consequence of the (symmetric) prior specification. For other hypotheses, it is not so easily seen what this proportion is, for example, for $H_0 : |\mu_{amn} - \mu_{pat}| < \delta$ (i.e., the shaded area in the right-hand plot in Figure 4.1). For such a hypothesis, $1/c_t$ can be estimated by taking a large sample from the unconstrained prior and counting how often the sampled values of μ_{amn} and μ_{pat} are in agreement with the constraint. This will be illustrated for the two groups in the DID data. The encompassing prior, specified according to the guidelines presented in Section 4.2.1 and further elaborated in Section 4.4.1, is $p(\mu_{amn}, \mu_{pat}) = \mathcal{N}(\mu_{amn}|\mu_0 = 3.75, \tau_0^2 = 3.42)\mathcal{N}(\mu_{pat}|\mu_0 = 3.75, \tau_0^2 = 3.42)$. For $H_0 : |\mu_{amn} - \mu_{pat}| < \delta$, the value $\delta = 0.3$ is used.

To estimate $1/c_t$ for each of the hypotheses, $10,000$ draws of μ_{amn} and μ_{pat} are sampled from the encompassing prior. In Table 4.2 the first 20 draws are listed, as well as the information about whether each of these draws is a "hit" for both H_0 and H_1 (a value 1 if the sampled values are in agreement with the constraint and 0 otherwise). For instance, the first draw delivers the sampled values $\mu_{amn} = 2.07$ and $\mu_{pat} = 7.35$. These values are therefore not in agreement with the constraint of $H_1 : \mu_1 > \mu_2$ nor with $H_0 : |\mu_1 - \mu_2| < 0.3$. Iteration 4, with $\mu_1 = 3.08$ and $\mu_2 = 3.25$, provides an example of a "hit" for H_0 but not H_1. The total number of hits in $10,000$ iterations is provided at the bottom of the table. The proportions $1/c_t$ for both H_1 and H_0 are simply computed by division of the obtained sum by the total number of iterations (i.e., $10,000$). Note that the value for $1/c_1$ obtained by estimation is indeed about .5.

The encompassing posterior distribution combines the information in the encompassing prior and the information in the data (i.e., the likelihood). One of the specification guidelines for the encompassing prior is that it is specified to be noninformative or low informative. Therefore, the encompassing posterior is dominated by the information in the data. The posterior provides objective information (i.e., without a subjective prior) about the model parameters after taking the observed data into account. A sample of μ_{amn} and μ_{pat} values is drawn from the encompassing posterior $p(\mu_{amn}, \mu_{pat}|\boldsymbol{y}, d_{amn}, d_{pat}, \sigma^2 = 3) \propto \mathcal{N}(\mu_{amn}|4.53, 0.12)\mathcal{N}(\mu_{pat}|3.14, 0.15)$. Note that without constrained means and with σ^2 known, the posterior distributions of μ_{amn} and μ_{pat} are independent. Therefore, the posterior means

Table 4.2. Estimation of $1/c_t$ and $1/d_t$ using sampled values for μ_{amn} and μ_{pat} from encompassing prior and posterior

	Prior				Posterior			
Iteration	μ_{amn}	μ_{pat}	H_1	H_0	μ_{amn}	μ_{pat}	H_1	H_0
1	2.07	7.35	0	0	4.48	3.28	1	0
2	3.68	4.25	0	0	4.25	2.05	1	0
3	3.24	4.14	0	0	4.89	3.74	1	0
4	3.08	3.25	0	1	4.26	2.18	1	0
5	0.32	4.66	0	0	5.03	3.69	1	0
6	0.48	6.79	0	0	4.91	2.73	1	0
7	3.16	2.72	1	0	4.38	2.67	1	0
8	6.77	5.47	1	0	4.40	3.70	1	0
9	3.40	4.29	0	0	5.40	2.71	1	0
10	3.16	3.78	0	0	3.83	3.98	0	1
11	3.12	7.67	0	0	4.08	3.84	1	1
12	1.60	4.25	0	0	5.28	3.65	1	0
13	6.39	2.70	1	0	4.20	3.31	1	0
14	2.14	5.36	0	0	4.35	3.54	1	0
15	2.92	2.63	1	1	4.70	4.17	1	0
16	0.45	3.56	0	0	4.28	3.37	1	0
17	0.93	4.80	0	0	4.59	3.08	1	0
18	5.62	0.99	1	0	4.56	3.07	1	0
19	5.11	4.09	1	0	4.22	3.01	1	0
20	4.83	7.07	0	0	4.37	2.39	1	0
⋮	⋮	⋮	⋮	⋮	⋮	⋮	⋮	⋮
Sum			4997	908			9963	159
Proportion			.500	.091			.996	.016

are

$$\frac{\mu_0/\tau_0^2 + M_j(N_j/3)}{1/\tau_0^2 + N_j/3}, \quad \text{for} \quad j = amn, pat,$$

and the posterior variances are

$$(1/\tau_0^2 + N_j/3)^{-1}, \quad \text{for} \quad j = amn, pat,$$

(see Chapter 3). Again, the total number of "hits" for both H_1 and H_0 in 10,000 iterations provide information about the amount of agreement with the constraints. The total number of hits out of 10,000 iterations provides the required proportions.

The resulting proportions for the DID illustration (last row of Table 4.2) provide the ingredients required to compute all Bayes factor values:

$$BF_{12} = 0.996/0.500 = 1.99,$$
$$BF_{02} = 0.016/0.091 = 0.18,$$
$$BF_{10} = 1.99/0.18 = 11.06.$$

The first Bayes factor shows that the data support the constraint of H_1 ($BF_{12} > 1$). Likewise, the Bayes factor of H_0 with H_2 shows that the null hypothesis is not supported by the data ($BF_{02} < 1$). It is therefore not surprising that a confrontation of H_1 with H_0 provides strong support for the former ($BF_{10} = BF_{12}/BF_{02} = 11.06$). This value shows that after observing the data and taking both fit and complexity of both hypotheses into account, H_1 is about 11 times more likely than H_0.

Summarizing, the Bayes factor for each of the constrained models with the encompassing model equals the ratio of the proportions $1/d_t$ and $1/c_t$, which can be estimated by taking large samples from the unconstrained prior and unconstrained posterior and counting the numbers of iterations in agreement with the constrained model at hand. In the case of more than one constrained hypothesis, all Bayes factors BF_{t2} can be estimated with just one prior and one posterior sample. The Bayes factor provides a measure of support where both fit and complexity are included and is therefore especially useful for inequality constrained modeling.

4.2.4 Posterior Model Probabilities

Bayes factors contain the information about the relative support each model receives after observing the data and taking both model fit and complexity into account. However, when a finite set of hypotheses is under consideration, it can be helpful to compute so-called posterior model probabilities for all the models within this set. A posterior model probability (PMP) combines the information represented by the Bayes factors with user-specified prior model probabilities. To obtain an as objective as possible model selection procedure, in this chapter each hypothesis will be considered equally likely a priori. For our example with three hypotheses, this implies that the prior probability of each hypothesis is specified as 1/3. Note that in the sequel, the results and the formulas used for computation of PMPs are based on the assumption of equal prior model probabilities.

The support found in the data as represented by the Bayes factor values updates the prior probabilities into posterior probabilities. A PMP has a value between zero and one, where a larger value implies more support. PMPs for a finite set of hypotheses are computed such that the sum of the probabilities equals one. Therefore, they represent the relative support for a hypothesis within a set. Note that, in contrast with the usual interpretation of p-values in null hypothesis testing, a PMP is *not* used to draw dichotomous decisions based on some arbitrary critical value. A PMP provides the amount of evidence for a model relative to the other models under consideration.

For the three hypotheses H_0, H_1, and H_2 the PMP for each hypothesis H_t is computed by

$$\text{PMP}(H_t) = \frac{BF_{t2}}{\sum_{t'=0}^{2} BF_{t'2}},$$

Table 4.3. Posterior model probabilities

Hypothesis	Full set	Set: H_0 and H_1
$H_0 : \mu_1 \approx \mu_2$	.06	.08
$H_1 : \mu_1 < \mu_2$	.63	.92
$H_2 : \mu_1, \mu_2$	.32	–

with $BF_{22} = 1$. The results for the example at hand are provided in Table 4.3. The inequality constrained hypothesis receives the largest support with a PMP of .63. Note that the unconstrained hypothesis is also part of the set that is evaluated. It may be more interesting to include only hypotheses that represent a theory (i.e., hypotheses with equality and/or inequality constraints). In this small example this implies the inclusion of just H_0 and H_1, leading to a PMP of 0.18/(0.18+1.99)=.08 for H_0 and 1.99/(0.18+1.99)=.92 for H_1.

4.2.5 Types of Constraints and Corresponding Prior Sensitivity

In Chapter 3, with respect to Bayesian *estimation*, we have seen that diffuse (low informative, vague) prior distributions can be used and lead to estimates that represent the information in the observed data and are *not* affected by the prior. Consider, for instance, the situation that a prior normal distribution with a variance of 10 can be considered not very informative for the data at hand. This means that the data dominate the prior and, consequently, posterior estimates are determined almost by the data alone. For the same data, a normal prior with a variance of say 50 is then uninformative as well and will lead to the same posterior estimates.

Unfortunately, this does not hold for Bayesian *model selection*. Even if two priors are both low informative compared to the data at hand (e.g., the normal priors with variances 10 and 50, respectively), the resulting Bayes factors and posterior model probabilities can differ substantially. Low or uninformative priors in the estimation context are therefore not necessarily objective priors in the context of model selection. Worse, in many applications low or uninformative priors lead to arbitrary results that are completely determined by the priors and not by the data [3].

In this chapter, a general approach that can be used for inequality and about equality constrained hypotheses was presented, in which the proposed specification of the encompassing prior is relatively uninformative and symmetric with respect to the parameters that are constrained in one or more of the hypotheses. These specification guidelines make sense from an intuitive point of view given that we aim for "objective" model selection; that is, we do not wish to favor any of the constrained hypotheses a priori. These specification guidelines also provide a procedure that is not sensitive for the exact specification of the (diffuse) encompassing prior for inequality constrained hypotheses. This does, however, not hold for hypotheses with about equality

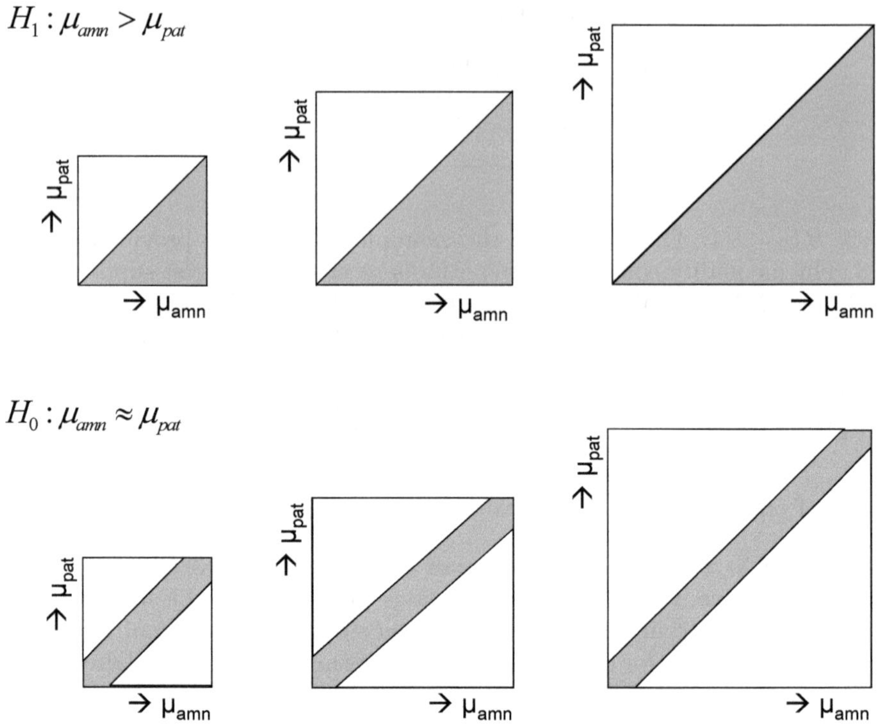

Fig. 4.3. Effect of the encompassing prior on $1/c_t$ for an inequality constrained hypothesis (first row) and an about equality constrained hypothesis (second row)

constraints. Prior sensitivity for both types of constraints will be elaborated in this section.

The simplified example with H_0, H_1, and H_2 provides illustrations of two basic types of constraints under consideration: the inequality constraint and the about equality constraint. For both constraints, sensitivity to the exact specification of the encompassing prior is discussed by examining $1/c_t$ and $1/d_t$ separately. Starting with the latter, it is well known that for relatively noninformative, diffuse prior distributions, the posterior is dominated by the data and thus not sensitive to the prior specification. The third criterion in the specification guidelines proposed in Section 4.2.1 states that the encompassing prior is low informative (i.e., relatively diffuse). As a consequence, the value $1/d_t$ is hardly affected by the exact specification of the encompassing prior, irrespective of the type of constraints used.

This is not always the case for the value $1/c_t$; however, it is for a subset of constraints. Let us start with the results for H_1 and BF_{12} derived from the results in Table 4.2. The proportion of sampled μ_{amn} and μ_{pat} values from the encompassing prior in agreement with H_1 is .50. This result is not surprising and shows a nice property of the encompassing prior approach and the criteria for specification of the encompassing prior. This is illustrated in Figure 4.3. In the top row, the three squares denote three different specifications of the encompassing prior distribution for μ_{amn} and μ_{pat}. The smallest square represents a prior with a relatively small value for τ_0^2, the middle square represents a prior with a larger value, and the largest square represents a very diffuse prior with the largest value for τ_0^2. The proportion of the prior in agreement with the constraint $\mu_{amn} > \mu_{pat}$ is the part of the square below the line $\mu_{amn} = \mu_{pat}$ (i.e., the shaded area). It can be seen that in each plot, $1/c_1 = .5$ irrespective of the size of the square (i.e., irrespective of the diffuseness of the encompassing prior). By criterion 1 and 2 of the specification guidelines for the encompassing prior approach, symmetry in the prior is obtained for all constrained parameters. As a consequence, the value for $1/c_t$ is a fixed constant that depends only on the number of inequality constraints and not on the specification of μ_0 and τ_0^2 of the prior distribution.

Since we also specified the prior to be relatively uninformative, leading to a value for $1/d_1$ that does not or hardly depend on the prior distribution, a model selection approach is formulated that is not sensitive to the prior specification for the model $\mu_{amn} > \mu_{pat}$. This result generalizes to models with more constrained means (e.g., $H_{1c} : \mu_{con} > \{\mu_{amn}, \mu_{pat}\} > \mu_{sim}$).

However, this result does *not* generalize to (about) equality hypotheses. This can be seen in the second row of Figure 4.3. Again, the three squares represent priors with increasing values for τ_0^2 (i.e., different levels of diffuseness). For a small but fixed value for δ, the shaded area contains parameter values in agreement with $H_0 : \mu_1 \approx \mu_2$. Therefore, the proportion of each of the squares that falls within these bounds must be evaluated to obtain the value $1/c_0$. The size of this proportion depends clearly on the size of the square. For the largest square, for instance, the proportion in agreement with H_0 becomes rather small. Stated formally, if τ_0^2 approaches ∞, it follows that the proportion $1/c_0$ approaches 0. Since $BF_{02} = (1/d_0)/(1/c_0)$, this implies that the Bayes factor becomes infinitely large in favor of the null model irrespective of the observed data. This is a phenomenon known as Bartlett's or Lindley's paradox [4, 20].

In the specification of the encompassing prior, a balance must be found between low informative, leading to values for $1/d_t$ that are virtually objective for all types of constraints, but not too diffuse to avoid the Lindley problem for about equality models. This explains and motivates the last criterion in the specification guidelines given earlier. The results for about equality constrained hypotheses can only be interpreted conditional on the prior specification. It seems a natural choice to specify the prior on a range that is relevant for the data at hand. In the encompassing prior approach, the range of the observed data is used for the specification of the encompassing prior. This is elaborated in the illustration in Section 4.4.1. The encompassing prior is therefore data-based, but only little information of the data is used, and the resulting prior distribution is still relatively uninformative. Simulation studies show satisfactory results for the encompassing prior approach based on these specification guidelines [15].

4.3 The Encompassing Prior Approach for Inequality Constrained ANOVA

The encompassing prior approach to model selection introduced in the previous section is elaborated here for the general ANOVA model. This section is more technical than the previous sections. Readers more interested in practical applications than equations can skip Section 4.3 and continue with the illustrations in Section 4.4.

The density of the data and the prior and posterior distributions for the general ANOVA model are provided in Section 4.3.1. They form the ingredients of the Bayes factor for a nested with the encompassing model within this context. The derivation of the Bayes factor is provided in Section 4.3.2. So far, we defined about equality hypotheses with a not too small user-specified relevance measure δ, but in Section 4.3.3 an adjustment is presented that can be used to let δ approach zero and therefore approach the (strict) equality constrained hypothesis. In Section 4.3.4, an elaboration on prior sensitivity in the encompassing prior approach will be given.

4.3.1 The Encompassing Prior for ANOVA Models

The general ANOVA model

$$y_i = \sum_{j=1}^{J} \mu_j d_{ji} + \varepsilon_i, \qquad \text{with } \varepsilon_i \sim \mathcal{N}(0, \sigma^2) \tag{4.4}$$

has the likelihood function

$$f(\boldsymbol{y}|\boldsymbol{\mu}, \sigma^2, D) = \prod_{i=1}^{N} \frac{1}{\sqrt{2\pi\sigma^2}} \exp\left\{-\frac{(y_i - \sum_{j=1}^{J} \mu_j d_{ji})^2}{2\sigma^2}\right\}. \tag{4.5}$$

The encompassing prior is the prior distribution specified for the model parameters for the unconstrained hypothesis. Since prior independence between model parameters is assumed, $p(\boldsymbol{\mu}, \sigma^2|H_2) = p(\sigma^2|H_2)\prod_{j=1}^{J} p(\mu_j|H_2)$. The conjugate prior for a mean parameter is a normal distribution, so $p(\mu_j|H_2) \sim \mathcal{N}(\mu_j|\mu_0, \tau_0^2)$. Note that in the encompassing prior approach, each μ_j is specified to have the same prior distribution (see the specification guidelines presented in Section 4.2.1). The normal prior is low informative (diffuse, vague) for large values of τ_0^2. The conjugate prior for the variance is a scaled inverse χ^2-distribution: $p(\sigma^2|H_2) \sim \text{Inv-}\chi^2(\sigma^2|\nu_0, \sigma_0^2)$. Loosely formulated, this distribution can be interpreted as adding ν_0 prior observations with a scale σ_0^2 to the analysis. Therefore, this distribution is low informative for small values of ν_0. We use $\nu_0 = 1$ in all analyzes. This leads to the encompassing prior

$$p(\boldsymbol{\mu}, \sigma^2 | H_2) \propto \text{Inv-}\chi^2(\sigma^2 | \nu_0, \sigma_0^2) \prod_{j=1}^{J} \mathcal{N}(\mu_j | \mu_0, \tau_0^2). \tag{4.6}$$

The specification of μ_0, τ_0^2, and σ_0^2 is data-based. To obtain values, a sample is drawn from the unconstrained posterior using a constant prior; that is, $p(\boldsymbol{\mu}, \sigma^2) \propto c$. Note, again, that although such a prior should not be used for the actual model selection, it does provide objective parameter estimates. For σ_0^2, the posterior mean of σ^2 is used. This provides a value that is reasonable for the data at hand and, since we always use $\nu_0 = 1$, a posterior that is hardly affected by the prior. For μ_0 and τ_0^2, the information about each of the μ's obtained by the posterior sample is combined as follows: Based on the posterior sample, the 99.7% credibility interval for each μ_j ($j = 1, \ldots, J$) is determined; the smallest lower bound and the largest upper bound of the J intervals define one broad interval containing all reasonable values for each of the μ's; μ_0 is defined to be the center of this interval (i.e., the average of the lower and upper bound) and τ_0^2 gets a value that is equal to the range of the interval divided by 2. An illustration is provided in Section 4.4.1, where the DID data and hypotheses are evaluated.

The product of the encompassing prior distribution and the likelihood provides the posterior of the unconstrained ANOVA model:

$$p(\boldsymbol{\mu}, \sigma^2 | \boldsymbol{y}, D, H_2) \propto f(\boldsymbol{y} | \boldsymbol{\mu}, \sigma^2, D) \text{Inv-}\chi^2(\sigma^2 | \nu_0, \sigma_0^2) \prod_{j=1}^{J} \mathcal{N}(\mu_j | \mu_0, \tau_0^2). \tag{4.7}$$

Note once more that all the constrained hypotheses H_t are nested in the unconstrained, encompassing model H_2. The constraints of a hypothesis are incorporated in the prior distribution by truncation of the parameter space through an indicator function $I_{\boldsymbol{\mu} \in H_t}$. This leads to the general prior distribution for any hypothesis H_t:

$$p(\boldsymbol{\mu}, \sigma^2 | H_t) \propto \text{Inv-}\chi^2(\sigma^2 | \nu_0, \sigma_0^2) \prod_{j=1}^{J} \mathcal{N}(\mu_j | \mu_0, \tau_0^2) I_{\boldsymbol{\mu} \in H_t}, \tag{4.8}$$

where $I_{\boldsymbol{\mu} \in H_t} = 1$ if the means μ_j are in agreement with the constraints of hypothesis H_t, and zero otherwise.

Subsequently, we also obtain the general posterior for a hypothesis H_t:

$$p(\boldsymbol{\mu}, \sigma^2 | \boldsymbol{y}, D, H_t) \propto$$

$$f(\boldsymbol{y} | \boldsymbol{\mu}, \sigma^2, D) \text{Inv-}\chi^2(\sigma^2 | \nu_0, \sigma_0^2) \prod_{j} \mathcal{N}(\mu_j | \mu_0, \tau_0^2) I_{\boldsymbol{\mu} \in H_t}. \tag{4.9}$$

The unconstrained prior (4.6) and the constrained prior (4.8) of any nested model are similar except for the truncation of the parameter space by the constraints. The same similarity exists in the encompassing (4.7) and constrained (4.9) posterior distributions. This aspect is used in the derivation of the Bayes factor for a nested with the encompassing hypothesis.

4.3.2 Deriving the Bayes Factor

The marginal likelihood $m(\boldsymbol{y}|H_t)$ for an ANOVA model with data $\boldsymbol{y}$, group membership denoted by D, and parameters $\boldsymbol{\mu}$ and σ^2 a priori assumed independent, can be expressed as

$$m(\boldsymbol{y}|H_t) = \frac{f(\boldsymbol{y}|\boldsymbol{\mu}, \sigma^2, D, H_t)p(\boldsymbol{\mu}, \sigma^2|H_t)}{p(\boldsymbol{\mu}, \sigma^2|\boldsymbol{y}, D, H_t)}; \tag{4.10}$$

see also [9]. A Bayes factor is the ratio of two marginal likelihoods; so, for two models H_t and H_2

$$BF_{t2} = \frac{f(\boldsymbol{y}|\boldsymbol{\mu}, \sigma^2, D, H_t)p(\boldsymbol{\mu}|H_t)p(\sigma^2|H_t)/p(\boldsymbol{\mu}, \sigma^2|\boldsymbol{y}, D, H_t)}{f(\boldsymbol{y}|\boldsymbol{\mu}, \sigma^2, D, H_2)p(\boldsymbol{\mu}|H_2)p(\sigma^2|H_2)/p(\boldsymbol{\mu}, \sigma^2|\boldsymbol{y}, D, H_2)}. \tag{4.11}$$

Since no constraints are imposed on σ^2, $p(\sigma^2|H_t) = p(\sigma^2|H_2)$ and therefore these terms cancel. Furthermore, due to the nesting of H_t in the encompassing model H_2, the models contain the same parameters; so, for a value of $\boldsymbol{\mu}$, say $\boldsymbol{\mu}'$, that exists in both models H_t and H_2, $f(\boldsymbol{y}|\boldsymbol{\mu}', \sigma^2, D, H_t) = f(\boldsymbol{y}|\boldsymbol{\mu}', \sigma^2, D, H_2)$ and therefore the Bayes factor as given in (4.11) reduces to

$$BF_{t2} = \frac{p(\boldsymbol{\mu}'|H_t)/p(\boldsymbol{\mu}', \sigma^2|\boldsymbol{y}, H_t)}{p(\boldsymbol{\mu}'|H_2)/p(\boldsymbol{\mu}', \sigma^2|\boldsymbol{y}, H_2)}. \tag{4.12}$$

Finally, the $\boldsymbol{\mu}$'s have the same prior distributions, except for the truncation of the prior for H_t by the constraints, formally denoted by

$$p(\boldsymbol{\mu}'|H_t) = \left(\frac{I_{\boldsymbol{\mu}' \in H_t}}{\int p(\boldsymbol{\mu}'|H_2)I_{\boldsymbol{\mu}' \in H_t}\delta\boldsymbol{\mu}'}\right)p(\boldsymbol{\mu}'|H_2) = c_t \cdot p(\boldsymbol{\mu}'|H_2). \tag{4.13}$$

Likewise, $p(\boldsymbol{\mu}', \sigma^2|\boldsymbol{y}, H_t) = d_t \cdot p(\boldsymbol{\mu}', \sigma^2|\boldsymbol{y}, H_2)$. This leads to

$$BF_{t2} = \frac{c_t \cdot p(\boldsymbol{\mu}'|H_2)/d_t \cdot p(\boldsymbol{\mu}', \sigma^2|\boldsymbol{y}, H_2)}{p(\boldsymbol{\mu}'|H_2)/p(\boldsymbol{\mu}', \sigma^2|\boldsymbol{y}, H_2)} = \frac{c_t}{d_t}. \tag{4.14}$$

Due to the nesting and the encompassing setup, the Bayes factor BF_{t2} reduces to the ratio of two constants. These constants are the inverse of the proportions of the unconstrained prior and posterior in agreement with the constraints of the nested model, respectively $1/c_t$ and $1/d_t$. This approach using (4.14) is previously presented and investigated in [15] and successfully applied in analysis of (co)variance models [17], in multilevel models [14], and in contingency tables [18]. Note that inequality constrained hypotheses in models other than the ANOVA are also presented in the last part of this book.

The proportions $1/c_t$ and $1/d_t$ are easily estimated by taking large samples from both the encompassing prior and posterior by application of the Gibbs sampler. If several constrained models are under consideration, the Bayes factor for each of these constrained hypotheses with the unconstrained hypothesis can be estimated with the same prior and posterior sample. Furthermore, for many hypotheses the value for $1/c_t$ needs not to be estimated because it is known exactly from the constraints, as we have seen before in $1/c_1$ for $H_1 : \mu_1 < \mu_2$. This value is known to be .5 (see Figure 4.2) and no prior sample is required. Therefore, the estimation approach is rather efficient.

However, efficiency may be a problem for constrained hypotheses that severely limit the parameter space. The value for $1/c_t$ for a fully ordered model with J groups is $\frac{1}{J!}$. So, for instance, $1/c_1$ for $H_1 : \mu_1 < \cdots < \mu_8$ is $\frac{1}{8!}$, meaning that in an unconstrained sample of 100,000 iterations about 2 or 3 will be in agreement with the constraint. More iterations will be required to get a stable estimate. Obviously, in this example with $1/c_1$ known, no sample is required at all. However, for a constraint of the type $H_1 : \mu_1 < cldots < \mu_6 < (\mu_7 \approx \mu_8)$, the value $1/c_1$ is not that straightforward, and the parameter space is again very small, leading to the requirement of large samples. Another example of a constrained model severely limiting the parameter space is $H_0 : \mu_1 \approx \mu_2$ (i.e., $H_0 : |\mu_1 - \mu_2| < \delta$ with a very small value for δ). So far, we have chosen δ to be a user-specified relevance measure large enough to be estimated with reasonable samples and thus rather efficiently. However, if the user-specified δ-value is very small or when researchers do prefer to approach the (strict) equality hypothesis, an adjustment of the estimation approach is required. This adjustment is presented for about equality constrained hypotheses in the next section but can also be applied to complex inequality constrained models.

4.3.3 ANOVA Models with (About) Equality Constraints

In this chapter, null hypotheses (i.e., equality constraints) are evaluated as "about equality" constraints. The motivation is twofold. The encompassing prior approach is based on the nesting of constrained models and the subsequent equal dimension of the parameter space for all hypotheses under consideration. A strict equality constrained hypothesis would change the dimension of the parameter space and therefore does not fit in the approach. The Bayes factor BF_{t2} for a nested model H_t also shows that strict equalities are problematic: The proportions of encompassing prior and posterior that are in agreement with the constraints of a strict equality hypothesis are both zero. This would lead to a Bayes factor that is not defined.

The second motivation for about equality constraints is that they are often more realistic. It seems more natural to hypothesize "no relevant effect" than an effect of exactly zero. See, for further discussion, [2, 10]. Therefore, we often do not hypothesize two or more means to be exactly equal, but restricted to differ less than a prespecified number δ. The number δ can be chosen such that it represents irrelevant differences between means from a substantive (psychological, not statistical) perspective.

However, for small values of δ the estimation may become very inefficient. For two means restricted to be about equal, the value for δ determines the size of the interval around the line $\mu_1 = \mu_2$. This interval can become very small compared to the unconstrained parameter space. To estimate the value for $1/c_0$ for such a hypothesis may require a very large sample from the unconstrained prior since a certain minimum number of hits is necessary to obtain a stable estimate. Similar efficiency problems may occur when estimating $1/d_0$ based on a sample from the unconstrained posterior. Therefore, an adjustment of the sampling and estimation procedure is suggested. See also [19] and Chapters 12 and 15 in this book. Consider the hypothesis $H_0 : \mu_1 \approx \mu_2$ with $\delta \to 0$. This is the situation where any difference between the means is considered relevant (i.e., the approximation of a strict null hypothesis).

The following procedure leads to an estimate of BF_{02}:

1. Choose a (not too) small value δ_1 and define $H_{01} : |\mu_1 - \mu_2| < \delta_1$.
2. Sample from the encompassing prior and posterior and compute $1/c_{01}$ and $1/d_{01}$ by counting for how many samples $|\mu_1 - \mu_2| < \delta_1$ holds. Compute $BF_{(01)2} = c_{01}/d_{01}$.
3. Define $\delta_2 < \delta_1$ and $H_{02} : |\mu_1 - \mu_2| < \delta_2$.
4. Sample from the constrained ($|\mu_1 - \mu_2| < \delta_1$) prior and posterior and compute $1/c_{02}$ and $1/d_{02}$ by counting for how many samples $|\mu_1 - \mu_2| < \delta_2$ holds. Compute $BF_{(02)(01)} = c_{02}/d_{02}$.

Repeating steps 3 and 4, with $\delta_1 > \delta_2 > \cdots > \delta_{r-1} > \delta_r$ leads to a sequence of Bayes factors $BF_{(01)2}, BF_{(02)(01)}, \ldots, BF_{(r)(r-1)}$. The estimate for BF_{02} follows from the product rule

$$BF_{(01)2} \times BF_{(02)(01)} \times \cdots \times BF_{r(r-1)}.$$

This procedure decreases the δ-value used for evaluation of the hypothesis in a stepwise procedure. In each step, it is calculated how the Bayes factor changes compared to the previously used δ-value. At a certain point (here denoted by r), a further decrease of the value for δ no longer changes the Bayes factor [2]. At this point, the value for $BF_{(r)(r-1)}$ will therefore be very close to one and convergence of $\delta \to 0$ is obtained.

Using this approach, BF_{02} can be estimated for very small values of δ with a still relatively efficient sampling procedure.

4.3.4 Prior Sensitivity for the ANOVA Model

In Section 4.2.5, we concluded that for constraints of the type $\mu_1 > \mu_2$, the model selection is not sensitive to the exact specification of the encompassing prior and that for constraints of the type $\mu_1 \approx \mu_2$, it is. In this section, the results will be elaborated for different types of inequality constraints (cf. [16, 17]).

In the previous section, it was already seen that $\pi(\sigma^2)$ does not influence $1/c_t$ for any hypothesis H_t. In BF_{t2} the prior for σ^2 cancels in numerator and denominator. This does not hold for the effect of $\pi(\sigma^2)$ on $1/d_t$, but since we specified $\pi(\sigma^2)$ to be low informative, the effect on the posterior will be negligible (the data will dominate the prior). What remains is the possible effect of the specification of the prior distribution for each of the μ's. Again, the effect on the posterior (i.e., $1/d_t$, can be neglected since we always use noninformative or low informative priors. This leaves only the possible influence of $\pi(\mu_j)$ on $1/c_t$. Several types of inequality constraints are discussed.

In Section 4.2.5, we concluded that $1/c_1 = .5$ for $H_1 : \mu_1 > \mu_2$ due to the symmetry in the encompassing prior. This was graphically illustrated in Figure 4.3. Before extending the result to more than two constrained parameters, another way of explaining the resulting value of .5 is helpful. Note that since $\pi(\mu_1) = \pi(\mu_2)$, sampling both μ_1 and μ_2 gives two random values from the same distribution. It is easily seen that the probability that the first is larger than the second is equal to .5.

Likewise for more than two means, all having the same prior distribution, the probability of each possible ordering of the means is equally likely. For instance, three means can have six different orderings and therefore the value for $1/c_t$ for any fully ordered hypothesis about three means (e.g., $\mu_1 < \mu_2 < \mu_3$) is 1/6. In general, for J means and fully ordered hypotheses, the value for $1/c_t = 1/(J!)$. The only assumption made to obtain this result is that each mean comes from the same (prior) distribution, which is one of the properties of the encompassing prior approach.

Based on similar arguments, a few more types of hypotheses lead to $1/c_t$ values that are not affected by the exact specification of the prior. They are presented for $J = 4$, but hold in general for any J. The first type of constraint that is useful has subgroups of parameters that are mutually constrained, but no constraints between parameters in different subgroups are formulated. An example is $H_1 : \{\mu_1 > \mu_2\}, \{\mu_3 > \mu_4\}$. Since the prior probability for both $\mu_1 > \mu_2$ and $\mu_3 > \mu_4$ is .5, the value for $1/c_1$ is $.5 \cdot .5 = .25$.

The value for $1/c_t$ for $\{\mu_1 + \mu_2\} > \{\mu_3 + \mu_4\}$ as well as for $\{3\mu_1\} > \{\mu_2 + \mu_3 + \mu_4\}$ is .5. In general, as long as the number of means on each side of the inequality constraint is equal, or the sum of the weights is equal, the value for $1/c_t$ is fixed (i.e., does not depend on the prior). Similar considerations hold for constraints of the form $\{\mu_1 \cdot \mu_2\} > \{\mu_3 \cdot \mu_4\}$. However, this type of constraint may be not very realistic in the context of analysis of variance models (i.e., for hypotheses about group means). In the context of cross-tabulated data and odds ratios, it turns out to be a valuable result though (see Chapter 12 for constrained hypotheses in contingency tables).

Table 4.4. Recognition: Mean (M), standard deviation (SD), and sample size (N) per subgroup

	M	SD	N
1. DID-patients	3.11	1.59	19
2. Controls	13.28	1.46	25
3. Simulators	1.88	1.59	25
4. True amnesiacs	4.56	1.83	25

Summarizing, we have seen that model selection using the encompassing prior approach and the specification guidelines as outlined provides virtually objective results for certain types of constraints. Essentially, these are the constraints that restrict certain parameters to be larger or smaller than other parameters. The insensitivity does not hold for about equality constraints. In that case, the results are conditional on the specification of the encompassing prior. In those cases, the observed data are used to obtain information about a reasonable range for the prior.

4.4 Illustrations

In this section, the three datasets introduced in Chapter 2 will be analyzed using the encompassing prior approach. A competing set of models is formulated, Bayes factors for comparison of each constrained model with the unconstrained (encompassing) model are computed, and, assuming equal prior model probabilities for each hypothesis under investigation, posterior model probabilities for a set of models follow from the Bayes factors. We refer the interested reader to `http://www.fss.uu.nl/ms/informativehypotheses` for software for Bayesian model selection using the encompassing prior approach in the context of inequality constrained analysis of variance models.

4.4.1 Dissociative Identity Disorder Data

The Data and Competing Theories

In Table 4.4, the sample means and standard deviations for the four groups in the DID illustration are presented. The data are analyzed twice to compare and illustrate the differences in results as a consequence of different specifications of hypotheses. In the first analysis, a set of four models contains the null hypothesis (H_0), the unconstrained, encompassing model (H_2), and the two competing informative alternatives (H_{1a} and H_{1b}):

$$\begin{aligned} H_0 &: \mu_{con} \approx \mu_{amn} \approx \mu_{pat} \approx \mu_{sim}, \\ H_{1a} &: \mu_{con} > \{\mu_{amn} \approx \mu_{pat}\} > \mu_{sim}, \\ H_{1b} &: \mu_{con} > \mu_{amn} > \{\mu_{pat} \approx \mu_{sim}\}, \\ H_2 &: \mu_{con}, \mu_{amn}, \mu_{pat}, \mu_{sim}. \end{aligned}$$

The second set of models that will be analyzed contains the same H_0 and H_2, but the informative alternative hypotheses are now formulated as

$$H_{1c} : \mu_{con} > \{\mu_{amn}, \mu_{pat}\} > \mu_{sim},$$
$$H_{1d} : \mu_{con} > \mu_{amn} > \{\mu_{pat}, \mu_{sim}\}.$$

The encompassing prior approach is used to evaluate each set of hypotheses. The observed data are used to specify the parameters of the encompassing prior. This is done such that the prior is low informative but with reasonable values for the data at hand. For $\mu_j (j = con, amn, pat, sim)$ this is done following the next steps:

- The 99.7% credibility intervals for each $\mu_j (j = con, amn, pat, sim)$ are estimated based on a sample from the unconstrained posterior using a constant prior (i.e., noninformative with a range from $-\infty$ to ∞). The resulting intervals were $\langle 0.9, 2.8\rangle$, $\langle 2.0, 4.2\rangle$, $\langle 3.5, 5.6\rangle$, and $\langle 12.3, 14.3\rangle$, respectively.
- The smallest of the lower bounds and the largest of the upper bounds are used to define one interval $l - u$, which led to $l = 0.9$ and $u = 14.3$.
- The prior distribution for each μ_j is chosen such that the mean of the prior minus one standard deviation equals l and the mean plus one standard deviation equals u. This leads to the normal prior distribution with a prior mean μ_0 of $(u + l)/2 = 7.6$ and a prior variance τ_0^2 of $((u - l)/2)^2 = 45.3$: $p(\mu_j) = \mathcal{N}(\mu_j | \mu_0 = 7.6, \tau_0^2 = 44.9)$.

This procedure ensures that, per group, the range for each μ_j corresponds to the range of values seen in $\boldsymbol{y}$, but excludes values that are highly unlikely. The prior for σ^2 is the scaled inverse χ^2-distribution with parameters ν_0 and σ_0^2. Specifying $\nu_0 = 1$ leads to a diffuse (low informative) prior distribution. The posterior mean of σ^2 (derived from the posterior sample just described) provides a reasonable value for σ_0^2. A similar approach for the specification of prior distributions is used in the illustrations in Sections 4.4.2 and 4.4.3.

Results and Conclusions

For the first set of hypotheses including H_{1a} and H_{1b}, the Bayes factor of the null hypothesis with the encompassing model is 0.0 confirming that the "nothing relevant is going on" hypothesis is not at all supported by the data. The Bayes factors for the informative alternative hypotheses with the unconstrained model are both larger than one (2.24 and 8.02, respectively), confirming that both theories are reasonably supported by the data. The posterior model probabilities for the four models (assuming equal prior model probabilities) are presented in the left panel of Table 4.5.

For this set of models, we would conclude that the informative hypothesis stating that DID-patients score equivalent to Simulators is mostly supported

Table 4.5. Posterior model probabilities for the DID data

Model	Full set	Set: 1a and 1b	Model	Full set	Set: 1c and 1d
H_0	.000	–	H_0	.000	–
H_{1a}	.199	.218	H_{1c}	.479	.499
H_{1b}	.712	.782	H_{1d}	.481	.501
H_2	.089	–	H_2	.040	–

by the data (PMP(H_{1b}) = .712). However, after seeing the data, the alternative hypothesis stating that DID-patients score equivalent to True amnesiacs cannot completely be ruled out, since it has a posterior probability of .199. Note that the two numbers provide the relative amount of evidence found for each hypothesis and that this information is not used to draw a dichotomous decision or reject one of the theories.

The posterior probability of the null hypothesis is $< .001$. Note, however, that this hypothesis does not reflect any theory of the researchers and is not at all helpful for deciding which of the two competing theories about DID-patients is more likely. A difference can certainly be expected between people who are asked to remember facts from a short story (Controls) and people whose answers are based on random guessing (True amnesiacs) and that difference alone is enough to reject the null. Also, the unconstrained, encompassing model is not very helpful for the question the researchers want to investigate. A reason to include it in the (initial) analysis is to evaluate the fit of each of the constrained hypotheses. To illustrate this, consider the following results for a set of three constrained hypotheses A, B, and, C: PMP(A) = .20, PMP(B) = .00, PMP(C) = .80. The conclusion based on these numbers would be that model B is not supported, model C is "the best model" with a posterior probability of .80, but model A also gets some support (.20). Note, however, that the unconstrained hypothesis is not included in the set. The Bayes factors of each of the three models with the unconstrained model could have been 0.01, 0.00, and 0.04, respectively, telling us that even the constraints imposed in model C reduces the fit with a factor 25 compared to the unconstrained model. In this case, the posterior model probabilities listed above just tell us that from three basically not supported theories, model C is "less wrong" than the other two. Including the unconstrained model (denoted D) in the set would lead to very different probabilities and conclusions: PMP(A) = .01, PMP(B) = .00, PMP(C) = .04, PMP(D) = .95.

It is up to the researcher which hypotheses to include in the set; that is, which hypotheses he or she wants to evaluate. The posterior model probabilities are, however, conditional on which other hypotheses are contained in the set. Adding or deleting a hypothesis can change the results in terms of PMPs considerably. For the example at hand, let us assume that the researchers most of all want to confront hypotheses H_{1a} and H_{1b} with each other. After having seen that the two Bayes factor values for each model with the encompassing model are larger than 1, the posterior model probabilities within a set containing just the two hypothesis of interest may be the most informative in this case. In Table 4.5, these probabilities are also provided. Although the numbers are slightly different, the conclusion remains the same. H_{1a} cannot completely be ruled out (with a probability of .22) but H_{1b} has a considerably stronger support (.78). Note, finally, that the Bayes factor value for comparison of H_{1a} and H_{1b} does *not* depend on the number of models included in the set: $BF_{1b,1a} = .712/.199 = .782/.218 = 3.58$.

For the second set of hypotheses (replacing H_{1a} and H_{1b} with H_{1c} and H_{1d}), the results for the null model are the same ($BF_{02} = 0.0$). The Bayes factors for both informative alternative hypotheses with the unconstrained model are again larger than one, confirming that also these two theories are reasonably supported by the data. The values for the two Bayes factors are larger than in the first set of models (for H_{1c} and H_{1d}, respectively 11.94 and 12.00) and virtually equal to each other. In the formulation of hypotheses where just order information is imposed on the group means, no preference for either of the alternative theories is found after observing the data. This is also reflected in the posterior model probabilities computed from the Bayes factors assuming equal prior model probabilities. They are presented in the right panel of Table 4.5.

4.4.2 Emotional Reactivity Data

The Data and Competing Theories

Reijntjes et al. studied the influence of depression severity on emotional reactivity after different types of peer evaluation feedback in preadolescent children between the age of 10 and 13 years [22]. For assessment of depressed mood, all participants filled in the Children's Depression Inventory (CDI). They were subsequently divided in three groups of about equal size (low, moderate, and high depression) based on their CDI scores.

All children were led to believe that they participated in an Internet version of a peer evaluation contest but in reality the contestants were fictitious. Social evaluation by peers was thus manipulated. Each child was randomly assigned to one of three conditions: success feedback, failure feedback, or neutral feedback. In the success (failure) feedback condition the child was told that he or she got the highest (lowest) "likeability" score. In the neutral condition, the child's name was not presented as highest or lowest score.

To assess changes in positive affect induced by the peer evaluation, the scores on the positive affect scale of the Positive And Negative Affect Schedule (PANAS-P) before and after the experimental phase were compared. The change in positive affect (i.e., the difference score of these two measurements) provides the outcome variable. The data are summarized in Table 4.6.

Three competing informative hypotheses were formulated for this research (note that group labels can also be found in Table 4.6). In the translation of the theoretical expectations, the neutral feedback group serves repeatedly as

Table 4.6. Emotional reactivity: Mean (M), standard deviation (SD), and sample size (N) per subgroup of depression by feedback condition

	Feedback condition											
	Positive				Neutral				Negative			
Depression		M	SD	N		M	SD	N		M	SD	N
Low	[1]	0.27	4.67	18	[2]	0.29	4.76	17	[3]	−9.33	8.77	12
Moderate	[4]	0.41	4.96	12	[5]	−1.50	5.99	14	[6]	−5.78	5.75	19
High	[7]	5.76	4.39	17	[8]	−0.56	4.25	16	[9]	−3.85	6.49	14

Note: Numbers in square brackets denote the group labelling as used in the hypotheses.

a reference for the results of both the positive and negative feedback groups. Expectations are translated in terms of the hypothesized difference between the *positive* and neutral condition within each of the depression groups ($\mu_1 - \mu_2, \mu_4 - \mu_5, \mu_7 - \mu_8$), as well as between the *negative* and neutral condition ($\mu_3-\mu_2, \mu_6-\mu_5, \mu_9-\mu_8$). The mood facilitation hypothesis, for instance, states that positive feedback will have a larger positive effect for lower depression groups. Imposing this expectation on the differences between means as just formulated gives $\{\mu_7 - \mu_8\} < \{\mu_4 - \mu_5\} < \{\mu_1 - \mu_2\}$. Similar translation of expectations into differences between negative and neutral feedback conditions leads to the remaining part $\{\mu_9-\mu_8\} < \{\mu_6-\mu_5\} < \{\mu_3-\mu_2\}$. Both orderings must be supported for the mood facilitation hypothesis to be supported. In a similar way, the emotion context insensitivity hypothesis (H_{1b}) and the discrepancy hypothesis (H_{1c}) were formulated:

$$\begin{aligned}
H_{1a} &: \{\mu_7-\mu_8\} < \{\mu_4-\mu_5\} < \{\mu_1-\mu_2\}, \{\mu_9-\mu_8\} < \{\mu_6-\mu_5\} < \{\mu_3-\mu_2\},\\
H_{1b} &: \{\mu_7-\mu_8\} < \{\mu_4-\mu_5\} < \{\mu_1-\mu_2\}, \{\mu_9-\mu_8\} > \{\mu_6-\mu_5\} > \{\mu_3-\mu_2\},\\
H_{1c} &: \{\mu_7-\mu_8\} > \{\mu_4-\mu_5\} > \{\mu_1-\mu_2\}, \{\mu_9-\mu_8\} > \{\mu_6-\mu_5\} > \{\mu_3-\mu_2\}.
\end{aligned}$$

See Chapter 2 for a more elaborate introduction of this research and the development of the hypotheses.

Results and Conclusions

Using the specification guidelines of the encompassing prior approach, the priors for each of the model parameters under the unconstrained hypothesis are derived from the observed ranges in the data. For each μ, a normal prior with $\mu_0 = -2.3$ and $\tau_0^2 = 145.1$ is used, and for the variance σ^2, the scaled inverse χ^2-distribution has $\nu_0 = 1$ degrees of freedom and a scale parameter $\sigma_0^2 = 31.7$. Note that this encompassing prior is low informative (relatively diffuse) for each of the model parameters, meaning that the posterior will be dominated by the data. Furthermore, the constraints in all three hypotheses are of a type that leads to model selection results that are not sensitive to the exact specification of the (low informative) encompassing prior. The model selection procedure can therefore be considered virtually objective.

Table 4.7. Posterior model probabilities for the emotional reactivity data

Model	Full set	Set: 1a, 1b, 1c
H_{1a}	.000	.000
H_{1b}	.003	.003
H_{1c}	.881	.997
H_2	.116	–

The results are presented in Table 4.7. In the first column with posterior model probabilities, the unconstrained model is included in the set of hypotheses; in the second column it is not. The latter seems to be more appropriate since the unconstrained model does not reflect a hypothesis of interest. The results are quite conclusive and in favor of the discrepancy hypothesis ($\text{PMP}(H_{1c}) = .997$). It is concluded that success feedback leads to a stronger increase in mood for children with higher depression and that negative feedback leads to a stronger decrease in mood for children with lower depression.

4.4.3 Complicated Grief Data

The Data and Competing Theories

The last illustration addresses gender differences in the development of complicated grief (CG) after the loss of partner or child [5, 6]. Eight subgroups of respondents are compared in a Gender (2) by Kinship (2) by Time from loss (2) design. The data are summarized in Table 4.8.

The following set of hypotheses is evaluated:

Table 4.8. Complicated grief: Mean (M), standard deviation (SD), and sample size (N) per subgroup of the Gender by Kinship by Time from loss design

		Time from loss							
		Recent				Remote			
Gender	Kinship		M	SD	N		M	SD	N
Men	Partner	[1]	84.91	21.59	106	[2]	78.60	20.31	131
	Child	[3]	79.77	21.88	26	[4]	77.79	22.37	52
Women	Partner	[5]	86.42	18.56	229	[6]	78.36	19.28	374
	Child	[7]	84.88	17.33	91	[8]	83.02	21.74	165

Note: Numbers in square brackets denote the group labelling as used in the hypotheses.

$$
\begin{aligned}
H_0 &: \mu_1 \approx \mu_2 \approx \mu_3 \approx \mu_4 \approx \mu_5 \approx \mu_6 \approx \mu_7 \approx \mu_8, \\
H_{1a} &: \{\mu_1 > \mu_2\}, \{\mu_3 > \mu_4\}, \{\mu_5 > \mu_6\}, \{\mu_7 > \mu_8\}, \\
H_{1b} &: \text{constraints of } H_{1a} \text{ and } \{\mu_5 > \mu_1\}, \{\mu_6 > \mu_2\}, \{\mu_7 > \mu_3\}, \{\mu_8 > \mu_4\}, \\
H_{1c} &: \text{constraints of } H_{1b} \text{ and } \{\mu_3 > \mu_1\}, \{\mu_4 > \mu_2\}, \{\mu_7 > \mu_5\}, \{\mu_8 > \mu_6\}, \\
H_{1d} &: \text{constraints of } H_{1a} \text{ and } \{\mu_7 > \mu_5\}, \{\mu_8 > \mu_6\}, \{\mu_7 > \mu_3\}, \{\mu_8 > \mu_4\}, \\
H_{1e} &: \text{constraints of } H_{1d} \text{ and } \{\mu_1 > \mu_3\}, \{\mu_2 > \mu_4\}, \{\mu_1 > \mu_5\}, \{\mu_2 > \mu_6\}, \\
H_2 &: \mu_1, \mu_2, \mu_3, \mu_4, \mu_5, \mu_6, \mu_7, \mu_8.
\end{aligned}
$$

In this illustration the informative hypotheses differ substantially in their amount of restrictiveness. For instance, H_{1a} just imposes directional simple main effects of Time from loss. In each subgroup formed by gender and kinship, it is expected that the average CG is larger for Recent than for Remote. Since previous studies have convincingly shown that CG levels are stronger in the early months of bereavement than later on, in all subsequent informative hypotheses this time effect is included. Hypotheses H_{1b} contains the directional simple main effects of both time and gender (CG larger for women than for men). Subsequently, H_{1c} contains the directional simple main effects of time, gender, and kinship (CG larger for child loss than for partner loss). Hypotheses H_{1d} and H_{1e} impose constraints that represent specific expected interaction effects of gender and kinship on CG. For an elaboration of the motivation for each of these theories, we refer to Chapter 2. In the set, also the unconstrained (H_2) and null model (H_0) are included. For the latter, a δ-value of 3.0 is used.

Results and Conclusions

The specification of the encompassing prior is again based on the observed data. For each μ, a normal prior with $\mu_0 = 79.9$ and $\tau_0^2 = 131.8$ is used, and for the variance σ^2, the scaled inverse χ^2-distribution has $\nu_0 = 1$ degrees of freedom and a scale parameter $\sigma_0^2 = 396.9$.

The results are presented in Table 4.9. The interpretation of the posterior model probabilities is a bit more complicated here than in the previous examples, since the hypotheses investigated are not really representing competing theories but are of increasing specificity. Let us first conclude from the results that the null model ("nothing relevant is going on") can be discarded. The results show hardly any support for this hypothesis ($BF_{02} = 1.06$). Subsequently, let us examine the results of hypotheses H_{1a}, H_{1b}, and H_{1c}. These are the hypotheses just representing directional simple main effects of respectively Time (H_{1a}), Time and Gender (H_{1b}), and Time, Gender, and Kinship (H_{1c}). Compared to the unconstrained model, the addition of the simple main effect of Time increases the support ($BF_{1a,2} = 7.9$; i.e., larger than one). Addition of Gender to this model further increases the support ($BF_{1b,2} = 20.8$, which is larger than $BF_{1a,2}$). This is, however, not the case when the directional simple main effect of Kinship is also added. Compared to the previous

Table 4.9. Posterior model probabilities for the complicated grief data

Model	BF_{t2}	PMP
H_0	1.06	.02
H_{1a}	7.87	.14
H_{1b}	20.78	.38
H_{1c}	3.81	.07
H_{1d}	13.44	.25
H_{1e}	6.79	.12
H_2	1.00	.02

hypothesis (H_{1b}), the support has decreased ($BF_{1c,2} = 3.8$). The theory stating three directional simple main effects seems not to be the best explanation of the complicated grief data.

Possible explanations could be that there is no substantial effect of Kinship, or the hypothesized direction of the effect of Kinship is not supported, or there are specific interaction effects of the different factors that need to be taken into account. The last explanation is conform two of the alternative hypotheses and is therefore already included in the analysis. Hypothesis H_{1d} represents the expectation that losing a child is especially devastating for women. This hypothesis receives considerable support ($BF_{1d,2} = 13.4$). However, compared to the hypothesis stating just simple main effects of Time and Gender, the support is lower. Hypothesis H_{1e} represents the expectation that losing a child is more devastating for women *and* losing a partner is more devastating for men. This extended hypothesis about interaction of Gender and Kinship receives less support than the previous one ($BF_{1e,2} = 6.8$).

In conclusion, it is not entirely clear which hypothesis is the best. All constrained theories have a Bayes factor value larger than one, stating that compared to the unconstrained model, (some) support is found for the constraints. However, this support can be due mainly to the directional simple main effect of Time. Therefore, the Bayes factors of more specific theories were compared to less restricted hypotheses (e.g., H_{1b} with H_{1a} and H_{1c} with H_{1b}). This tells us that examining main effects only (in the form of directional simple main effects), the expected effect of Time and Gender were supported but the expected effect of Kinship was not. Furthermore, within different specifications of the expected interaction of Gender and Kinship, the part stating that losing a child is more devastating for women than for men was supported, but the part stating that losing a partner is more devastating for men than for women was not.

In Chapter 5 in this book, we will further reflect on the results and the interpretation of the three illustrations. In that chapter, inequality constrained analysis of variance using Bayesian model selection will be evaluated from a statistical point of view, including a comparison with a classic null hypothesis testing approach and corresponding advantages and disadvantages.

References

[1] Akaike, H.: Factor analysis and AIC. Psychometrika, **52**, 317–332 (1987)
[2] Berger, J.O., Delempady, M.: Testing precise hypotheses. Statistical Science, **2**, 317–352 (1987)
[3] Berger, J., Pericchi, L.: Objective Bayesian methods for model selection: Introduction and comparison. In: Lahiri, P. (ed) Model Selection. Monograph Series, Lecture Notes Vol. 38 pp. 135–207. Beachwood OH (2001)
[4] Bernardo, J.M., Smith, A.F.M.: Bayesian Theory. Chichester, Wiley (1994)
[5] Boelen, P.A., Bout, J. van den: Complicated grief, depression, and anxiety as distinct post-loss syndromes: A confirmatory factor analysis study. American Journal of Psychiatry, **162**, 2175–2177 (2005)
[6] Boelen, P.A., Keijser, J. de, Bout, J. van den: Psychometrische eigenschappen van de Rouw VragenLijst (RVL) [Psychometric properties of the Inventory of Traumatic Grief]. Gedrag & Gezondheid, **29**, 172–185 (2001)
[7] Bozdogan, H.: Model selection and Akaike's information criterion (AIC): The general theory and its analytic extensions. Psychometrika, **52**, 345–370 (1987)
[8] Carlin, B.P., Chib, S.: Bayesian model choice via Markov Chain Monte Carlo methods. Journal of the Royal Statistical Society, B, **57**, 473–484 (1995)
[9] Chib, S.: Marginal likelihood from the Gibbs output. Journal of the American Statistical Association, **90**, 1313–1321 (1995)
[10] Cohen, J.: The earth is round ($p < .05$). American Psychologist, **12**, 997–1003 (1994)
[11] Huntjens, R.J.C., Peters, M.L., Woertman, L., Bovenschen, L.M., Martin, R.C., Postma, A.: Inter-identity amnesia in dissociative identity disorder: A simulated memory impairment? Psychological Medicine, **36**, 857–863 (2006)
[12] Jefferys, W., Berger, J.: Ockham's razor and Bayesian analysis. American Scientist, **80**, 64–72 (1992)
[13] Kass, R.E., Raftery, A.E.: Bayes factors. Journal of the American Statistical Association, **90**, 773–795 (1995)
[14] Kato, B.S., Hoijtink, H.: A Bayesian approach to inequality constrained hierarchical models: Estimation and model selection. Statistical Modelling, **6**, 231–249 (2006)
[15] Klugkist, I., Hoijtink, H.: The Bayes Factor for inequality and about equality constrained models. Computational Statistics and Data Analysis, **51**, 6367–6379 (2007)
[16] Klugkist, I., Kato, B., Hoijtink, H.: Bayesian model selection using encompassing priors. Statistica Neerlandica, **59**, 57–69, (2005)
[17] Klugkist, I., Laudy, O., Hoijtink, H.: Inequality constrained analysis of variance: A Bayesian approach. Psychological Methods, **10**, 477–493 (2005)
[18] Laudy, O., Hoijtink, H.: Bayesian methods for the analysis of inequality constrained contingency tables. Statistical Methods in Medical Research, **16**, 123–138 (2007)

[19] Laudy, O.: Bayesian Evaluation of Equality and Inequality Constrained Hypotheses for Contingency Tables. Ph.D. thesis, Utrecht University (2006)

[20] Lindley, D.V.: A statistical paradox. Biometrika, **44**, 187–192 (1957)

[21] Newton, M.A., Raftery, A.E.: Approximate Bayesian inference by the weighted likelihood bootstrap. Journal of the Royal Statistical Society, B, **56**, 3–48 (1994)

[22] Reijntjes, A., Dekovic, M., Vermande, M., Telch, M.: The role of depressive symptoms in early adolescents online emotional responding to a peer evaluation challenge. Depression and Anxiety (in press)

[23] Smith, A.F.M., Spiegelhalter, D.J.: Bayes factors and choice criteria for linear models. Journal of the Royal Statistical Society, Series B, **42**, 213–220 (1980)

5

An Evaluation of Bayesian Inequality Constrained Analysis of Variance

Herbert Hoijtink[1], Rafaele Huntjens[2], Albert Reijntjes[3], Rebecca Kuiper[1], and Paul A. Boelen[4]

[1] Department of Methodology and Statistics, Utrecht University, P.O. Box 80140, 3508 TC Utrecht, the Netherlands h.hoijtink@uu.nl and r.m.kuiper@uu.nl
[2] Department of Experimental Psychopathology, Groningen University, Grote Kruisstraat 2/1, 9712 TS Groningen, the Netherlands r.j.c.huntjens@rug.nl
[3] Department of Pedagogical and Educational Sciences, Utrecht University, P.O. Box 80140 , 3508 TC Utrecht, the Netherlands a.h.a.reijntjes@uu.nl
[4] Department of Clinical and Health Psychology, Utrecht University, P.O. Box 80140, 3508 TC Utrecht, the Netherlands p.a.boelen@uu.nl

5.1 Introduction

In Chapters 2, 3, and 4 inequality constrained analysis of variance was introduced and illustrated. This chapter contains an evaluation of inequality constrained analysis of variance. Section 5.2 contains an evaluation from the perspective of psychologists on the use of inequality constrained analysis of variance. The questions raised will be discussed in Sections 5.3 and 5.4. Among other things, the interpretation of posterior model probabilities and the sensitivity of Bayesian model selection with respect to the choice of the prior distribution will be discussed.

5.2 An Evaluation from a Psychological Perspective

5.2.1 The DID Data

Certain controversies in clinical psychology are long-standing and heated, like the mechanisms underlying the technique of Eye Movement Desensitisation and Reprocessing (EMDR) and the advantages versus the risks of hypnotic techniques in the recovery of memories of the past, to name but a few. As noted in Chapter 2, the controversy surrounding Dissociative Identity Disorder (DID) certainly classifies as heated, with proponents and opponents not even agreeing on the basic phenomena that are at the core of this condition (i.e., the existence of multiple identities and the corresponding amnesia between identities). Such a controversy sometimes results in articles in which opposing claims are systematically evaluated [10, 16]. At the same time, controversies such as these sometimes lead those involved to use arguments by authority or arguments based on emotion rather than reason. Moreover, they can cause researchers to focus solely on research that is in accord with their own line of thinking, ignoring other theoretical and empirical contributions and, important in the current context, alternative hypotheses.

It is precisely the area of hypothesis formulation in which Bayesian analyses seems to have additional value over and above traditional hypothesis testing. By allowing for a test of informative hypotheses, the Bayesian approach encourages researchers to explicitly formulate expected and alternative hypotheses. In the context of DID research, proponents of the so-called "trauma model" [12] argue that dissociative amnesia is a defensive reaction to overwhelming trauma, resulting in painful memories becoming fragmented or separated from other parts of autobiographical memory, rendering them inaccessible to normal consciousness. Proponents of the opposing model, the so-called "sociocognitive model" [16], argue that people adopt the role of DID-patient, acting according to their perception of how a DID-patient should behave, which is shaped by therapist education, television documentaries, and self-help books. In this model, reported amnesia is regarded an element of the immersed role enactment. Informative hypothesis formulation results in explicitly juxtaposing "the patients resembling amnesiacs hypothesis" and "the patients resembling simulators hypothesis." Hopefully this approach will encourage researchers from both theoretical stances to use designs incorporating the hypotheses generated by both the trauma model and sociocognitive model, with past research in the sociocognitive tradition only focusing on the ability of normal controls to simulate symptoms of DID and most studies in the trauma model tradition testing patients without including DID simulators and/or objective memory tasks.

Bayesian statistics seems specifically worthwhile in studies including multiple hypothesis testing (i.e., designs with more than two comparison groups and/or multiple measures). In the DID study, besides a recognition test, we also included a recall test, and besides a normal control group and a simulator group, an additional control group was included to assess the unusualness of each answer alternative selected by patients (i.e., choosing an answer like "animal" when the correct answer was "woman" to the question "was the story about a man, woman, or animal?"). This entailed a lot of (post hoc) testing, while actually, we were interested in the same two critical hypotheses in all these tests. The multiple post hoc testing needed to reach a conclusion regarding these measures in the classical approach could have been reduced by using the informative hypothesis testing procedure that is part of the Bayesian approach.

Not only does the informative hypotheses formulation in the Bayesian tradition encourage the explicit formulation of competing hypothesis, but it also enables the systematic evaluation of these competing hypotheses by directly comparing hypotheses against each other instead of evaluating them one by one against the traditional null hypothesis. The calculated posterior probabilities and Bayes factor provide a clear-cut quantitative index of hypothesis support. If there are any disadvantages to be mentioned, it should be that in some situations, it may be difficult for the researcher to decide when the difference between probabilities is "big enough" to favor one hypothesis over another.

One way to guard against indecisive Bayes factors is to specify more informative hypotheses. In the DID example, DID-patients were expected to perform either comparable to Simulators (as hypothesized by the sociocognitive model) or comparable to True amnesiacs (as hypothesized by the trauma model). The equivalence of DID-patients to either True amnesiacs or Simulators can be formulated using (about) equality constraints, as in the following hypotheses:

$$H_{1a} : \mu_{con} > \{\mu_{amn} = \mu_{pat}\} > \mu_{sim},$$

$$H_{1b} : \mu_{con} > \mu_{amn} > \{\mu_{pat} = \mu_{sim}\}.$$

Alternatively, a less restrictive translation of the competing expectations is using an imposed ordering on the group's means without restricting some means to be (about) equal, as in the hypotheses

$$H_{1c} : \mu_{con} > \{\mu_{amn}, \mu_{pat}\} > \mu_{sim},$$

$$H_{1d} : \mu_{con} > \mu_{amn} > \{\mu_{pat}, \mu_{sim}\}.$$

If one assumes the mean of DID-patients to be about equal to either the True amnesiacs or the Simulator mean (H_{1a} and H_{1b}), this allows for more specific evaluation of hypotheses compared to only assuming that the means of these three groups are ordered (H_{1c} and H_{1d}). As can be seen in Table 4.5 in Chapter 4, the more specific formulation using (about) equality constraints yielded a clear preference for the patients equal simulators hypothesis (with a posterior model probability of .782 for H_{1a} compared to .218 for H_{1b}). The less restrictive translation yielded a less informative "no preference" for either of the alternative theories (with PMPs of .499 and .501 for H_{1c} and H_{1d}, respectively).

Additionally in the formulation of hypotheses, there is another important way of incorporating previous theoretical knowledge in Bayesian analyses, and that is in specifying the prior distribution of the parameters. While psychologists have experience in theoretical hypothesis formulation (e.g., in specifying group comparisons) be it sometimes with less restrictive hypotheses (cf. hypothesis (2.5) and (2.6) in Chapter 2), it is harder to come up with hypotheses about the expected values of test parameters like means or variances, or even more so, expectations about the distribution of those parameters. This is especially true in the case of a new patient group for which comparable test results are not available or in the situation where large differences in test performance have been found due to group differences in, for example, age, education, and symptom severity. Moreover, knowing that different prior distributions may lead to different conclusions may render colleagues not to become enthusiastic about using Bayesian statistics in their research. However, as is explained in previous chapters, in most examples the unconstrained prior is chosen to be low or noninformative, in the sense that parameters are assumed to be normally distributed and the range determined by the minimal and maximal scores possible on the test or the highest and lowest scores found in the data. These kind of constraints do not pose problems in most instances of psychological research, be it either experimental or survey research, and lead to "objective" results determined by the observed data, not overly restricted or influenced by "subjective" priors.

5.2.2 The Emotional Reactivity Data

It has been noted for a long time that proficiency in emotion regulation is a fundamental prerequisite for adaptive daily functioning, including feelings of general well-being and the capacity to work and to relate to others (cf. [9]). However, people can experience serious difficulties in modulating their emotions in response to contextual demands.

Depression in particular has been increasingly conceptualized as a disorder of emotion regulation (cf. [11, 15, 19]). Indeed, the core emotional symptoms of depression – persistent sad mood and the diminished capacity to experience pleasure and enjoyment (i.e., anhedonia) – strongly allude to difficulties with emotion regulation.

Most researchers agree that depression involves abnormalities in the temporal course of an emotional response as it unfolds over time (i.e., emotion dysregulation), such as difficulty sustaining or enhancing positive affect, difficulty terminating sadness, or both (cf. [8, 18]). However, as noted in Chapter 2, the linkage between depression and the magnitude of emotional reactivity (i.e., emotion activation) when faced with a salient stimulus event is less undisputed. Specifically, according to the proponents of the "mood-facilitation" hypothesis, enduring mood states potentiate matching emotions. They thus predict that those higher in depression experience higher levels of emotional distress (e.g., sadness) when faced with a negative event like peer rejection and weaker increases in happiness/joy after being faced with a "success" event like peer praise. In contrast, based on research among adults suffering from a major depressive disorder (MDD), the "emotional context insensitivity" hypothesis has been advanced, positing that depressed mood is linked to attenuated emotional reactivity to both positive and negative stimuli and hence predicts that those higher in depression react in a more flattened way to both peer praise and peer rejection (i.e., with relatively low levels of emotional reactivity). Finally, it may be that emotional reactivity in depression is mainly governed by the extent to which the peer evaluation outcome is (in)consistent with individuals' expectations and/or self-views. If this "discrepancy" hypothesis is correct, children high in depression will display higher increases in state mood after (unexpected) success and weaker decreases in mood after (presumably less unexpected) failure, both relative to children low in depression.

To test these three competing hypotheses (i.e., mood facilitation, emotion-context insensitivity, and discrepancy), it stands to reason to compare mood changes after success and failure relative to mood changes following a neutral/reference condition (in the neutral condition, emotional reactivity may also differ across different levels of depression; see Chapter 2). A traditional frequentist approach typically entails a 3×3 ANOVA analysis, with depression status (e.g., low, medium, high) and feedback condition (succes, neutral, and failure) serving as the between-subjects factors and changes in state mood from prefeedback to postfeedback serving as the dependent variable. This approach will yield results that are informative with regard to the (non)significance of depression status, feedback outcome, and their interaction.

However, an important drawback of this approach is that the obtained results do not directly speak to the major question of interest; namely "which of these three competing hypotheses should be preferred because they fit the data best?" Rather, designs with multiple comparison groups or multiple measures typically require post hoc testing procedures and pairwise comparisons that may exert detrimental effects on either type I or type II error. In fact, the typical (somewhat cumbersome) procedure is to test each of the hypotheses separately against the traditional null hypothesis.

It has been suggested that in these instances the informative hypothesis testing procedure that is part of the Bayesian approach may prove useful because it allows for directly comparing the relative merit of competing hypotheses. Indeed, as noted earlier in this chapter, one major advantage of this alternative approach is the possibility to directly test competing hypotheses against each other in the context of a single analysis. Moreover, it appears that by forcing researchers to be explicit about all their hypotheses (both expected and alternative) and allowing for the possibility to incorporate theoretical knowledge in the specification of the prior distribution of the parameters, scientific progress may be more rapidly brought about.

Taken together, when conducting studies that involve testing multiple competing hypotheses, it appears that the Bayesian approach offers important strengths and may outperform more traditional approaches (e.g., ANOVA).

However, it should be noted that the Bayesian approach is mute with regard to one pivotal issue, namely based on what criterion can we decide that one hypothesis is "better" than the other? Sure, one can easily say that a hypothesis with a PMP of .80 is more strongly confirmed than a hypothesis with a PMP of .01. However, how should we judge the difference between the fit of competing hypotheses if the differences between the two PMPs would be, say, .05, rather than .79?

Moreover, formulation of inequality constrained hypotheses with respect to means may prove difficult, especially when research on a certain topic is still in its infancy. It may therefore be that the added value of the Bayesian approach increases as a function of the amount of similar prior research from which competing hypotheses can be derived.

5.2.3 The Grief Data; Everybody Has Won?

Gender differences in psychological functioning is an intriguing topic that lends itself to nice discussions. As noted in Chapter 2, gender differences is also a popular topic in psychological research. Among other areas, gender differences have been studied in the field of coping with trauma and loss. There are several theoretical reasons why it is important to study gender differences in coping with stressors. Generally stated, it is known that there are differences between men and women in the way they cope with stress. Men are generally more inclined to engage in problem-focused coping strategies (What is the problem? How can I solve it?), whereas women have a tendency to engage in emotion-focused coping (How can I deal with my feelings concerning this problem?). In addition, men and women have different roles and are treated differently in our Western society. Noteworthy too is that there are obviously biological differences between men and women (e.g., differences in sex hormones and immune system functioning). With this in mind, it seems fair to say that knowledge of the differential impact of stress on the psychological and emotional well-being of men and women sheds light on coping styles, societal influences, and biological factors that potentially contribute to this differential impact.

Although scientifically important, the question "who suffers more?" in the context of loss and trauma is all but decided. From the viewpoint of scientific research, there is no final answer. For instance, in literature on coping with loss, some studies have found no differences [5], others found that men suffer more [4], and still others concluded that loss has a more profound emotional impact on women [22].

Although we seem to continue to have an urge to look for a simple answer, the question "who suffers more?" is a difficult one, simply because there is whole range of variables that moderate the impact of gender on problems after negative life events. One of the things that we stressed in Chapter 2 is that we cannot ignore variables such as time since the death occurred and the kinship relationship with the deceased when studying differences between men and women in coping with loss. But that is where it easily gets complicated.

As noted in Chapter 2, we had a few hypotheses concerning the interplay of gender, time from loss, and kinship in contributing to grief symptomatology. Specifically, we hypothesized that (a) recent losses coincide with stronger grief reactions than remote losses, (b) women suffer more than do men, irrespective of time and kinship, and (c) the loss of a child is always more devastating than the loss of a partner, irrespective of time and gender. Apart from the expectation of such one-sided simple main effects, there were two additional, more complex hypotheses: (d) one stating that losing a child is more devastating for women than for men, and more devastating for women than losing a partner and (e) a final hypothesis stating that for men, but not women, partner loss results in more severe grief, whereas for women, but not men, the loss of a child is more disrupting.

From a conventional viewpoint, one could consider exploring the impact of these variables on grief, in a 2 (men, women) $\times$ 2 (recent, remote loss) $\times$ 2 (partner, child death) ANOVA design. Yet, again, this easily becomes complex. The complexity of such a design starts when setting up a study to test it, given that the inclusion of this number of variables requires a fairly large sample size. Moreover, when performing such a three-way ANOVA, one is immediately faced with findings that are difficult to interpret. For instance, in itself, the finding of a significant interaction among gender, time, and kinship is difficult to interpret and requires a whole set of additional dismantling steps and pairwise comparisons. Moreover, if one does engage in running such an ANOVA and exploring the outcomes of pairwise comparisons, as was done in Chapter 2, one only gets two statistically significant findings. Together, these suggest that recent losses coincide with stronger grief reactions than remote losses among women confronted with the death of a partner. Although such a conclusion tells us something about gender differences in coping with loss, it leaves us wondering if this is really the only thing that can be said based on such an extensive set of analyses and to what extent limitations of the sample size are responsible for the fact that so little is found. So, all in all, a conventional approach to the examination of the contribution of gender, time from loss, and kinship to grief severity would leave us with a whole bunch of outcomes that are difficult to interpret and with an unsatisfactory evaluation of our a priori hypotheses.

It was exactly this observation that formed the starting point for the Bayesian evaluation of the competing hypotheses about the grief data, in Chapter 4. The analyses in that chapter nicely illustrated one of the key points of this book that were also noted in the earlier sections of this chapter: that the Bayesian approach to inequality constrained hypothesis testing allows for a relatively easy examination of the appropriateness or fit of different competing hypotheses. More or less explicitly, Huntjens et al. [14] had competing hypotheses about the relative performance of different groups on a memory task and the Bayesian approach allowed for a direct comparison of these hypotheses (see Chapters 2 and 4). Likewise, the research on gender differences in grief included similar alternative hypotheses, with one important additional feature and that is that the hypotheses in that research gradually included more group comparisons and more specific competing ideas about interactions. The analyses in Chapter 4 showed that the Bayesian approach allowed for a relatively easy evaluation of these hypotheses. In the end, these analyses suggested that gender and time, but not kinship had an impact on grief severity. Moreover, findings indicated that the hypothesis that women (but not men) suffered more after losing their child than after losing their partner was supported by the data more than the hypotheses that women suffer more after losing a child and men more after losing a partner. Although one can argue that these findings leave room for alternative interpretations, they seem more informative than the conclusion that grief only varies as a function of time, in women who lost a partner based on the conventional ANOVA.

So, this summary further underscores the pros of the Bayesian approach that have been pointed out at several places in this chapter and other parts of this book. However, one can also be more critical and point at the potential drawbacks of this approach. One pitfall is the construction and selection of informative hypotheses. It seems that the usefulness of applying the Bayesian versus the traditional, frequentist approach to the test of group differences varies as a function of the degree to which one is able to formulate informative competing hypotheses.

In a related vein, when looking at the matter through the eyes of a "traditional Null Hypothesis Significance Testing (NHST) psychologist," one might well continue to have a need for "rules of thumb": When do we know for sure that one of our competing hypotheses fit best – a Bayes factor of 20.78 is higher than a Bayes factor of 13.44, how exactly should we interpret this difference in magnitude? Moreover, it is conceivable that some researchers prefer to have clear rules of thumb for determining an appropriate subjective prior (but see Section 5.4 in which it is elaborated that priors that may look subjective are not really subjective). An additional question is: What particular findings of earlier research are worthy of being included in the determination of priors that have the form of inequality constrained hypotheses and what findings should be excluded in such a process? Generally, the issue of how to translate a theory into inequality constrained hypotheses and which constraints should and should not be included in a set of alternative hypotheses is a largely unexplored territory that needs further scrutiny.

But then, when looking at the matter through the eyes of a "Bayesian psychologist," one can easily argue that "NHST psychologists" think that they have more sound rules of thumb than their Bayesian counterparts, but this is an incorrect assumption; that is, as will be elaborated further on in this chapter, why is $p = .049$ but not $p = .051$ a significant finding?

Nevertheless, the frequentist approach is still more popular than the Bayesian approach. One can argue that if Bayesians have the ambition of becoming more popular, they should perhaps develop rules of thumb (although, by their very nature, they seem inclined to avoid doing so at all costs) or – more realistically – should try to make a stronger case in showing that the existence of rules of thumb in NHST is based on illusions. Or perhaps the Bayesians should make a stronger case in defending the parts of the statistical realm that they are much better at handling than are the frequentists (e.g., directional hypotheses such as the ones formulated for the grief data). This might lead to a fruitful struggle that eventually could result in the situation in which "everybody has won, and all must have prices."

5.3 Classical Hypotheses Testing Versus Bayesian Evaluation of Informative Hypotheses

In all the three research projects introduced in Chapter 2 the research questions could be represented by a set of informative hypotheses. A few of these hypotheses are the following

$$H_{1b} : \mu_{con} > \mu_{amn} > \{\mu_{pat} = \mu_{sim}\} \tag{5.1}$$

from the DID project, which states that controls perform better than true amnesiacs, who, in turn, perform better than patients and simulators (who perform equally well);

$$H_{1a} : \{\mu_7 - \mu_8\} < \{\mu_4 - \mu_5\} < \{\mu_1 - \mu_2\},$$
$$\{\mu_9 - \mu_8\} < \{\mu_6 - \mu_5\} < \{\mu_3 - \mu_2\} \tag{5.2}$$

from the emotional reactivity in children project, which states (a) that relative to the neutral condition, children in the success condition are increasingly emotionally reactive moving from low via moderate to highly depressed children and (b) that controlled for the neutral condition, children in the failure condition are increasingly emotionally reactive moving from low via moderate to highly depressed children; and

$$H_{1a} : \mu_1 > \mu_2, \mu_3 > \mu_4, \mu_5 > \mu_6, \mu_7 > \mu_8 \tag{5.3}$$

from the gender differences in coping with loss project, which states that recent loss leads to more grief than remote loss for each combination of gender of the respondent and kinship to the deceased (partner or child).

For each of the three research projects, Chapter 2 also presented traditional data analyses based on null hypothesis testing. For the DID data, for example, the null hypothesis

$$H_0 : \mu_{con} = \mu_{amn} = \mu_{pat} = \mu_{sim} \tag{5.4}$$

was tested against the alternative

$$H_2 : \mu_{con}, \mu_{amn}, \mu_{pat}, \mu_{sim}. \tag{5.5}$$

Since the p-value was rather small (.00), H_0 was rejected in favor of H_2. Subsequently, a pairwise comparison of means analysis was executed to determine which means differ from each other. All pairs of means were significantly different with the exception of μ_{pat} and μ_{sim}. Since this is in accordance with (5.1), in a final step the sample means were inspected to determine if their order was in agreement with (5.1). This inspection revealed that the order of the means and (5.1) were in agreement.

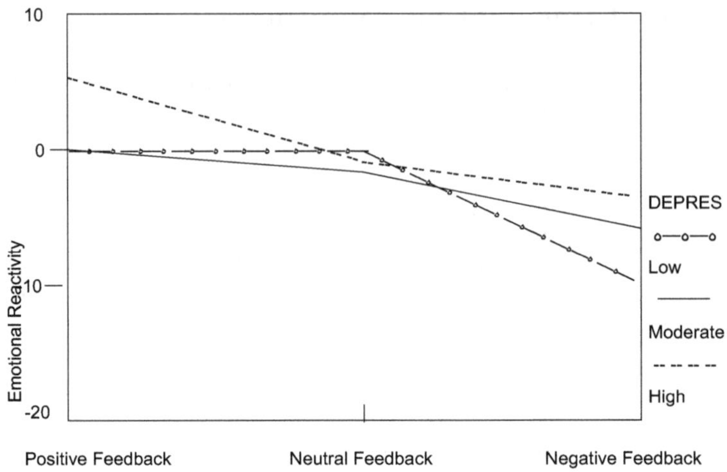

Fig. 5.1. Visual display of average emotional reactivity in the nine experimental groups

Another example is the traditional analysis executed on the emotional reactivity in children data. A two-way analysis of variance rendered two significant main effects (depression level and feedback condition) and a non-significant interaction effect. Since this rendered insufficient information to determine whether the data are in accordance with (5.2) and the other informative hypotheses, Figure 5.1 was inspected, which displays the means for each combination of depression level and feedback condition. Since from Figure 5.1 it is hard to determine which differences between means are significant, it was concluded that further testing of differences between means would be necessary to avoid overinterpretation of the figure. In the end, no real conclusion with respect to the informative hypotheses under investigation was obtained. As was illustrated in Chapter 2, classical hypothesis testing was also not helpful for the evaluation of the informative hypotheses formulated for the gender differences in grief project.

5.3.1 Classical Hypothesis Testing

The disadvantages of traditional data analysis based on hypothesis testing if the goal is to evaluate a set of informative hypotheses were also discussed in Chapter 2. The main issues will now be summarized.

The Null Hypothesis

In none of the three examples is the traditional null hypothesis of importance. Stated otherwise, in none of the three research projects did the null hypothesis seem to represent a reasonable theory. Since statistical inference should provide information with respect to the hypotheses that are of interest to a researcher, this renders NHST an insignificant tool for the evaluation of informative hypotheses.

Furthermore, what if the null hypothesis is not rejected? Not many social scientists think that the null hypothesis gives a reasonable description of the population in which they are interested [7, 23]. Scientifically interesting populations where truly nothing is going on are probably very rare. Furthermore, many statisticians ([2, 3, 20]; see also Chapter 9) consider effects or differences that are exactly null unrealistic and consequently share the opinion of the social scientists. Only one answer to the question posed in the first line of this paragraph seems possible: There is not enough power [6] to reject the null; that is, the sample size is too small to reject the null hypothesis.

The Alternative Hypothesis

Loosely formulated, alternative hypotheses like (5.5) state "something is going on, but I don't know what." Consequently, if the null is rejected in favor of the alternative, it is still not clear what is going on. A common course of action is to continue with further testing (e.g., pairwise comparison of means) followed by a visual inspection of the data (e.g., tables of means or figures displaying means) to determine "what is going on." Two examples of this process were given above for the amnesia data and the emotional reactivity data.

This procedure has several disadvantages if the goal is to evaluate a set of informative hypotheses. Consider again the hypothesis (5.1). What should be done if all pairwise comparisons would have been significant? In that case, neither (5.1) nor the competing hypothesis $H_{1a} : \mu_{con} > \{\mu_{amn} = \mu_{pat}\} > \mu_{sim}$ would be supported by the data. Should it be concluded that both hypotheses are wrong? To give another example, suppose that all but one of the differences in Figure 5.1 are in agreement with a hypothesis like (5.2). Does this imply that the hypothesis is strongly supported by the data, or does it imply that the hypothesis should be rejected? What if all but two of the differences are in agreement with a hypothesis? What if the violation of the hypothesis is minor; that is, the data are almost in accordance with the restrictions?

There are two reasons for further testing. First, for the amnesia data it is clear that the alternative hypothesis (5.5) has a relation with the informative hypotheses under consideration since (5.1) is just a further specification of (5.5). However, looking at the sample means to check if their values are in accordance with an informative hypothesis like (5.1) may lead to overinterpretation. A rejection of the null hypothesis does not imply that all pairwise differences between means are different from zero; it implies that at least one of the pairwise differences is different from zero. Overinterpretation – that is interpretation of nonsignificant differences between means – can only be avoided if each pairwise difference is tested before checking whether the values of the means involved are in agreement with the informative hypothesis under investigation. In the next section the disadvantages of such multiple hypothesis testing will be discussed.

Second, there may not be a straightforward relation between the alternative hypothesis and the informative hypotheses under investigation. The hypotheses tested for the emotional reactivity in children data (main and interaction effects of depression level and feedback condition) do not have a straightforward relation with the informative hypotheses under investigation. The hypotheses tested for the coping with loss data (main and interaction effects in a three-way analysis of variance) also do not have a straightforward relationship with the directional simple effects that are used to construct informative hypotheses like (5.3). This problem can be solved by testing more specific hypotheses like the two-way analysis of variance executed for pairs of groups from both conditions in the emotional reactivity in children data.

Multiple Hypothesis Testing

Classical statistical inference can handle the testing of *one* hypothesis. The probability of a type I error (also called the alpha-level) is under control. A usual value is .05; that is, if the null hypothesis is true, the probability that it is incorrectly rejected is 5%. Once the sample size is known and a researcher has specified the smallest effect size (e.g., the smallest difference between two means) that he is interested in, the probability of a type II error (also called the beta-level) can be computed. A usual value is .20; that is, the probability that the null hypothesis is accepted if it is false is 20%. Perhaps better known is the counterpart of the type II error, the power [6]. Power is the probability to reject the null hypothesis if it is false. A usual value for the power is .80.

However, classical statistical inference is in trouble when more than one hypothesis is tested. For example, pairwise comparison of means for the amnesia data involves the evaluation of six null hypotheses (one for each pair of means). This implies that there are six opportunities to obtain a type I error. Stated otherwise, the probability of one or more type I errors will be (much) larger than .05. To give an example, for six independent tests this probability is $1 - .95^6 = .265$. If the fact that multiple hypotheses are tested is ignored, data analysis will render many effects that do not exist in the population from which the data were sampled.

There are procedures that can be used to control the number of type I errors when multiple hypotheses are tested. Perhaps the best known procedure is the Bonferroni correction [13]. It can be proven that the probability of one or more type I errors is smaller or equal to .05 if the p-value of each hypothesis is compared to $.05/K$ (where K denotes the number of hypotheses tested). Stated otherwise, if a p-value is smaller than $.05/K$, reject the corresponding hypothesis; otherwise, accept the hypothesis. More refined procedures than the Bonferroni correction exist [1], but the basic principles involved are similar. However, corrections for multiple hypothesis testing come at a price: they lead to an increase in the number of type II errors. There is a relation between type I errors and power: smaller probabilities (e.g., $.05/K$ instead of .05) of incorrectly rejecting the null hypothesis imply smaller probabilities of correctly rejecting the null hypothesis. If the power was computed to be .80 for an alpha-level of .05, it will be smaller than .80 for an alpha-level of $.05/K$. Stated otherwise, if many hypotheses are tested and the researcher chooses to control the number of type I errors, this implies an increase in the number of type II errors; that is, data analysis will not render many effects that do exist in the population from which the data are sampled.

Another disadvantage of multiple hypothesis testing, as we applied it to the amnesia, the emotional reactivity, and the grief data, is the fact that the inequality constraints in the alternative hypothesis are ignored. It would be much better if the null hypothesis is tested directly against hypotheses like (5.1), (5.2), and (5.3). This would avoid the multiple testing problem. Furthermore, such a test would also be more powerful because hypotheses like (5.1), (5.2), and (5.3) are more specific than the unconstrained alternative (5.5). A nice overview of such testing procedure is given in [21]. However, even these procedures have disadvantages. Tests like

$$H_0 : \mu_{con} = \mu_{amn} = \mu_{pat} = \mu_{sim} \tag{5.6}$$

against

$$H_{1b} : \mu_{con} > \mu_{amn} > \{\mu_{pat} = \mu_{sim}\} \tag{5.7}$$

exist only for a limited class of models and hypotheses and cannot generally be applied. Furthermore, suppose that besides (5.7) a researcher is also interested in

$$H_{1a} : \mu_{con} > \{\mu_{amn} = \mu_{pat}\} > \mu_{sim} \tag{5.8}$$

and that both the test of H_0 against H_{1b} and the test of H_0 against H_{1a} render a p-value smaller than .05. What is then to be concluded? Furthermore, what is to be concluded if both p-values are larger than .05. In that case is the null hypothesis the best hypothesis? But we were not interested in the null to begin with! With which we have gone full circle, and have returned to the beginning of the subsection "The Null Hypothesis" that can be found above.

5.3.2 Bayesian Model Selection

Chapter 4 introduced Bayesian model selection as a method for the evaluation of informative hypotheses. In this section it will be argued that Bayesian model selection is better suited for the evaluation of informative hypotheses than classical hypothesis testing. The structure of the argument will be similar to the structure of our criticism of classical hypothesis testing in the previous section. Bayesian model selection also has an Achilles' heel (the specification of the prior distribution). This will be discussed in the next section.

As was explained in Chapter 4, Bayesian evaluation of informative hypotheses consists of three steps. In the first step a researcher has to translate his theories into a set of competing hypotheses. An example of a set of competing hypotheses is $H_{1a} : \mu_{con} > \{\mu_{amn} = \mu_{pat}\} > \mu_{sim}$ and $H_{1b} : \mu_{con} > \mu_{amn} > \{\mu_{pat} = \mu_{sim}\}$ from the amnesia example. If there is a theory that $H_0 : \mu_{con} = \mu_{amn} = \mu_{pat} = \mu_{sim}$ could actually represent an interesting state of the population, it can be added to the set of competing hypotheses. In the second step a researcher has to compute the posterior model probability (PMP) of each hypothesis under consideration. In Chapter 4 it can be found that these probabilities were .22 and .78 for H_{1a} and H_{1b}, respectively. This implies that after observing the data H_{1b} was about $3.5 \approx .78/.22$ as likely as H_{1a}. In the third step the researcher has to decide how to evaluate the PMPs. In contrast to classical statistics, clear decision rules do not exist. The ".05-rule" that is so abundantly used for the evaluation of p-values does not have a counterpart that can be used for the evaluation of posterior model probabilities. The reason for this is that such rules are often more a nuisance than a help. What if a researcher finds a p-value of .051? Most researchers would have preferred to find .049. But, paraphrasing [17], "surely god loves the [.051] almost as much as the [.049]." To give just another example. Suppose a researcher tests six hypotheses and finds the following p-values: .051, .051, .051, .051, .051, and .051. Should he conclude that nothing is going on (strict adherence to the ".05-rule")? Or should he conclude that six such small p-values are very unlikely if nothing is going on, so something must be going on? Thus, it is likely that strict guidelines for the evaluation of posterior model probabilities are bound to have the same flaws as the ".05-rule." Furthermore, since posterior model probabilities are straightforward quantifications of the support in the data for the hypotheses under investigation, such guidelines are not necessary. Consequently, posterior probabilities of .22 and .78 for H_{1a} and H_{1b}, respectively, do not have to be transformed into a dichotomous "it is H_{1a}" or "it is H_{1b}" decision. The message is clear: H_{1b} receives more support from the data than H_{1a}, but the latter cannot yet be ruled out completely.

The Null Hypothesis

For a Bayesian evaluation of informative hypotheses the null hypothesis is not needed. Stated otherwise, if the null hypothesis does not represent a plausible theory, it does not have to be considered. The Bayesian approach is not tied to the null hypothesis (and null hypothesis related problems) in a way that classical hypothesis testing is.

The Alternative Hypothesis

The Bayesian approach directly evaluates the hypotheses of interest. This avoids the problems that occur when classical hypothesis testing is used for the evaluation of informative hypotheses. PMPs directly quantify the support in the data for each hypothesis under consideration. This avoids the problem that it can be hard or impossible to determine to which degree analysis results support each of the hypotheses under consideration. This removes the necessity to inspect tables of means or figures to determine the support for each informative hypothesis. This also avoids the problem of a vague relation between the hypotheses tested and the hypotheses of interest, because in the Bayesian approach there is no difference between both sets of hypotheses.

What can happen in a Bayesian analysis is that none of the informative hypotheses under consideration appropriately reflects the state of affairs in the population of interest. Note that this can only happen if the researchers who formulate the hypotheses are completely ignorant with respect to the state of affairs in the population. The latter is probably as unlikely as the possible truth of a classical null hypothesis. Nevertheless, it can happen. This problem can be detected if the unconstrained hypothesis (i.e., the classical alternative hypothesis) is added to the set of competing hypotheses. The amnesia example can be used to illustrate this. The informative hypotheses considered there were $H_{1a} : \mu_{con} > \{\mu_{amn} = \mu_{pat}\} > \mu_{sim}$ and $H_{1b} : \mu_{con} > \mu_{amn} > \{\mu_{pat} = \mu_{sim}\}$ with PMPs of .22 and .78, respectively. Although it is clear that H_{1b} is preferred to H_{1a}, it may still be a bad representation of the population of interest. However, if H_{1b} is also a better hypothesis than $H_2 : \mu_{con}, \mu_{amn}, \mu_{pat}, \mu_{sim}$, this means that the constraints in H_{1b} receive more support from the data than the absence of constraints in H_2. As can be read in Chapter 4, the PMPs for H_{1a}, H_{1b}, and H_2 were .20, .71, and .09, respectively. This shows that H_{1b} is preferred to H_{1a} and that a hypothesis with constraints is a better hypothesis than a hypothesis without constraints. Had the PMPs been, .20, .09, and .71, respectively, it would have been concluded that neither H_{1a} nor H_{1b} is preferred to H_2 and that the constraints in both hypotheses are not supported by the data.

Multiple Hypothesis Testing

Since Bayesian model selection does not use p-values, multiple testing problems like too many false positives if the number of type I errors is not controlled or too many false negatives if the number of type I errors is controlled do not occur. However, it is worthwhile to shortly consider a related problem that can occur.

Suppose a researcher has formulated two competing hypotheses. The resulting PMPs could be .92 and .08 for hypotheses 1 and 2, respectively. The support in the data is .92/.08=11.5 times stronger for hypothesis 1 than for hypothesis 2. From these results many researchers would conclude that it is relatively safe to discard hypothesis 2. Suppose now the same researcher did not start with a set of two hypotheses, but with a set of eight competing hypotheses (including the original two). The resulting PMPs could now be .69, .06, .05, .05, .04, .04, .04, and .03, respectively. The support in the data for hypothesis 1 versus the combination of the other seven hypothesis is now

$.69/.31 \approx 2$. Although hypothesis 1 is still clearly the best hypothesis, the certainty with which this can be claimed has been reduced because of the extension of the set of hypotheses under investigation.

If only a limited number of competing theories/hypotheses are under investigation, a researcher has either excluded all other theories through the design of his study or is very knowledgeable with respect to his area of research, which allows him to exclude all other theories prior to looking at the data. This holds, for example, for each of the three research projects that were discussed in Chapter 2. Larger sets of hypotheses can occur if the design is more complicated and/or the amount of prior knowledge is limited. Consider, for example, a researcher who wants to determine which of four variables can be used to predict a fifth variable using multiple regression. This researcher has to consider one hypothesis with four predictors, four with three predictors, six with two predictors, four with one predictor, and one hypothesis without predictors; that is, a total of 16 hypotheses. Model selection with only a few well-chosen competing hypotheses is confirmatory of nature. The example with 16 hypotheses is completely exploratory: Which of all possible hypotheses is the best? The latter could also be described with the phrase "fishing for the best hypothesis." As illustrated in the previous paragraph, the larger the number of hypotheses under consideration, the larger the probability of an erroneous decision.

This once more illustrates one of the main messages we try to bring across in this book: Use the existing knowledge, ideas, experience and previous research to formulate a set of plausible competing hypotheses. This ensures that a research project does not start from scratch, but that all irrelevant hypotheses are a priori excluded from the analysis and, consequently, that the current state of knowledge is appropriately accounted for in the set of hypotheses that are under consideration. Assuming that researchers are knowledgeable with respect to their research field, this will strongly decrease the probability of erroneous decisions and thus lead to a faster advancement of scientific knowledge.

However, Bayesian model selection has an Achilles' heel: the specification of the prior distribution. The next section will show that in the context of selecting the best of a set of (in)equality constrained hypotheses, three different ways to specify the prior distribution renders rather similar results. Stated otherwise, for (in)equality constrained hypotheses, Bayesian model selection is rather robust with respect to the precise specification of the prior distribution of the model parameters.

5.4 Bayes' Achilles' Heel: The Prior Distribution

As shown and discussed in Chapter 4, selecting the best from a set of hypotheses is sensitive to the prior specification, if equality constraints are used to specify one or more of the hypotheses. Therefore, the prior distribution should be chosen carefully; that is, those aspects of prior distributions that hypotheses like

$$\begin{aligned} H_0 &: \ \mu_1 = \mu_2, \\ H_1 &: \ \mu_1 > \mu_2, \\ H_2 &: \ \mu_1, \ \mu_2 \end{aligned} \tag{5.9}$$

have in common must have little impact on the support in the data for each hypothesis. As was shown in Chapter 4, this can be achieved if the the prior distribution for constrained hypotheses like H_0 and H_1 is derived from the prior distribution of the corresponding unconstrained hypothesis (in this simple example, H_2). Differences between the hypotheses (unconstrained, inequality constrained, or equality constrained) must have a large impact on the support in the data for each of the hypotheses. In what follows, three different prior specifications that attempt to achieve this will be discussed and compared:

- The specification of the encompassing prior (described in Chapter 4)
- The specification of a (normalized) conditional prior (to be described in Chapter 8)
- The specification of a prior distribution for the noncentrality parameter of a relevant test statistic (to be described in Chapter 6).

In this chapter the main features of each prior specification will be summarized. Subsequently, using the three examples from Chapter 2, it will be shown that Bayesian model selection using PMPs, renders rather similar evaluations of the hypotheses under consideration for each of the three different prior specifications.

5.4.1 The Encompassing Prior

In the encompassing prior approach described in Chapter 4, the prior distribution of constrained hypotheses (like $H_0 : \mu_1 \approx \mu_2$ and $H_1 : \mu_1 > \mu_2$) can be derived from the prior distribution of the corresponding unconstrained/encompassing hypothesis (like $H_2 : \mu_1, \ \mu_2$). Consequently, only the (encompassing) prior distribution for H_2 needs to be specified. Note that in Chapter 4, $\approx$ is defined as $|\mu_1 - \mu_2| < d$. In the tables that follow, the value of d used in the computation of PMPs will be reported.

The encompassing prior is specified using the following guidelines:

- All model parameters are a priori independent.
- The prior distributions for all parameters that are constrained in one or more of the hypotheses are equal.
- The encompassing prior is a relatively noninformative, conjugate, data-based distribution.

According to the first specification guideline, μ_1 and μ_2 are independent, so it holds that $p(\mu_1, \mu_2|H_2) = p(\mu_1|H_2)p(\mu_2|H_2)$. The second guideline indicates

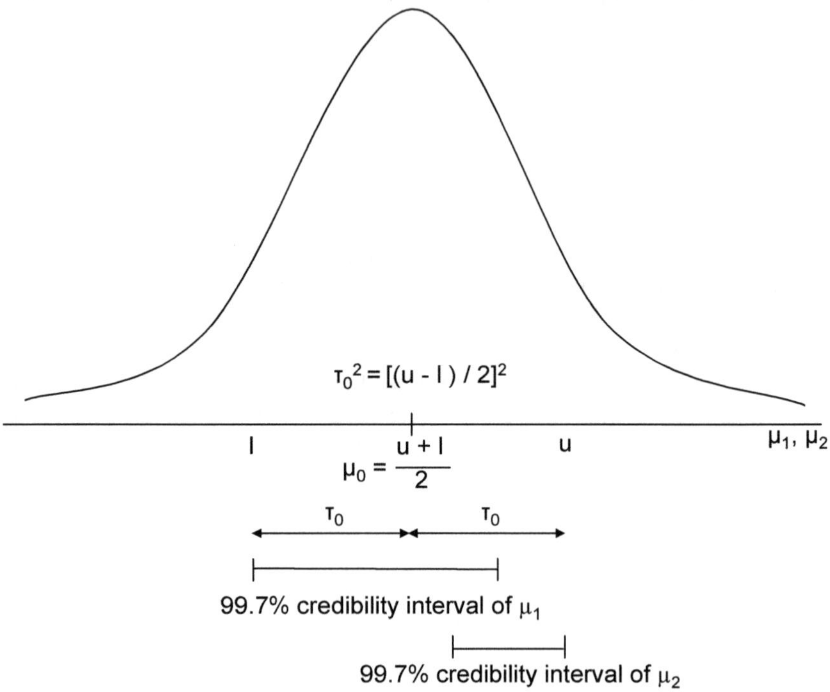

Fig. 5.2. Prior distribution of μ_1 and μ_2

that the prior distributions for μ_1 and μ_2 are the same: $p(\mu_1|H_2) = p(\mu_2|H_2)$. The conjugate prior distribution for a mean parameter is a normal distribution with mean μ_0 and variance τ_0^2. From these three requirements it follows that $p(\mu_1|H_2) = p(\mu_2|H_2) = \mathcal{N}(\mu_0, \tau_0^2)$.

The parameters μ_0 and τ_0^2 are data based, the details can be found in Chapter 4. Besides the means, there is also a variance term (σ^2). The encompassing (conjugate) prior for σ^2 is a scaled inverse χ^2-distribution, with data-based parameters $\nu_0 = 1$ and σ_0^2. In this section we will focus on the prior distribution for the means, because the prior distribution of σ^2 is of little importance when the goal is to select the best of a set of hypotheses like (5.9).

The prior distribution for μ_1 and μ_2 is shown in Figure 5.2. The data-based mean (μ_0) and the data based variance (τ_0^2) of the distribution are based on the 99.7% credibility intervals for both μ_1 and μ_2. The smallest value of the two lower bounds (l) and the largest value of the two upper bounds (u) are used to determine $\mu_0 = \frac{u+l}{2}$ and $\tau_0^2 = \left(\frac{u-l}{2}\right)^2$.

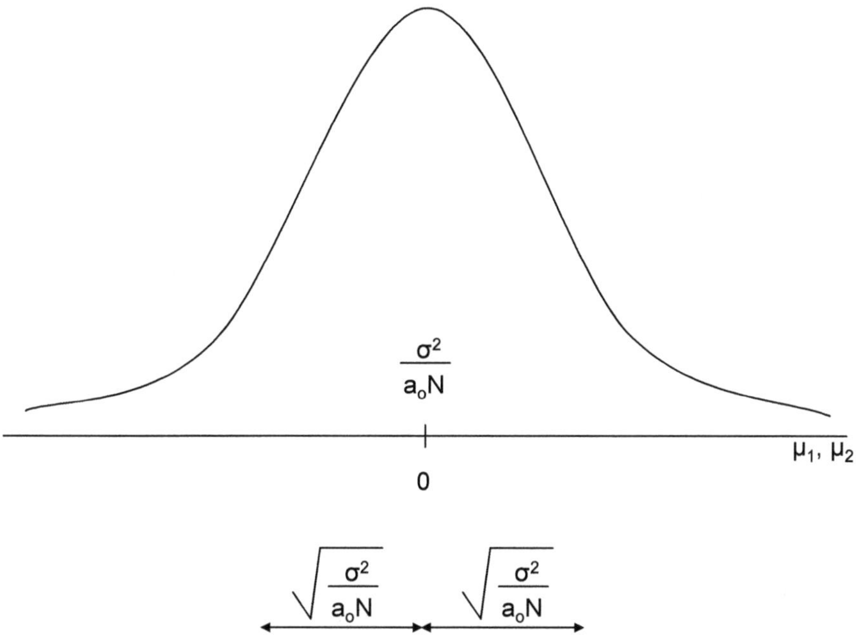

Fig. 5.3. Prior distribution of μ_1 and μ_2 (in case $N_1 = N_2 = N$)

5.4.2 Specification of a (Normalized) Conditional Prior

The (normalized) conditional prior described in Chapter 8 resembles the specification of the encompassing prior in Chapter 4. The adjective "normalized" is used because the normalization constant is taken into account such that this prior integrates to 1.0 not only for the unconstrained but also for the constrained hypotheses. A difference with respect to the encompassing prior is that the prior for μ is now specified conditional on the value of σ^2. For the unconstrained/encompassing hypothesis the (normalized) conditional prior for μ_1 given σ^2 equals $p(\mu_1|H_2, \sigma^2) = \mathcal{N}(0, \frac{\sigma^2}{a_0 N_1})$ and for μ_2 given σ^2: $p(\mu_2|H_2, \sigma^2) = \mathcal{N}(0, \frac{\sigma^2}{a_0 N_2})$. The prior distribution for μ_1 and μ_2, in case $N_1 = N_2 = N$ (i.e., $\mathcal{N}(0, \frac{\sigma^2}{a_0 N})$) is shown in Figure 5.3. In this method the variation in the prior is determined by a_0. In Chapter 8 it is elaborated how the data can be used to determine a value for a_0. In the examples that follow we will use the "best" values determined in Chapter 8. The (conjugate) prior for the residual variance σ^2 is an inverse gamma distribution, with hyperparameters α_0 and β_0. Since the prior for σ^2 is of minor importance, it will not be further discussed in this chapter.

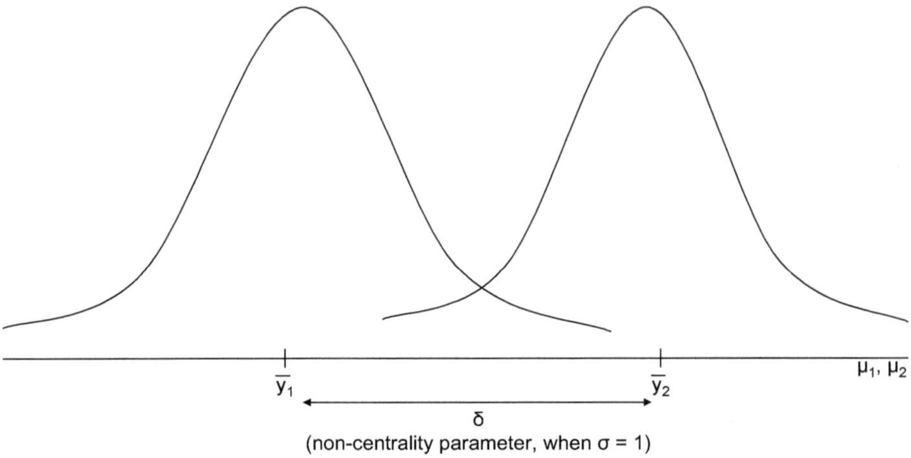

Fig. 5.4. Density of data

The resemblance between this approach and the encompassing prior approach is that every μ has the same prior if $N_1 = N_2$ and that the variance of the prior distribution of each μ is data based (via a_0 for the (normalized) conditional prior) for both approaches. The difference between the two approaches is that the (normalized) conditional prior approach assumes that the prior distribution of each μ has a mean of zero and the encompassing prior approach has a data based mean.

5.4.3 Specification of Prior Distributions for the Noncentrality Parameter of a Relevant Test Statistic

The approach by which the Bayes factor is computed in Chapter 6 is quite different from the approaches described in Chapters 4 and 8. Computation of Bayes factors requires the evaluation of the marginal likelihood of the data under each hypothesis. As was elaborated in Chapter 4, in order to be able to compute the marginal likelihood, a prior distribution has to be specified for the hypothesis under investigation. In the previous two sections, priors were determined both for the means and the residual variance.

An alternative is to compute the marginal likelihood of a test statistic instead of the marginal likelihood of the data. If the test statistic is well chosen, it contains almost the same information as the data. The T statistic used in Chapter 6 applied to the hypotheses in (5.9) contains information with respect to the differences between the means of group 1 and group 2. As shown in Chapter 6, the T statistic can be generalized to a situation where more than two groups are involved.

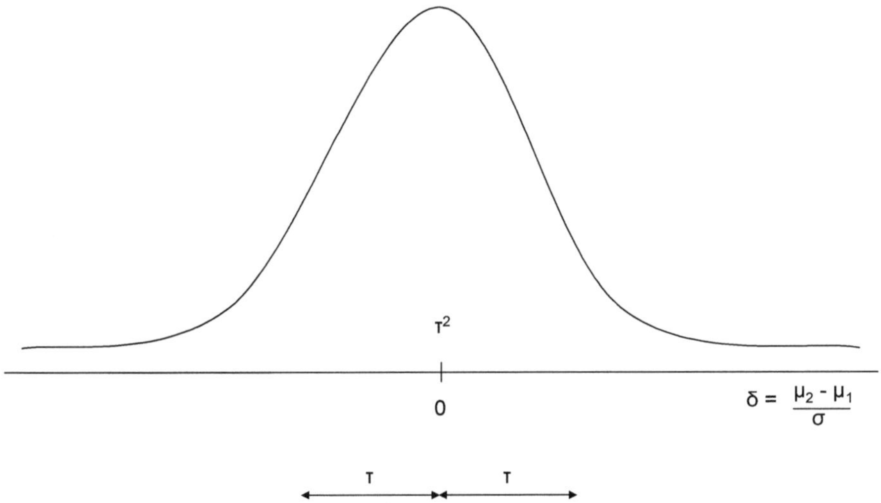

Fig. 5.5. Prior distribution of δ

The T statistic contains one unknown parameter, namely the noncentrality parameter δ. As is visually displayed in Figure 5.4, in the case of two means $\delta = \frac{\mu_2 - \mu_1}{\sigma}$. In order to be able to compute the marginal likelihood of T, a prior has to be specified for δ. In Chapter 6 the prior for δ equals $p(\delta \mid H_2) = \mathcal{N}(0, \tau^2)$ (Figure 5.5). In the sequel, τ^2 is set to 0.5 because a priori it is expected that "the standardized differences between means δ are not large" (Chapter 6). Stated otherwise, a priori it is expected that the standardized difference between the means is, on average, 0, with a standard deviation of $\sqrt{.5} = .707$.

"To implement this method, one chooses a test statistic and then computes the density of the test statistic under each hypothesis. Bayes factors are obtained as ratios of these densities" (Chapter 6). This method greatly reduces associated computations, because the specification of prior distributions is only required for the noncentrality parameter δ and not for all model parameters $\boldsymbol{\mu}$ and σ^2 as in the encompassing prior and (normalized) conditional prior approaches.

5.4.4 Posterior Model Probabilities Resulting from Three Different Prior Specifications

Four of the hypotheses specified in Chapter 2 for the DID data are

Table 5.1. PMPs obtained for the DID data using the prior distributions from Chapters 4, 8, and 6

Ch. 4 PMP ($d = 0.3$)		Ch. 8 PMP ($a_0 = 0.001$)		Ch. 6 PMP ($\tau = 0.5$)	
H_{1b}	0.71	H_{1b}	0.71	H_{1b}	0.67
H_{1a}	0.20	H_{1a}	0.22	H_{1a}	0.21
H_2	0.09	H_2	0.08	H_2	0.13
H_0	0.00	H_0	0.00	H_0	0.00

$$
\begin{aligned}
H_0 &: \mu_{con} \approx \mu_{amn} \approx \mu_{pat} \approx \mu_{sim}, \\
H_{1a} &: \mu_{con} > \{\mu_{amn} \approx \mu_{pat}\} > \mu_{sim}, \\
H_{1b} &: \mu_{con} > \mu_{amn} > \{\mu_{pat} \approx \mu_{sim}\}, \\
H_2 &: \mu_{con}, \mu_{amn}, \mu_{pat}, \mu_{sim}.
\end{aligned}
$$

In Table 5.1 the PMPs obtained using the three different prior specifications discussed in the previous sections are displayed for the DID data. Note that the hypotheses in Chapters 6 and 8 are based on equality restrictions ("=") and order restrictions ("<", ">"), and in Chapter 4 on about equality restrictions ("$\approx$") and order restrictions ("<", ">"). As explained above, $\approx$ is defined as $|\mu_1 - \mu_2| < d$. For the DID data, $d = 0.3$. As can be seen, the three approaches are very much in agreement with respect to the support in the data for each of the hypotheses under investigation. In Table 5.2 the PMPs obtained using two of the three prior specifications discussed in the previous sections are displayed for the hypotheses that were specified in Chapter 2 for the emotional reactivity data. As can be seen, the results are again very similar for the hypotheses that were evaluated by both approaches (H_0 was not evaluated in Chapter 4).

In Table 5.3 the PMPs obtained using two of the three prior specifications discussed in the previous sections are displayed for the hypotheses that were specified in Chapter 2 for the grief data. As can be seen the results are again rather similar (both in size and order of the PMPs), with the exception of H_{1e}.

The overall conclusion is that Bayesian model selection of (in)equality constrained hypotheses using Bayes factors or PMPs is rather robust with respect

Table 5.2. PMPs obtained for the emotional reactivity data using the prior distributions from Chapters 4 and 8

Ch. 4 PMP		Ch. 8 PMP ($a_0 = 1$)	
H_{1c}	0.88	H_{1c}	0.74
H_2	0.12	H_2	0.13
		H_0	0.09
H_{1b}	0.00	H_{1b}	0.03
H_{1a}	0.00	H_{1a}	0.00

Table 5.3. PMPs obtained for the grief data using the prior distributions from Chapters 4 and 6

Ch. 4 PMP ($d = 3$)		Ch. 6 PMP ($\tau = 0.5$)	
H_{1b}	0.40	H_{1b}	0.30
H_{1d}	0.26	H_{1e}	0.26
H_{1a}	0.15	H_{1d}	0.20
H_{1e}	0.13	H_{1a}	0.15
H_{1c}	0.07	H_{1c}	0.08

to the specification of the prior distribution. This conclusion is based on two results. First of all, as was shown in this chapter, three sensible but different ways to specify the prior distributions render similar PMPs for the hypotheses specified for the empirical examples introduced in Chapter 2. Second, as was shown in Chapter 4, if only inequality constraints are used to specify hypotheses, prior distributions can be specified such (see, for example, the encompassing prior discussed in Section 5.4.1) that the resulting PMPs are independent of the mean and variance of the prior distribution.

References

[1] Benjamini, Y., Hochberg, Y: Controlling the false discovery rate, a practical and powerful approach to multiple testing. Journal of the Royal Statistical Society, Series B, **57**, 289–300 (1995)

[2] Berger, J.O., Delempady, M.: Testing precise hypotheses. Statistical Science, **2**, 317–352 (1987)

[3] Berger, J.O., Sellke, T.: Testing a point null hypothesis, the irreconcilability of p-values and evidence. Journal of the American Statistical Association, **82**. 112–122 (1987)

[4] Bierhals, A.J., Prigerson, H.G., Fasiczka, A., Frank, E., Miller, M., Reynolds III, C.: Gender differences in complicated grief among the elderly. Omega: Journal of Death and Dying, **32**, 303–317 (1996)

[5] Boelen, P.A., Bout, J. van den: Gender differences in traumatic grief symptom severity after the loss of a spouse. Omega: Journal of Death and Dying, **46**, 183–198 (2003)

[6] Cohen, J.: Statistical Power Analysis for the Social Sciences. Mahway, NJ, Erlbaum (1988)

[7] Cohen, J.: The earth is round ($p < .05$). American Psychologist, **49**, 997–1003 (1994)

[8] Forbes, E.E., Dahl, R.E.: Neural systems of positive affect: Relevance to understanding child and adolescent depression? Development and Psychopathology, **17**, 827–850 (2005)

[9] Freud, S.: Inhibitions, symptoms, anxiety (A. Strachey, Trans. and J. Strachey, eds.). New York, Norton (1959) (Original work published 1926)

[10] Gleaves, D.H.: The sociocognitive model of dissociative identity disorder: A reexamination of the evidence. Psychological Bulletin, **120**, 42–59 (1996)

[11] Gross, J.J., Munoz, R.F.: Emotion regulation and mental health. Clinical Psychology: Science and Practice, **2**, 151–164 (1995)

[12] Hart, O. van der, Nijenhuis, E.: Generalized dissociative amnesia: Episodic, semantic, and procedural memories lost and found. Australian and New Zealand Journal of Psychiatry, **35**, 589–600 (2001)

[13] Holm, S.: A simple sequentially rejective multiple test procedure. Scandinavian Journal of Statistics, **6**, 65–70 (1979)

[14] Huntjens, R.J.C., Peters, M.L., Woertman, L., Bovenschen, L., Martin, R.C., Postma, A.: Inter-identity amnesia in dissociative identity disorder: A simulated memory impairment? Psychological Medicine, **36**, 857–863 (2006)

[15] Kring, A.M., Bachorowski, J.: Emotions and psychopathology. Special Issue: Functional accounts of emotion. Cognition and Emotion, **13**, 575–599 (1999)

[16] Lilienfeld, S.O., Lynn, S.J., Kirsch, I., Chaves, J.F., Sarbin, T.R., Ganaway, G.K., Powell, R.A.: Dissociative identity disorder and the sociocognitive model: Recalling the lessons of the past. Psychological Bulletin, **125**, 507–523 (1999)

[17] Rosnow, R.L.,Rosenthal, R.: Statistical procedures and the justification of knowledge in psychological science. American Psychologist, **44**, 1276–1284 (1989)

[18] Rottenberg, J., Gross, J.J.: When emotion goes wrong: Realizing the promise of affective science. Clinical Psychology: Science and Practice, **10**, 227–232 (2003)

[19] Rottenberg. J., Kasch, K.L., Gross, J.J., Gotlib, I.H.: Sadness and amusement reactivity differentially predict concurrent and prospective functioning in major depressive disorder. Emotion, **2**, 135–146 (2002)

[20] Sellke, T., Bayarri, M.J., Berger, J.O.: Calibration of p values for testing precise null hypotheses. The American Statistician, **55**, 62–71 (2001)

[21] Silvapulle, M.J., Sen, P.K.: Constrained Statistical Inference. New York, Wiley (2004)

[22] Wijngaards-de Meij, L., Stroebe, M., Schut, H., Stroebe, W., Bout, J. van den, Heijden, P. van der, Dijkstra, I.C.: Couples at risk following the death of their child: Predictors of grief versus depression. Journal of Consulting and Clinical Psychology, **73**, 617–623 (2005)

[23] Wilkinson, L. and the Task Force on Statistical Inference: Statistical methods in psychology journals. American Psychologist, **54**, 594–604 (1999)

Part II

A Further Study of Prior Distributions and the Bayes Factor

6

Bayes Factors Based on Test Statistics Under Order Restrictions

David Rossell[1], Veerabhadran Baladandayuthapani[2], and Valen E. Johnson[2]

[1] Bioinformatics and Biostatistics Unit, Institute for Research in Biomedicine of Barcelona, Josep Samitier 1-5, 08028 Barcelona, Spain rosselldavid@gmail.com
[2] Department of Biostatistics, The University of Texas M. D. Anderson Cancer Center, 1515 Holcombe Blvd., 77030 TX, Houston, TX, USA
veera@mdanderson.org and vejohnson@mdanderson.org

6.1 Introduction

The subject of statistical inference under order restrictions has been studied extensively since Bartholomew's likelihood-ratio test for means under restricted alternatives [1]. Order restrictions explicitly introduce scientific knowledge into the mathematical formulation of the problem, which can improve inference.

Here we restrict our attention to the comparison of means of normally distributed random variables under ordered alternatives; that is, we assume $y_i = \sum_{j=1}^{J} \mu_j d_{ji} + \varepsilon_i$ with $\varepsilon_i \sim \mathcal{N}(0, \sigma^2)$ with independence for $i = 1, \ldots, N$ and where d_{ji} is 1 if individual i is in group j and 0 otherwise. There is abundant literature addressing this problem from a frequentist standpoint. Most authors consider simple ordering (i.e., $\mu_1 \leq \mu_2 \leq \ldots \leq \mu_J$). Classical reviews can be found in [2, 19, 20]. In some situations it is sensible to consider other orderings. In [21] it was suggested that experimenters occasionally know which mean is smallest and which is largest, but do not know the relative ordering of the remaining means. In [17] unimodal or umbrella orderings for which $\mu_1 \leq \ldots \leq \mu_{j^*-1} \leq \mu_{j^*} \geq \mu_{j^*+1} \geq \ldots \geq \mu_J$ for some j^* were considered.

From a Bayesian standpoint, inference for a wide variety of models using Gibbs sampling was considered in [8]. In [6] posterior draws from an unconstrained generalized linear model were obtained, and then an isotonic transformation that enforces monotonicity was applied. In [14] it was argued to use encompassing prior distributions, which condition an overall prior to the range of allowable parameter values under each hypothesis, and they discussed an intuitive interpretation of the resulting Bayes factors. In [13] encompassing priors were used for estimation and hypothesis testing of ordered group means in general linear models. More details about their approach can be found in Chapter 4. All these Bayesian approaches test either inequality or approximate equality constraints, but they do not extend to testing an exact equality constraint. For instance, they cannot be used to test the null hypothesis that all group means are exactly equal.

A difficulty that may be encountered with fully Bayesian approaches like those just described is that they require the specification of prior distributions on all the model parameters. This can be a challenging task in situations where there is not much prior knowledge, since formal Bayesian hypothesis tests cannot generally be performed with improper priors. In [12] an alternative approach for computing Bayes factors was proposed, which only requires specifying prior distributions on noncentrality parameters and which greatly reduces associated computations. To implement this method, one chooses a test statistic and then computes the density of the test statistic under each hypothesis. Bayes factors are obtained as ratios of these densities.

In this chapter we propose the use of methodology from [12] in the context of problems with order restrictions. In Section 6.2 we formulate our approach, and in Section 6.3 we assess its behavior by applying it to real data. Finally, in Section 6.4 we present some concluding remarks.

R functions to implement the methodology described in this chapter are in the R library `isoregbf`, see `http://rosselldavid.googlepages.com`.

6.2 Methods

Normally, computing Bayes factors requires evaluating the marginal density of the data under each hypothesis. In this chapter, we instead define Bayes factors by evaluating the density of a relevant test statistic under each hypothesis. The choice of test statistic is described in Section 6.2.1. In Section 6.2.2 we show how the sampling density of the test statistic depends on a noncentrality parameter vector $\boldsymbol{\delta}$, and in Section 6.2.3 we discuss the specification of a prior distribution for $\boldsymbol{\delta}$. The computation of Bayes factors and posterior probabilities is addressed in Section 6.2.4.

6.2.1 Choice of Test Statistic

As a first step, we summarize the data with the statistic $(\bar{y}_1, \ldots, \bar{y}_J, s_p^2)$, where $\bar{y}_j$ is the mean of the jth group and s_p^2 is the pooled estimate of the variance. Note that this is the sufficient statistic under the unrestricted hypothesis that all group means are different. We then define $\mathbf{T} = (T_1, \ldots, T_{J-1})'$ to be the following function of the sufficient statistic:

$$T_j = \frac{\bar{y}_{j+1} - \bar{y}_j}{s_p \sqrt{\frac{1}{N_j} + \frac{1}{N_{j+1}}}}, \qquad j = 1, \ldots, J-1; \tag{6.1}$$

that is, we construct pairwise statistics that compare each pair of consecutive groups. This may look like an arbitrary decision at first, since one could also consider other pairwise comparisons; for example, one could compare each category to the first. In Appendix 6.1 we prove that our approach is essentially invariant to this choice in the sense that we obtain the same result no matter which pairwise comparisons are used to define the test statistic.

It is important to emphasize that our goal is to make inferences about differences in group means. Since $\mathbf{T}$ contains the information of all possible pairwise comparisons, most of the information in the sufficient statistic that is relevant for model comparisons is preserved in $\mathbf{T}$. We note that other authors have also argued for the use of pairwise t-test statistics in an ANOVA setup with order restrictions [9, 10].

Finally, note that the test statistic in (6.1) remains a reasonable choice when a hypothesis specifies a unimodal ordering or some other form of non-simple ordering.

6.2.2 Distribution of the Test Statistic

We first introduce some notation. Let N_j be the number of individuals in group j and let $\boldsymbol{\delta} = (\delta_1, \ldots, \delta_{J-1})'$, where $\delta_j = (\mu_{j+1} - \mu_j)/\sigma$ for $j = 1, \ldots, J-1$ are standarized mean differences. We use the notation $\mathbf{X} \sim \mathcal{N}(\boldsymbol{\mu}, \Sigma)$ to indicate that a random vector $\mathbf{X}$ follows a multivariate normal distribution and $\mathbf{X} \sim \mathcal{T}_\nu(\boldsymbol{\mu}, \Sigma)$ to indicate that a random vector $\mathbf{X}$ follows a multivariate $\mathcal{T}$ with ν degrees of freedom, where $\boldsymbol{\mu}$ and Σ denote the location parameter and the scale matrix, respectively. For $\nu > 2$ the covariance matrix is defined and it is equal to $\frac{\nu}{\nu-2}\Sigma$.

Proposition 1 states that the sampling distribution of $\mathbf{T}$ is a noncentral multivariate $\mathcal{T}$. Here the noncentral $\mathcal{T}$ is defined as in [18], i.e. by dividing a normal with nonzero mean by the square root of an independent chi-square divided by its degrees of freedom. A difficulty with this definition is that its density function cannot be evaluated explicitly, but it is possible to approximate it with the more usual definition of the non-central $\mathcal{T}$, which is obtained by shifting a central $\mathcal{T}$ [7, 15]. See Appendix 6.2 for more details and an outline of the derivation of this result.

Proposition 1. *Given $\boldsymbol{\delta}$, $\mathbf{T} \sim \mathcal{T}_{N-J}(B\boldsymbol{\delta}, \Sigma_T)$, where B and Σ_T are $(J-1) \times (J-1)$ symmetric matrices with elements $B_{j,j'}$ and $\Sigma_{j,j'}$ given by*

$$B_{j,j} = \sqrt{\frac{N-J}{2}} \frac{\Gamma\left(\frac{N-J-1}{2}\right)}{\Gamma\left(\frac{N-J}{2}\right)} \left(\sqrt{\frac{1}{N_j} + \frac{1}{N_{j+1}}}\right)^{-1},$$

$$B_{j,j'} = 0,$$

$$\Sigma_{j,j'} = \mathcal{I}(j=j') + \sqrt{c_j c_{j'}}\delta_j\delta_{j'} - \frac{\mathcal{I}(|j-j'|=1)}{N_{j+1}}\left(\frac{1}{N_j}+\frac{1}{N_{j+1}}\right)^{-1/2}\left(\frac{1}{N_{j'}}+\frac{1}{N_{j'+1}}\right)^{-1/2}, \quad \textit{for } j \le j',$$

$$c_j = \left(\frac{1}{N_j}+\frac{1}{N_{j+1}}\right)^{-1}\left[1-\frac{N-J-2}{2}\left(\frac{\Gamma\left(\frac{N-J-1}{2}\right)}{\Gamma\left(\frac{N-J}{2}\right)}\right)^2\right], \tag{6.2}$$

where $\mathcal{I}(\cdot)$ is the indicator function. Notice that for $\delta_j = 0$ we get $E(T_j|\delta_j = 0) = 0$ and $\Sigma_{jj} = 1$, i.e. a standard Student-t distribution with $N-J$ degrees of freedom. If $\delta_j = 0$ for all i, then $\Sigma_{j,j+1} = -\frac{1}{N_{j+1}}\left(\frac{1}{N_j}+\frac{1}{N_{j+1}}\right)^{-1/2}\left(\frac{1}{N_{j+1}}+\frac{1}{N_{j+2}}\right)^{-1/2}$ *and* $\Sigma_{j,j'} = 0$ *for* $|j-j'| \neq 1$.

6.2.3 Prior Specification

In Section 6.2.2 we demonstrated that the distribution of the test statistic $\mathbf{T}$ depends solely on a non-centrality parameter vector $\boldsymbol{\delta}$ having components $\delta_j = (\mu_{j+1} - \mu_j)/\sigma$. We complete the model by specifying a prior distribution on $\boldsymbol{\delta}$ under each of the hypotheses that we wish to compare. We focus the ensuing discussion on hypotheses that specify monotonic orderings, although extensions to more complicated orderings are straightforward.

By monotonic ordering of hypotheses we mean that increments between two consecutive hypothesized means are either always positive (negative) or exactly 0. Denote by $\boldsymbol{\Delta}$ the set of all standardized increments and $\boldsymbol{\Delta}^+$ the subset of those that are non-zero. It is important to notice that the ordering specified by each of the hypotheses will, in general, not coincide with the original ordering $\mu_1, \ldots, \mu_J$, which is the ordering that is used to compute the test statistic $\mathbf{T}$; that is, to find $\boldsymbol{\delta}$ from $\boldsymbol{\Delta}$, one generally needs to apply a linear transformation, say $\boldsymbol{\Delta} = A\boldsymbol{\delta}$, where A is a square matrix. For example, if we have three groups and the current hypothesis specifies that $\mu_2 < \mu_1 = \mu_3$, we have that $\Delta_1 = (\mu_1 - \mu_2)/\sigma$ and $\Delta_2 = (\mu_3 - \mu_1)/\sigma = 0$ (hence $\boldsymbol{\Delta}^+ = \Delta_1$). We then define $\delta_1 = (\mu_2 - \mu_1)/\sigma = -\Delta_1$ and $\delta_2 = (\mu_3 - \mu_2)/\sigma = \Delta_2 + \Delta_1$.

We first define a prior specification under the full alternative hypothesis that does not impose any restrictions on the group means. We specify a multivariate $\mathcal{T}$ centered on the null hypothesis of homogeneity of means,

$$\boldsymbol{\Delta} \sim \mathcal{T}_1(\mathbf{0}, \tau_0^2 \Sigma^{\Delta}), \tag{6.3}$$

where $\mathbf{0}$ is a J-dimensional vector of zeroes and $\tau_0 > 0$ is the only prior parameter that needs to be specified. The (i,j) element of the scale matrix is given by $\Sigma^{\Delta}_{j,j'} = 1$ if $j = j'$, $\Sigma^{\Delta}_{j,j'} = -0.5$ if $|j-j'| = 1$, and 0 elsewhere. The use of a multivariate $\mathcal{T}$ as a default prior has been advocated by several authors and has been shown to have desirable properties [3, 4]. The choice of Σ^{Δ} is motivated by an invariance property that is explained in Appendix 6.1. Here we note that (6.3) is the implied prior on $\boldsymbol{\Delta}$ when $\boldsymbol{\mu} \sim \mathcal{N}(\mathbf{0}, 0.5\tau_0^2 I)$, where $\mathbf{0}$ is a vector of zeroes and I is the identity matrix, and σ arises independently from a chi-square distribution with 1 degree of freedom.

We now consider hypotheses with order constraints (i.e., a prior for $\boldsymbol{\Delta}^+$). We condition the prior in (6.3) to the region of allowable parameter values under the current hypothesis. This results in a truncated multivariate $\mathcal{T}$ prior for $\boldsymbol{\Delta}^+$, possibly of lower dimensionality than (6.3) to account for some elements of $\boldsymbol{\Delta}$ being exactly equal to zero. We denote a multivariate truncated $\mathcal{T}$ with ν degrees of freedom by $\text{Trunc}\mathcal{T}_\nu(\boldsymbol{\mu}, \Sigma, R)$, where $\boldsymbol{\mu}$ is the location parameter, Σ is the scale matrix, and R indicates the region in which the variable is allowed to take values (typically the positive axis or quadrant, according to the ordering).

$$\boldsymbol{\Delta}^+ \sim \text{Trunc}\mathcal{T}_1(\mathbf{0}^+, \tau_0^2 \Sigma^{\Delta^+}, \boldsymbol{\Delta}^+ > \mathbf{0}), \tag{6.4}$$

where $\mathbf{0}^+$ and Σ^{Δ^+} are the appropriate subvector and submatrix of $\mathbf{0}$ and Σ^{Δ} in (6.3), respectively.

Our approach thus implies a prior density on the standardized differences in means $\boldsymbol{\delta}$, which is the natural scale on which to specify hypotheses since it represents signal-to-noise ratios or, in the social sciences, the standardized effect size. The parameter τ_0 is easily interpretable, since (6.3) places .5 prior probability on the univariate interval $(-\tau_0, \tau_0)$ and (6.4) on the interval $(0, \tau_0)$. The use of $\tau_0 = 1$ is advocated in [4] and [3], which would assign .5 prior probability that the signal-to-noise ratios $\boldsymbol{\delta}$ are between -1 and 1. In general, when reliable prior information is not available, we recommend setting τ_0 between 0.1 and 2, since signal-to-noise ratios much smaller than 0.1 or larger than 2 are not common in practice.

6.2.4 Bayes Factors and Posterior Probabilities

The Bayes factor to compare hypotheses H_p and $H_{p'}$ is computed as the ratio of the marginal densities that they imply for $\mathbf{T}$, given by

$$BF_{pp'} = \frac{m(\mathbf{T}|H_p)}{m(\mathbf{T}|H_{p'})}, \tag{6.5}$$

where

$$m(\mathbf{T}|H_p) = \int f(\mathbf{T}|\boldsymbol{\Delta}, H_p) dP(\boldsymbol{\Delta}|H_p) \tag{6.6}$$

and $P(\boldsymbol{\Delta}|H_p)$ denotes the prior distribution of $\boldsymbol{\Delta}$ under H_p. The term $f(\mathbf{T}|\boldsymbol{\Delta}, H_p)$ denotes the sampling distribution of $\mathbf{T}$ given in (6.2) and integration is with respect to the prior distribution on $\boldsymbol{\Delta}$ under H_p, as described in Section 6.2.3. Note that integration is really only with respect to $\boldsymbol{\Delta}^+$ since the remaining elements of $\boldsymbol{\Delta}$ are set to be exactly zero.

Once the marginal densities in (6.6) have been obtained, it is straightforward to compute the posterior probability of each hypothesis as $P(H_p|\mathbf{T}) \propto m(\mathbf{T}|H_p)\pi_p$, where π_p is the prior probability assigned to hypothesis H_p.

In summary, to compute Bayes factors and posterior probabilities, we need only to evaluate the integrals in (6.6). When $\boldsymbol{\Delta}^+$ is one dimensional, there are numerous numerical integration routines that can be used for this purpose. This task is more challenging in higher dimensions, however. In such cases we have adopted an importance sampling approach; more sophisticated methods might also be used. Details concerning integration of $\boldsymbol{\Delta}^+$ are provided in Appendix 6.3.

If $\boldsymbol{\Delta}$ has the same number of nonzero elements under all hypotheses, then $BF_{pp'}$ simplifies to $P(R_p|\mathbf{T})P(R_{p'})/(P(R_{p'}|\mathbf{T})P(R_p))$, where $P(R_p|\mathbf{T})$ and $P(R_p)$ are the proportions of the posterior and prior distribution for which $\boldsymbol{\delta}$ falls in the region specified by H_p [14]. Therefore, it is enough to sample from the prior and posterior of $\boldsymbol{\Delta}$ under the unrestricted model.

6.3 Results

To illustrate our approach, we analyze data from the dissociative identity disorder study (DID) [11] and the grief study detailed in Chapter 2. We also simulate data with the same structure as the DID study.

6.3.1 DID Data

Our goal is to compare the mean values of a recognition score, which measures the recognition of text and pictures of subjects categorized into one of the following four groups: persons simulating to suffer from DID (Simulators, *sim*), DID-patients (*pat*), persons who guessed the answers to the questions (True amnesiacs, *amn*), and a group with healthy control patients (Controls, *con*). We wish to test the following hypotheses:

$$\begin{aligned} &H_0 : \mu_{sim} = \mu_{pat} = \mu_{amn} = \mu_{con}, \\ &H_{1a} : \mu_{sim} < \{\mu_{pat} = \mu_{amn}\} < \mu_{con}, \\ &H_{1b} : \{\mu_{sim} = \mu_{pat}\} < \mu_{amn} < \mu_{con}, \\ &H_2 : \text{not } H_0; \end{aligned} \tag{6.7}$$

that is, we expect a priori that the Simulators will have the lowest scores and the Controls the highest, but it is not clear what the relative ordering of the remaining two groups should be. To address this issue, one could begin by excluding H_0 and H_2 and compare the remaining two hypotheses and then

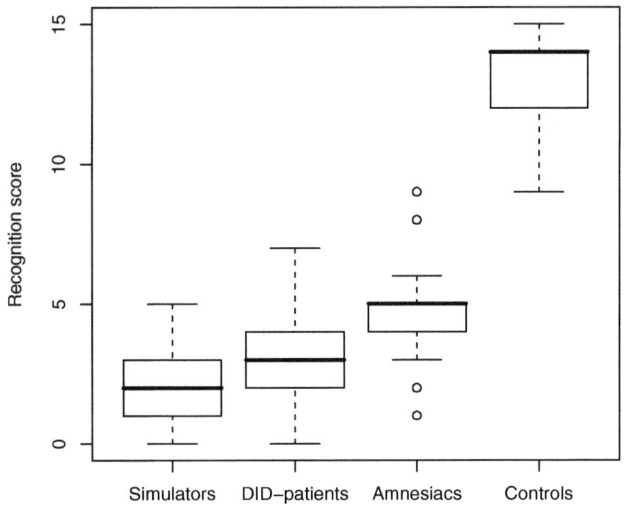

Fig. 6.1. Recognition score for each group

conduct a second comparison that includes all four hypotheses. However, in a Bayesian framework it seems more natural to start by computing the posterior probabilities for all four hypotheses and basing conclusions on these values.

We start with a descriptive analysis. Figure 6.1 provides a boxplot of the recognition score for the four groups. The Controls present the highest scores, the True amnesiacs have the second highest, and the Simulators tended to have slightly lower scores than the DID-patients. The pairwise t-test statistics for this dataset are $T_1 = 2.48$ to compare μ_{pat} versus μ_{sim}, $T_2 = 2.94$ for μ_{amn} versus μ_{pat} and $T_3 = 18.97$ for μ_{con} versus μ_{amn}. The associated two-sided p-values under the assumption of normality and common variance are .0151, .0042, and <.0001, respectively. These results qualitatively suggest that H_0 does not hold; they provide some evidence against H_{1a} and weaker evidence against H_{1b} (the Bonferroni correction for multiple comparisons results in a p-value cutoff of .017).

To compute Bayes factors based on the observed value of the test statistic we only need to specify a single prior parameter, namely the prior scale parameter τ_0 that was defined in Section 6.2.3. We obtain results for τ_0 ranging from 0.1 to 1 in increments of 0.1 and from 1 to 10 in increments of 1, so that we can assess the conclusions that would be reached under different prior specifications. In most datasets one observes standardized differences between means smaller than 1 in absolute value. A value of $\tau_0 = 1$ assigns .5 prior probability for $\delta_j < 1 \ \forall j = 1, \ldots, J$; that is, we deliberately obtain results for values of τ_0 larger than what most practitioners would typically use, so that we can assess the performance of our approach under these conditions.

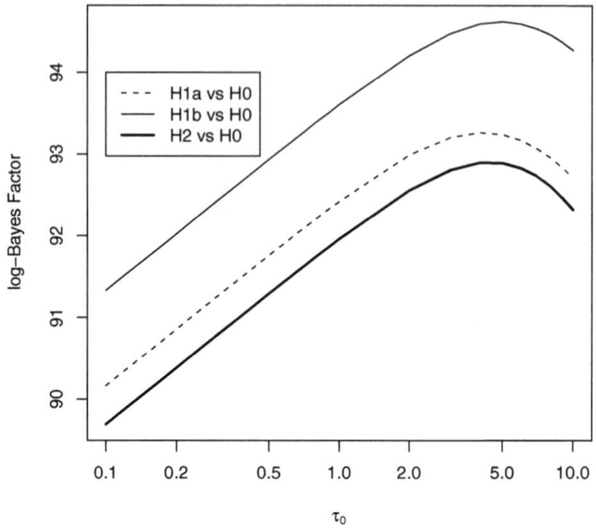

Fig. 6.2. Bayes factors as a function of the prior scale parameter τ_0

To evaluate the integral in (6.6) we use an importance sampling scheme with 1,000,000 Monte Carlo samples per dimension of the integral.

Figure 6.2 provides the Bayes factors (in log-scale) comparing H_0 with all the other hypotheses. For all the considered values of τ_0 there is overwhelming evidence against the null hypothesis. The preferred hypothesis is H_{1b}, followed by H_{1a} and H_2; that is, the Bayes factors favor the more parsimonious model H_{1b} and even H_{1a} over the full alternative H_2, even though the p-values comparing μ_{sim} versus μ_{pat} and μ_{pat} versus μ_{amn} were significant. We consider this to be an attractive feature of the Bayesian approach.

To obtain posterior probabilities, we assume that all hypothesis are equally likely a priori. Table 6.1 summarizes results obtained under this assumption. We see that H_{1b} has the largest posterior probability for all values of τ_0 considered, with probabilities ranging from .665 for $\tau_0 = 0.1$ to .740 for $\tau_0 =$

Table 6.1. Posterior probabilities for DID data and selected values of τ_0

τ_0	H_0	H_{1a}	H_{1b}	H_2
0.1	<.001	.206	.665	.129
0.5	<.001	.205	.666	.129
1.0	<.001	.204	.667	.129
10.0	<.001	.155	.740	.105

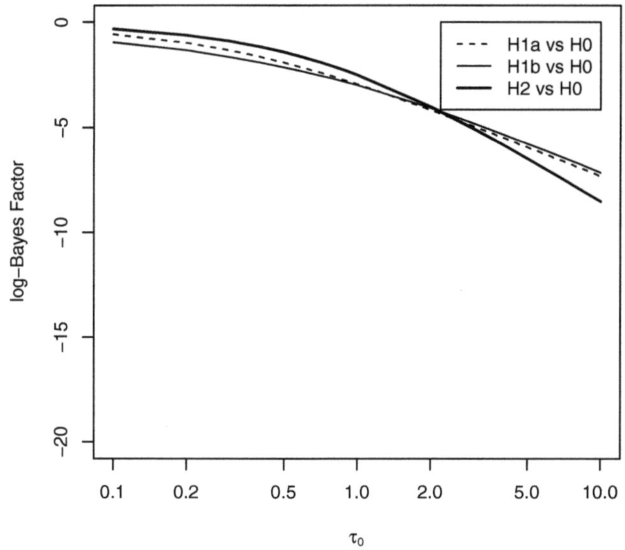

Fig. 6.3. Bayes factors as a function of the prior scale parameter τ_0 for simulated dataset with $N_j = 10$

10. The second most probable hypothesis is H_{1a}, whereas H_0 has a negligible posterior probability for all the selected values of τ_0. The results appear to be quite robust with respect to the prior specification.

6.3.2 Simulated Data

We now consider a study with the same structure as the DID study from Section 6.3.1; that is, we assume that there are four groups and the hypotheses are as defined in (6.7). We generate both a small and a large dataset under the null hypothesis H_0 (i.e., $y_i \sim \mathcal{N}(0, \sigma^2)\ \forall i$), and obtain results for values of the prior parameter τ_0 ranging between 0.01 and 10. The small and large datasets contain 10 and 10,000 observations per group, respectively, and in both cases the observational variance is set to $\sigma^2 = 1.625$, the value estimated from the DID dataset.

Figures 6.3 and 6.4 plot log-Bayes factors as a function of τ_0 for $N_j = 10$ and $N_j = 10,000$, respectively. Qualitatively we observe similar results for both datasets, even though for any fixed value of τ_0 the evidence in favor of H_0 is stronger in the large dataset. The results are positive in the sense that for all values of τ_0 there is moderate evidence in favor of H_0 in the small dataset and strong evidence in the large dataset. Also, H_{1a} and H_{1b} are favored over H_2, since they assume that some of the group means are equal to each other.

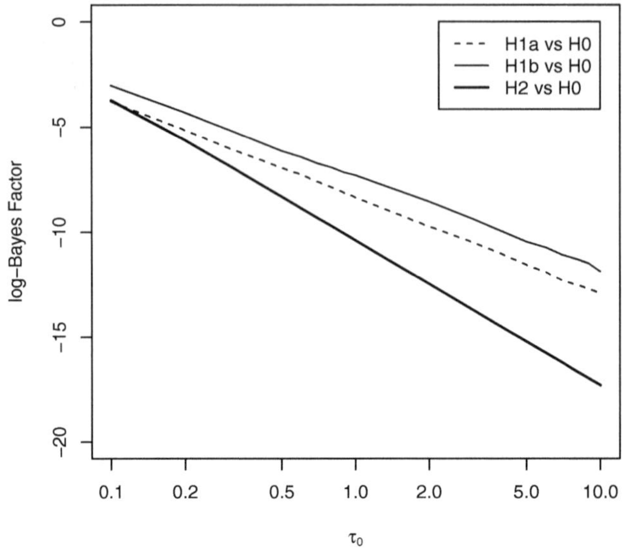

Fig. 6.4. Bayes factors as a function of the prior scale parameter τ_0 for simulated dataset with $N_j = 10,000$

The smaller the value of τ_0, the more mass that the alternative hypotheses concentrate around $\boldsymbol{\Delta} = 0$ and the more similar to H_0 that they become, resulting in log-Bayes factors closer to 0. As τ_0 increases evidence favoring H_0 becomes stronger.

Table 6.2 provides the posterior probability of H_0 for several values of τ_0. In the large dataset the results are robust with respect to τ_0. In the small dataset, H_0 is always favored, but for $\tau_0 = 0.1$, the other hypotheses also have relatively high posterior probabilities. As argued in Section 6.2.3, $\tau_0 = 1$ is a reasonable default choice. In both simulated datasets $\tau_0 = 1$ results in a high posterior probability for the correct hypothesis.

Table 6.2. Posterior probabilities for simulated data and selected values of τ_0

	Large data (N_j=10,000)				Small data ($N_j = 10$)			
τ_0	H_0	H_{1a}	H_{1b}	H_2	H_0	H_{1a}	H_{1b}	H_2
0.1	.913	.044	.021	.022	.374	.211	.144	.271
0.5	.997	.002	.001	<.001	.665	.098	.078	.159
1.0	.999	.001	<.001	<.001	.844	.045	.043	.069
10.0	1.000	<.001	<.001	<.001	.998	.001	.001	<.001

6.3.3 Grief Data

We next consider a study that examined a measure of complicated grief (CG) resulting from the loss of a loved one. The study recorded the CG score for 1179 patients along with their gender, whether a child or spouse had died, and whether the loss was recent or not. Thus, patients were divided in eight groups according to their sex, kinship, and time from loss. In Chapter 2 each group was labeled by a number between 1 and 8. Our goal is to assess the following statements:

H_0: there are no differences between groups,
H_{1a}: grief from a recent loss is greater than from a remote loss,
H_{1b}: additionally to H_{1a}, women grieve more than men,
H_{1c}: additionally to H_{1b}, losing a child causes more grief than losing a spouse,
H_{1d}: additionally to H_{1a}, losing a child is more severe for women than losing a partner, and losing a child is more severe for women than for men,
H_{1e}: additionally to H_{1d}, losing a partner is more severe for men than losing a child, and losing a partner is more severe for men than for women,
H_2: otherwise (i.e., neither H_0 nor H_{1a} hold).

These statements are made precise in Chapter 2 by stating inequalities between pairs of group means. We define H_2 in a slightly different manner than it is done in Chapter 2. Chapter 2 defines H_2 as an unrestricted hypothesis, whereas here we restrict H_2 so that it only contains the parameter values that are not allowed by any of the other hypotheses. The inequalities can be trivially re-expressed in terms of $\boldsymbol{\delta}$. For instance, $\mu_2 > \mu_1$ if and only if $\delta_1 = (\mu_2 - \mu_1)/\sigma > 0$. See Appendix 6.4 for a detailed elicitation of the hypotheses in terms of $\boldsymbol{\delta}$. As before, we use the vector of t-test statistics $\mathbf{T}$ to find Bayes factors and posterior probabilities for the above-mentioned hypotheses. For these data we find $\mathbf{T} = (-2.43, 0.27, -0.41, 2.82, -4.83, 2.80, -0.72)$. The large values of some of the components of $\mathbf{T}$ suggest that the null hypothesis does not hold.

From the results of previous studies (see Chapter 2), a priori we expect that the standardized differences between means $\boldsymbol{\delta}$ are not large. Therefore, it is unreasonable that τ_0 is larger than 1. We set $\tau_0 = 0.5$ as our primary choice, although we also use $\tau_0 = 0.1$ and $\tau_0 = 1$ to assess the sensitivity of the results. We assign prior probability 1/2 to the existence of differences between groups (i.e., $P(H_0) = 1/2$). This results in a posterior probability $P(H_0|\mathbf{T}) < .001$ for the three values of τ. Since H_0 has negligible posterior probability, we do not consider it further.

To compute Bayes factors for the remaining hypotheses, we generated 100,000 samples from the posterior distribution of $\boldsymbol{\delta}$ given $\mathbf{T}$ using Metropolis-Hastings sampling as implemented in the function `MCMCmetrop1R` from the R library `MCMCpack` [16]. As explained in Section 6.2.4, $BF_{pp'}$ in (6.5) can be approximated by $(\widehat{P}(R_p|\mathbf{T})\widehat{P}(R_{p'})/(\widehat{P}(R_{p'}|\mathbf{T})\widehat{P}(R_p))$, where R_p is the region of allowable parameter values under H_p. $P(R_p|\mathbf{T})$ and $P(R_p)$ are the posterior

Table 6.3. Bayes factors and posterior probabilities for the grief data

	Bayes factor vs. H_2					Posterior probability					
τ_0	H_{1a}	H_{1b}	H_{1c}	H_{1d}	H_{1e}	H_{1a}	H_{1b}	H_{1c}	H_{1d}	H_{1e}	H_2
0.1	15.00	23.14	7.08	17.12	21.32	.891	.347	.044	.449	.136	.109
0.5	16.76	32.64	9.12	22.03	28.00	.908	.409	.045	.481	.141	.092
1.0	16.38	36.26	9.92	22.35	29.94	.913	.441	.047	.484	.145	.087

and prior proportions of $\boldsymbol{\delta}$ values, respectively, in R_p under the unrestricted prior (6.3).

Table 6.3 provides the Bayes factors for each of the hypotheses versus H_2. For $\tau_0 = 0.5$ the data provides very strong evidence in favor of H_{1b} when compared to H_2, strong evidence for H_{1e}, H_{1d}, and H_{1a}, and substantial evidence for H_{1c}. Similar results are observed for $\tau_0 = 1$ and for $\tau_0 = 0.1$, although in the later case the evidence in favor of all these hypotheses is somewhat weaker.

We now complete our analysis by obtaining the posterior probabilities $P(H_p|\mathbf{T})$ for each hypothesis. It is important to distinguish these from $P(R_p|\mathbf{T})$, which is the proportion of the posterior distribution falling in the region specified by H_p. $P(H_p|\mathbf{T})$ depends on the prior probability π_p, whereas $P(R_p|\mathbf{T})$ depends on the prior proportion $P(R_p)$ (i.e., the volume of the multivariate $\mathcal{T}$ prior (6.3) that falls in R_p). For instance, under (6.3), the prior probability for R_{1a} is $P(\delta_1 < 0, \delta_3 < 0, \delta_5 < 0, \delta_7 < 0) = .5^4 = .0625$ and the prior probabilities for $R_{1b}, \ldots, R_{1e}$ are substantially smaller. Since $H_{1a}, \ldots, H_{1e}$ arise from psychological considerations and the results of previous studies, setting $\pi_p = P(R_p)$ would make $H_{1a}, \ldots, H_{1e}$ too unlikely a priori. Instead, we consider that each statement made in $H_{1a}, \ldots, H_{1e}$ is equally likely. As an illustration, suppose we only considered H_{1a}, H_{1b}, and H_2. In this case we would assign .5 probability that grief from a recent loss is greater and .5 probability that women grief more than men, with independence. H_{1b} holds if both statements are true simultaneously, which occurs with prior probability $.5 \times .5$, whereas $\pi_{1a} = .5$ and $\pi_2 = 1 - \pi_{1a} = .5$. Since H_{1b} is nested within H_{1a} (i.e., H_{1b} is true $\Rightarrow H_{1a}$ is true), we have that $\pi_{1a} + \pi_2 = 1$ instead of $\pi_{1a} + \pi_{1b} + \pi_2 = 1$.

When considering all hypotheses, we arrive at the following prior probabilities (see Appendix 6.4 for a detailed explanation): $\pi_{1a} = .5$, $\pi_{1b} = .167$, $\pi_{1c} = .056$, $\pi_{1d} = .278$, $\pi_{1e} = .056$, and $\pi_2 = 1 - \pi_{1a} = .5$. Of course, we use these prior probabilities as an illustration; other values are also possible.

The posterior probabilities must be computed carefully, for the hypotheses are not mutually exclusive. For instance, if H_{1c} is true, then H_{1b} and H_{1a} are also true, and hence we should have that $P(H_{1c}|\mathbf{T}) \leq P(H_{1b}|\mathbf{T}) \leq P(H_{1a}|\mathbf{T})$. As a consequence, the posterior probabilities of $H_{1a}, \ldots, H_{1e}, H_2$ should not add up to 1. It is the posterior probabilities of H_{1a} and H_2 that should add up to 1, since they are mutually exclusive and complementary. To deal with this issue, we divide the parameter space into mutually exclusive regions and then add their posterior probabilities (Appendix 6.4).

Table 6.3 provides the posterior probability of each hypothesis. For $\tau_0 = 0.5$ there is .908 posterior probability that recent loss leads to more grief than remote loss (H_{1a}). There is .409 probability that, additionally, women grieve more than men (H_{1b}) and .481 probability that women grieve more for the loss of a child than men (H_{1d}). Hypotheses H_{1c} and H_{1e} have low posterior probability, despite their large Bayes factor. This is due to their a priori probability being small.

To summarize, we confirm that recent loss causes more grief. We find evidence that women grieve more than men in general and that they grieve more after the loss of a child. Since this contradicts most earlier studies, this evidence is not conclusive. It is unlikely that losing a child causes more grief both for men and women, or that men are more attached to their spouse than to their children.

6.4 Conclusions

We have presented an approach for computing Bayes factors and posterior probabilities that can be used to compare several competing probability models or hypotheses under order restrictions. Because the resulting Bayes factors are based on test statistics rather than on all the observed data, there is a possibility that some information is lost. However the selected test statistic is closely related to the sufficient statistic under the model that specifies no order restrictions, which suggests that the loss of information will be negligible unless there is substantial prior information about attributes of the data not reflected in the test statistic. In addition, we have proven some invariance properties that make this choice of test statistic more compelling.

The main advantage of our approach is that it requires only the specification of prior distributions for the parameters that govern the sampling distribution of the test statistic, which can be substantially simpler than specifying priors for all the parameters that index the distribution of the original data.

The sampling distribution of our proposed test statistic is fully determined by a noncentrality parameter vector. We have proposed a prior distribution that only requires setting the value of a single scalar parameter τ_0. Based on practical considerations we have argued that reasonable default values for τ_0 may be between 0.1 and 2, and in analyzing both real and simulated data, this range of values seems to perform satisfactorily. For example, in simulated data with only 10 observations per group, our approach chose the correct model for any value of τ_0 within this range, albeit the posterior probability of competing models was not negligible. Indeed, we have observed that in larger datasets the approach can be quite robust to the specification of τ_0.

With minor modifications, our methodology can be extended beyond the case of simple ordering. For example, the so-called umbrella ordering can be addressed simply by truncating certain parameters to be negative instead of positive. Another possible extension would be to consider multiple regression models in a generalized linear models setup; that is, while one is still mainly interested in studying possible orderings between groups, the response may be non-normal and one may wish to include some additional explanatory variables in the model. In such a setup, one could define the test statistic using estimated regression coefficients instead of sample means, as we have done here. Note that the dimensionality of this statistic does not increase as more variables are added to the model, which results in considerable computational savings.

In summary, we have defined an approach that substantially simplifies prior specification and that appears to be reasonably robust with respect to this specification. We believe that this makes our approach an appealing choice when informative prior knowledge is not available.

References

[1] Bartholomew, D.: A test of homogeneity of means under restricted alternatives. Journal of the Royal Statistical Society B, **23**, 239–281 (1961)

[2] Barlow, R.E., Bartholomew, D.J., Bremner, J.M., Brunk, H.D.: Statistical Inference under Order Restrictions. The Theory and Application of Isotonic Regression. New York, Wiley (1972)

[3] Bayarri, M., Garcia-Donato, G.: Extending conventional priors for testing general hypotheses in linear models. Biometrika, **94**, 135–152 (2007)

[4] Berger, J., Pericchi, R.: Objective Bayesian methods for model selection: Introduction and comparison. In: Lahiri, P. (ed) Model Selection. Beachwood, OH, Institute of Mathematical Statistics Lecture Notes Monograph Series **38**, 135–207 (2001)

[5] Box, G.E.P., Tiao, G.C.: Bayesian Inference in Statistical Analysis. New York, Wiley (1992)

[6] Dunson, D., Neelon, B.: Bayesian inference on order-constrained parameters in generalized linear models. Biometrics, **59**, 286–295 (2003)

[7] Gelman, A., Carlin, J., Stern, H., Rubin, D.: Bayesian Data Analysis (2nd ed.). Chapman & Hall (2004)

[8] Gelfand, A., Smith, A., Lee, T.: Bayesian analysis of constrained parameter and truncated data problems using gibbs sampling. Journal of the American Statistical Association, **87**, 523–532 (1992)

[9] Hayter, A.: A one-sided studentized range test for testing against a simple ordered alternative. Journal of the American Statistical Association, **85**, 778–785 (1990)

[10] Hayter, A., Liu, W.: Exact calculations for the one-sided studentized range test for testing against a simple ordered alternative. Computational Statistics and Data Analysis, **22**, 17–25 (1996)

[11] Huntjens, R.J.C., Peters, M.L., Woertman, L., Bovenschen, L.M., Martin, R.C., Postma, A.: Inter-identity amnesia in dissociative identity disorder: A simulated memory impairment? Psychological Medicine, **36**, 857–863 (2006)

[12] Johnson, V.: Bayes factors based on test statistics. Journal of the Royal Statistical Society B, **67**, 689–701 (2005)

[13] Klugkist, I., Kato, B., Hoijtink, H.: Bayesian model selection using encompassing priors. Statistica Neerlandica, **59**, 57–69 (2005)

[14] Klugkist, I., Laudy, O., Hoijtink, H.: Inequality constrained analysis of variance: A Bayesian approach. Psychological Methods, **10**, 477–493 (2005)

[15] Kraemer, H., Paik, M.: A central t approximation to the noncentral t distribution. Technometrics, **21**, 357–360 (1979)

[16] Martin, A.D., Quinn, K.M., Park, J.H.: MCMCpack: Markov chain Monte Carlo (MCMC) Package (2007). R package version 0.9-2

[17] Mack, G., Wolfe, D.: K-sample rank test for umbrella alternatives. Journal of the American Statistical Association, **76**, 175–181 (1981)

[18] Owen, D.: A special case of a bivariate non-central t-distribution. Biometrika, **52**, 437–446 (1965)

[19] Robertson, T., Wright, F., Dykstra, R.: Order Restricted Statistical Inference. New York, Wiley (1988)

[20] Silvapulle, M., Sen, P.: Constrained Statistical Inference. New York, Wiley (2004)

[21] Singh, B., Wright, F.: The level probabilities for a simple loop ordering. Annals of the Institute of Statistical Mathematics, **45**, 279–292 (1993)

Appendix 6.1

We now prove that our approach delivers the same results no matter what pairwise comparisons are chosen to construct the test statistic $\mathbf{T}$ defined in Section 6.2.1, provided that the pairs are chosen in such a way that the test statistic has a nondegenerate distribution; that is, the only requisite is that the covariance matrix be of full rank.

Since Bayes factors depend uniquely on integrals of the form presented in (6.6), we will show invariance of this integral both for ordered hypotheses and the full alternative hypothesis.

First, consider an arbitrary ordered hypothesis. Since $\boldsymbol{\Delta}$ measures differences between groups that are adjacent under the order specified by the hypothesis, its prior distribution $f(\boldsymbol{\Delta}|H_p)$ is the same for any choice of pairwise comparisons. Therefore, we only need to establish the equivalence of the probability density function $f(\mathbf{T}|\boldsymbol{\Delta}, H_p)$. Recall that $\boldsymbol{\delta}$ is a one-to-one linear function of $\boldsymbol{\Delta}$ and therefore we can equivalently write $f(\mathbf{T}|\boldsymbol{\delta}, H_p)$. Using (6.2), we have that

$$f(\mathbf{T}, H_p|\boldsymbol{\delta}) \propto \left(1 + \frac{1}{N-J}(\mathbf{T} - B\boldsymbol{\delta})'\Sigma_T^{-1}(\mathbf{T} - B\boldsymbol{\delta})\right)^{(N-1)/2}. \tag{6.8}$$

Let $\mathbf{T}^*$ be another test statistic obtained through another set of $J-1$ pairwise comparisons. Since the elements in $\mathbf{T}^*$ are linear combinations of those in $\mathbf{T}$, we can write $\mathbf{T}^* = D\mathbf{T}$, where D is a $(J-1) \times (J-1)$ matrix. Since $\mathbf{T}|\boldsymbol{\delta} \sim \mathcal{T}_{N-J}(B\boldsymbol{\delta}, \Sigma_T)$ and D is assumed to be of full rank, it is a well-known result that $\mathbf{T}^*|\boldsymbol{\delta} \sim \mathcal{T}_{N-J}(DB\boldsymbol{\delta}, D\Sigma_T D')$. Therefore, the density of $\mathbf{T}^*$ given $\boldsymbol{\delta}$ is proportional to

$$\begin{aligned} f(\mathbf{T}^*|\boldsymbol{\delta}, H_p) &\propto \left(1 + \frac{1}{N-J}(\mathbf{T}^* - DB\boldsymbol{\delta})'(D\Sigma_T D')^{-1}(\mathbf{T}^* - DB\boldsymbol{\delta})\right)^{(N-1)/2} \\ &= \left(1 + \frac{1}{N-J}(\mathbf{T} - B\boldsymbol{\delta})'D'D'^{-1}\Sigma_T D^{-1}D(\mathbf{T} - B\boldsymbol{\delta})\right)^{(N-1)/2} \\ &= \left(1 + \frac{1}{N-J}(\mathbf{T} - B\boldsymbol{\delta})'\Sigma_T^{-1}(\mathbf{T} - B\boldsymbol{\delta})\right)^{(N-1)/2}, \end{aligned}$$

which is equivalent to (6.8). This proves the result for any ordered hypothesis.

Now consider the full alternative hypothesis specifying that all group means are different. Using the same argument as that for ordered hypotheses proves the invariance of $f(\mathbf{T}|\boldsymbol{\Delta}, H_p)$. However, note that for the full alternative the groups could have been given in any order; that is, the definition of what categories are adjacent is arbitrary. Hence, we need to prove that the prior is invariant with respect to reorderings of the groups.

To prove this it is convenient to consider the prior in terms of $(\boldsymbol{\mu}, \sigma)$ and then derive the implied prior for $\boldsymbol{\Delta}$. Let $\boldsymbol{\mu} \sim \mathcal{N}(\mathbf{0}, 0.5\tau_0^2 I)$, where $\mathbf{0}$ is a vector of zeroes and I is the identity matrix, and let σ arise independently from a chi-square distribution with 1 degree of freedom. Then $\boldsymbol{\delta}$ can be computed as $\boldsymbol{\delta} = P\boldsymbol{\mu}/\sigma$, where P is a $(J-1) \times J$ contrast matrix taking the differences between μ_{j+1}/σ and μ_j/σ for $j = 1, \ldots, J-1$. Then by definition of the multivariate $\mathcal{T}$ distribution we get that

$$\boldsymbol{\delta} \sim \mathcal{T}_1\left(\mathbf{0}, 0.5\tau_0^2 PP'\right), \tag{6.9}$$

where $0.5PP'$ is a matrix with ones in the diagonal, -0.5 in the upper and lower subdiagonals, and 0 elsewhere.

Suppose that we reorder the elements of $\boldsymbol{\mu}$, say by defining $\boldsymbol{\mu}^* = Q\boldsymbol{\mu}$ where Q is a matrix with canonical unit vectors as rows (i.e., $QQ' = I$). Then if we define $\boldsymbol{\delta}^* = P\boldsymbol{\mu}^* = PQ\boldsymbol{\mu}/\sigma$, again by definition we get that

$$\boldsymbol{\delta}^* \sim \mathcal{T}_1\left(\mathbf{0}, 0.5\tau_0^2 PQQ'P'\right). \tag{6.10}$$

But since $QQ' = I$, this is equivalent to (6.9); that is, any group reordering results in the same prior distribution for the noncentrality parameters, which proves the result.

Appendix 6.2

To find the distribution of $\mathbf{T}$, we start by considering $\mathbf{Z} = (Z_1, \ldots, Z_J)'$, where $Z_j = \sqrt{N_j}\bar{y}_j/s_p$ for $j = 1, \ldots, J$. It is well known that $\sqrt{N_j}\bar{y}_j/\sigma \sim \mathcal{N}(\sqrt{N_j}\mu_j/\sigma, 1)$ and $(N-J)s_p^2/\sigma^2 \sim \chi^2_{N-J}$ with mutual independence, from which it follows that $\mathbf{Z} \sim \mathcal{T}_{N-J}(\boldsymbol{\mu}_Z, \Sigma_Z)$ in the sense defined by [18].

This definition of the noncentral $\mathcal{T}$ is obtained by dividing normal variates with nonzero mean by the square root of an independent scaled chi-square; that is, $T_j = \frac{\mathcal{N}(\sqrt{N_j}\mu_j/\sigma, 1)}{\sqrt{\chi^2_{N-J}/(N-J)}}$. A difficulty with this definition is that its probability density function cannot be evaluated explicitly. For this reason, we approximate it with the more common definition of the non-central $\mathcal{T}$, obtained with a location shift of a central $\mathcal{T}$ [7, 15]; that is, $T_j = \frac{\mathcal{N}(0,1)}{\sqrt{\chi^2_{N-J}/(N-J)}} + \sqrt{N_j}\mu_j/\sigma$. It is worthwhile noticing that for the central multivariate $\mathcal{T}$ case, both definitions are equivalent and that as $N - J \to \infty$, both approximate the same limiting normal distribution. Also, for both definitions, any linear combination or subvector remain within the family.

After some algebra, one finds that

$$
\begin{aligned}
E(Z_j) &= a_n\sqrt{N_j}\frac{\mu_j}{\sigma}, \\
V(Z_j) &= \frac{N-J}{N-J-2}(1 + N_j\mu_j^2/\sigma^2) - E(Z_j)^2, \\
\mathrm{Cov}(Z_j, Z_{j'}) &= \sqrt{N_j N_{j'}}\frac{\mu_j\mu_{j'}}{\sigma^2}\left(\frac{N-J}{N-J-2} - a_N^2\right), \\
a_N &= \sqrt{\frac{N-J}{2}}\frac{\Gamma(0.5(N-J-1))}{\Gamma(0.5(N-J))}.
\end{aligned}
\tag{6.11}
$$

Now, note that we can obtain $\mathbf{T} = D\mathbf{Z}$ where D is a matrix with elements

$$
\begin{aligned}
d_{j,j} &= \frac{-1}{\sqrt{N_j}\sqrt{\frac{1}{N_j} + \frac{1}{N_{j+1}}}}, \\
d_{j,j+1} &= \frac{1}{\sqrt{N_{j+1}}\sqrt{\frac{1}{N_j} + \frac{1}{N_{j+1}}}}, \\
d_{j,j'} &= 0, \text{ elsewhere.}
\end{aligned}
\tag{6.12}
$$

Following standard results for the multivariate $\mathcal{T}$ (see, e.g., [5]) we get that $\mathbf{T} \sim \mathcal{T}_{N-J}(\boldsymbol{\mu}_T, \Sigma_T)$, where $\boldsymbol{\mu}_T = D\boldsymbol{\mu}_Z$ and $\Sigma_T = D\Sigma_Z D'$. Working out the algebra gives the expressions for $\boldsymbol{\mu}$ and Σ_T given in (6.2).

Appendix 6.3

We discuss the choice of importance sampling distribution in the numerical evaluation of (6.6). Ideally, this distribution should give high probability to the regions in which the integrand takes large values.

Since there are two functions contributing to the integrand, namely the likelihood $f(\mathbf{T}|\boldsymbol{\Delta}, H_p)$ and the prior $f(\boldsymbol{\Delta}|H_p)$, we use a mixture of two truncated multivariate normal distributions: one centered at the likelihood and the other at the prior. Both multivariate normals have independent components that are left-truncated at zero, so that the mixture has the same support as the integrand. We use independent components out of convenience, since under independence one only needs to deal with univariate truncation regions.

To find the normal component centered at the likelihood, recall that the likelihood increases with $(\mathbf{T} - B\boldsymbol{\delta})'\Sigma_T^{-1}(\mathbf{T} - B\boldsymbol{\delta})$ and that $\boldsymbol{\Delta} = A\boldsymbol{\delta}$. This can be rewritten as $(\boldsymbol{\delta} - B^{-1}\mathbf{T})'B\Sigma_T^{-1}B(\boldsymbol{\delta} - B^{-1}\mathbf{T})$; that is, the maximum is achieved at $\boldsymbol{\delta} = B^{-1}\mathbf{T}$. Therefore, we set the mean of the proposal to $\mathbf{m} = AB^{-1}\mathbf{T}$. For the variance we use the diagonal elements of $\widehat{\Sigma}$, where $\widehat{\Sigma}$ is obtained by plugging in $\mathbf{m}$ for $\boldsymbol{\delta}$ in (6.2). To center the other multivariate normal on the prior we simply take a normal with zero mean and variance equal to τ_0^2 for all its components.

All that is left to set is the mixing weights. We set the weight for the component centered at the likelihood to be proportional to the value of the integrand at $\mathbf{m}$ relative to its value at $\mathbf{0}$, that is,

$$\frac{f(\mathbf{T}|\mathbf{m}, \widehat{\Sigma}, H_p)f(\mathbf{m}|H_p)}{f(\mathbf{T}|\mathbf{m}, \widehat{\Sigma}, H_p)f(\mathbf{m}|H_p) + f(\mathbf{T}|\mathbf{0}, \Sigma_0, H_p)f(\mathbf{0}|H_p)}, \tag{6.13}$$

where Σ_0 is obtained by plugging in $\mathbf{0}$ for $\boldsymbol{\delta}$ in (6.2); that is, the smaller the value of this function at the origin the proportion of observations that will be generated from the multivariate normal centered at zero will be smaller.

Appendix 6.4

Here we specify the hypotheses for the grief data in terms of $\boldsymbol{\delta}$ (Chapter 2 states them in terms of the group means $\mu_1, \ldots, \mu_8$), and we assign prior probabilities to them. Since $\delta_j = (\mu_{j+1} - \mu_j)/\sigma$, we have that $(\mu_j - \mu_{j'})/\sigma = (\mu_j - \mu_{j-1} + \mu_{j-1} - \ldots + \mu_{j'+1} - \mu_{j'})/\sigma = \sum_{l=j'}^{j} \delta_l$. The five conditions used to define $H_{1a}, \ldots, H_{1e}$, which we denote $C_1, \ldots, C_5$, are as follows:

C_1: $\mu_1 > \mu_2$, $\mu_3 > \mu_4$, $\mu_5 > \mu_6$, $\mu_7 > \mu_8$ if and only if

$$\delta_1 < 0,\ \delta_3 < 0,\ \delta_5 < 0,\ \delta_7 < 0,$$

C_2: $\mu_5 > \mu_1$, $\mu_6 > \mu_2$, $\mu_7 > \mu_3$, $\mu_8 > \mu_4$ if and only if

$$\sum_{j=1}^{4} \delta_j > 0,\ \sum_{j=2}^{5} \delta_j > 0,\ \sum_{j=3}^{6} \delta_j > 0,\ \sum_{j=4}^{7} \delta_j > 0,$$

C_3: $\mu_3 > \mu_1$, $\mu_4 > \mu_2$, $\mu_7 > \mu_5$, $\mu_8 > \mu_6$ if and only if

$$\delta_1 + \delta_2 > 0,\ \delta_2 + \delta_3 > 0,\ \delta_5 + \delta_6 > 0,\ \delta_6 + \delta_7 > 0,$$

C_4: $\mu_7 > \mu_5$, $\mu_8 > \mu_6$, $\mu_7 > \mu_3$, $\mu_8 > \mu_4$ if and only if

$$\delta_5 + \delta_6 > 0,\ \delta_6 + \delta_7 > 0,\ \sum_{j=3}^{6} \delta_j > 0,\ \sum_{j=4}^{7} \delta_j > 0,$$

C_5: $\mu_1 > \mu_3$, $\mu_2 > \mu_4$, $\mu_1 > \mu_5$, $\mu_2 > \mu_6$ if and only if

$$\delta_1 + \delta_2 < 0,\ \delta_2 + \delta_3 < 0,\ \sum_{j=1}^{4} \delta_j < 0,\ \sum_{j=2}^{5} \delta_j < 0.$$

The hypotheses can thus be stated as

H_{1a}: C_1 holds,
H_{1b}: C_1, and C_2 hold,
H_{1c}: C_1, C_2 and C_3 hold,
H_{1d}: C_1, and C_4 hold,
H_{1e}: C_1, C_4 and C_5 hold,
H_2: C_1 does not hold.

We assign a prior probability of .5 that C_1 holds (i.e., $\pi_{1a} = \pi_2 = .5$). For the remaining hypotheses we consider that all the possible combinations of $C_2, \ldots, C_5$ are equally likely, after excluding the combinations that are not possible. For instance, C_2 and C_5 require a different sign for $\sum_{j=1}^{4} \delta_j$ and, hence, they are mutually exclusive. Similarly, C_3 and C_5 are also mutually exclusive. After careful examination, we find that only the following 9 out of the 16 possible combinations are feasible:

1. C_2, C_3, C_4 hold, C_5 does not hold,
2. C_2, C_4 hold, C_3, C_5 do not hold,
3. C_2 holds, C_3, C_4, C_5 do not hold,
4. C_3, C_4 hold, C_2, C_5 do not hold,
5. C_3 holds, C_2, C_4, C_5 do not hold,
6. C_4, C_5 hold, C_2, C_3 do not hold,
7. C_4 holds, C_2, C_3, C_5 do not hold,
8. C_5 holds, C_2, C_3, C_4 do not hold,
9. C_2, C_3, C_4, C_5 do not hold.

To compute the prior and posterior probability of each hypothesis, we sum the appropriate combinations. Consider, for example, H_{1b}, which requires both C_1 and C_2 to hold. C_2 holds in combinations 1, 2, and 3, and hence a priori it has probability 3/9. C_1 holds with probability 1/2 (with independence), so the prior probability of H_{1b} is $\frac{1}{2}\frac{3}{9} = .167$. For the probability a posteriori, we compute Bayes factors for each of the nine combinations, find their posterior probabilities and then sum the appropriate combinations.

The prior probabilities presented here should be considered illustrative. In particular, it would also be reasonable to consider some degree of subjectivity or dependence between hypotheses.

7

Objective Bayes Factors for Informative Hypotheses: "Completing" the Informative Hypothesis and "Splitting" the Bayes Factors

Luis Raúl Pericchi Guerra, Guimei Liu, and David Torres Núñez

Department of Mathematics, University of Puerto Rico at Rio Piedras Campus, P.O. Box 23355, San Juan 00931-3355, Puerto Rico lrpericchi@uprrp.edu and luarpr@gmail.com and liugm@zucc.edu.cn and david.torres.math@gmail.com

7.1 Introduction

Informative hypotheses are particular hypotheses about the vector of means μ in the classical normal ANOVA model:

$$\mathbf{Y} = X\mu + \epsilon,$$

where ϵ is the vector of uncorrelated normal errors with constant variance σ^2. Informative hypotheses involve a complex ordering of means like, say,

$$H_{1a} : \{\mu_1 = \mu_2\} < \mu_3 < \{\mu_4 = \mu_5\}$$

or

$$H_{1b} : \mu_1 < \{\mu_2 = \mu_3\} < \mu_4 < \mu_5,$$

among other possibilities. In order to compare models, in the canonical Bayesian way (i.e., in terms of Bayes factors) and upon specification of a prior under, say, H_{1a}, $\pi(\mu_1, \mu_2, \mu_3, \mu_4, \mu_5, \sigma | H_{1a})$, we need to calculate the marginal densities:

$$m(\mathbf{y}|H_{1a}) = \int_{H_{1a}} f(\mathbf{y}|\mu, \sigma, H_{1a}) \pi(\mu_1, \mu_2, \mu_3, \mu_4, \mu_5, \sigma | H_{1a}) \, d\mu \, d\sigma. \quad (7.1)$$

There are two difficulties with (7.1):

1. The difficulty of assessment of the prior $\pi(.|H_{1a})$, which is potentially highly influential.
2. The difficulty in the computation of integral in (7.1), due to the awkward region of integration, which is in the very definition of an informative hypothesis.

A direct approach to deal with inequality constraints is in Berger and Mortera, who studied in depth the simplest examples with inequality constraints [3]. However their detailed and incisive direct approach seems bound to be restricted to very simple situations, unsuited to the kind of examples encountered for example in psychology, particularly for dimensions greater than 2. Our own approach however, is conceptually closest to Berger and Mortera's approach. Another general approach is presented in Chapter 4, which suggests an encompassing approach plus the replacement of equality constraints by constraints that the means are close. Other relevant references are [15], [16], [17], and Chapter 8. The ideas in the present chapter have been developed independently of these cited works.

Our own approach, which we call "completing and splitting" the informative hypothesis, circumvents both difficulties with (7.1), permits dealing with complex hypotheses, and does not involve the assessment of extra parameters of "proximity" in hypotheses of equality. The basic idea is to make a natural completion of an informative hypothesis. Take, for instance, H_{1a} with a completed hypothesis

$$H_{1a}^{*} : \{\mu_1 = \mu_2\} \neq \mu_3 \neq \{\mu_4 = \mu_5\},$$

with a corresponding completion of H_{1b} denoted by H_{1b}^{*}, in which all order constraints are replaced by inequalities. These completions allow one to split the problem in (7.1) in two factors: One of a usual objective marginal density and (objective) Bayes factor and the other factor as a probability of a region with positive measure, thus, the name "completing and splitting" approach. In the sequel, it will be shown that

$$m(\mathbf{y}|H_{1a}) = m(\mathbf{y}|H_{1a}^{*}) \times Pr(H_{1a}|\mathbf{y}, H_{1a}^{*}), \tag{7.2}$$

which is the pivotal identity of the completion and splitting approach. Equation (7.2) entails a substantive methodological simplification of (7.1). The first factor can be dealt with using objective Bayes factor theory and the second through Markov chain Monte Carlo (MCMC) methods.

In the next sections we briefly outline some theory of objective Bayes factors, after which the development of our approach is presented. Several examples of different objective factors are exposed that lead to different versions of our general approach. A numerical illustration is presented and, finally, conclusions are given.

7.2 Preliminaries: General Coherent Approach of Posterior Model Probabilities

We start the general framework on posterior model probabilities and objective Bayes factors presenting the general, coherent, and elegant Bayesian formalism to tackle the problems of hypotheses testing and model selection. Notice, however, that it is harder than it looks at first sight.

7.2.1 Bayes Factors and Posterior Model Probabilities

Suppose that we are comparing q models for the data $\mathbf{y}$,

$$H_i : \mathbf{Y} \text{ has density } f_i(\mathbf{y}|\theta_i), \quad i = 1, \ldots, q,$$

where the $\theta_i = (\mu_i, \sigma^2)$ are the unknown model parameters. Suppose that we have available prior distributions, $\pi_i(\theta_i)$, $i = 1, \ldots, q$, for the unknown parameters. Define the marginal or predictive densities of $\mathbf{Y}$,

$$m_i(\mathbf{y}) = \int f_i(\mathbf{y}|\theta_i)\pi_i(\theta_i)\, d\theta_i.$$

A central quantity for comparing models, the Bayes factor of H_j to H_i, is given by

$$BF_{ji} = \frac{m_j(\mathbf{y})}{m_i(\mathbf{y})} = \frac{\int f_j(\mathbf{y}|\theta_j)\pi_j(\theta_j)\, d\theta_j}{\int f_i(\mathbf{y}|\theta_i)\pi_i(\theta_i)\, d\theta_i}.$$

The Bayes factor is often interpreted as the "evidence provided by the data in favor or against model H_j versus the alternative model H_i," but it should be remembered that Bayes factors depend also on the priors. In fact, assuming an objective Bayesian viewpoint, the priors may be thought of "weighting measures" and the Bayes factor as the "weighted averaged likelihood ratio," which are in principle more comparable than maximized likelihood ratios, as measures of relative fit. Notice that if H_i is nested in H_j, then, logically, the maximized likelihood of H_j is a fortiori larger than the maximized likelihood of H_i. On the other hand, Bayes factors introduce an automatic penalty for overparametrization.

If prior probabilities $P(H_i)$, $i = 1, \ldots, q$, of the models are available, then one can compute the posterior probabilities of the models from the Bayes factors. Using Bayes' rule, it is easy to see that posterior probability of H_i, given the data $\mathbf{y}$, is

$$\mathrm{PMP}(H_i \,|\, \mathbf{y}) = \frac{P(H_i)m_i(\mathbf{y})}{\sum_{j=1}^{q} P(H_j)m_j(\mathbf{y})} = \left[\sum_{j=1}^{q} \frac{P(H_j)}{P(H_i)} BF_{ji}\right]^{-1}. \tag{7.3}$$

A particularly common choice of the prior model probabilities is $P(H_i) = 1/q$, so that each model has the same initial probability, but there are other possible choices; see, for example, [4].

7.2.2 Comparisons and Connections Frequentist and Bayesian Evidences

In the frequentist literature, the different methods for model selection (as Akaike Information Criterion, AIC, [2]) can be written as

Likelihood Ratio × *Correction Factor.*

A Bayesian version for grouping the different methods is

Un-normalized Bayes Factor × *Correction Factor,*

or in symbols,

$$BF_{ij} = \frac{m_i^N(\mathbf{y})}{m_j^N(\mathbf{y})} \times CF_{ji} = BF_{ij}^N \times CF_{ji},$$

where BF_{ij}^N is the ratio of marginal densities, calculated with noninformative, and typically improper (do not integrate one in the parameter space), prior densities $\pi^N(\theta_k), k = i, j$. A very important question is whether this correction of BF_{ij}^N is actually equivalent with the use of proper and sensible priors, to calculate proper Bayes factors (i.e., scaled Bayes factors calculated with proper priors). The next general principle addresses this issue.

Principle 1. *Testing and model selection methods should correspond, in some sense, to actual Bayes factors, arising from reasonable default prior distributions.*

This principle was first stated in [4]. It is natural for a Bayesian to accept that the best discriminator between procedures is the study of the prior distribution (if any) that give rise to the procedure or that is implied by it. Other properties, like large sample size consistency, are rough as compared with the incisiveness of Principle 1. In [5] it is shown that intrinsic Bayes factors (see below) actually correspond to the use of (conditionally) proper priors. Also, fractional Bayes factors often correspond to conditionally proper priors. Training samples are important in several of the methods for objective Bayes factors, like intrinsic and fractional Bayes factors.

Definition 1. *Deterministic (minimal) training samples, are simply subsets (of the sample* $\mathbf{y}$*)* $\mathbf{y}(l)$, $l = 1, \ldots, L$, *of size* m, *as small as possible so that, starting with improper priors* π^N, *using the training sample as the sample in Bayes' theorem, all the updated posteriors under all models become proper.*

A general discussion about different kinds of training samples can be found in [8].

Consider two models, H_i and H_j, and take a minimal (or larger) training sample $\mathbf{y}(l)$, which is a subset of the whole sample $\mathbf{y}(l) \subseteq \mathbf{y}$. Denote by $\mathbf{y}(-l)$ the complement of the training sample. Now use $\mathbf{y}(l)$ to make the priors proper and use $\mathbf{y}(-l)$ to form the (proper) Bayes factor, and use Bayes' theorem to get the following identity:

$$\begin{aligned} BF_{ij}(\mathbf{y}(-l)|\mathbf{y}(l)) &= \frac{\int f(\mathbf{y}(-l)|\theta_i, \mathbf{y}(l))\pi^N(\theta_i|\mathbf{y}(l))\, d\theta_i}{\int f(\mathbf{y}(-l)|\theta_j, \mathbf{y}(l))\pi^N(\theta_j|\mathbf{y}(l))\, d\theta_j} \\ &= BF_{ij}^N(\mathbf{y}) \times BF_{ji}^N(\mathbf{y}(l)), \end{aligned}$$

where

$$BF_{ji}^N(\mathbf{y}(l)) = \frac{\int f(\mathbf{y}(l)|\theta_j)\pi^N(\theta_j)\, d\theta_j}{\int f(\mathbf{y}(l)|\theta_i)\pi^N(\theta_i)\, d\theta_i},$$

and $BF_{ij}^N(\mathbf{y})$ is the same but replacing $\mathbf{y}(l)$ by $\mathbf{y}$. Use of training samples to scale Bayes factors seemed to have been pioneered in [18]. This general reference is an important and often overlooked milestone, seminal in the area of training samples and model selection. It is timely that this book rescues this important monograph.

7.3 Different Approaches for Objective Bayes Factors

In this section a brief review is given to different approaches to objective Bayes factors, which are of direct relevance to the completing and splitting approach of informative hypothesis, as presented in (7.2) in the introduction. Application of these methods to informative hypotheses will be presented in Sections 7.4 and 7.5.

7.3.1 Conventional Prior Approach

It was Jeffreys [14] who recognized the problem of arbitrary constants arising in hypotheses testing problems, implied by the use of "Jeffreys' Rule" for choosing objective-invariant priors for estimation problems. For testing problems then, a convention has to be established. His approach is based on (i) using non-informative priors only for common parameters in the models, so that the arbitrary multiplicative constant for the priors would cancel in all Bayes factors and (ii) using default proper priors for orthogonal parameters that would occur in one model but not the other. These priors are neither vague nor overinformative, but correspond to a definite but limited amount of information.

Example 1: Normal Mean, Jeffreys' Conventional Prior
Suppose the data is $\mathbf{Y} = (Y_1, \ldots, Y_n)$, where the Y_l are iid $\mathcal{N}(\mu, \sigma_2^2)$ under H_2. Under H_1, the Y_l are $\mathcal{N}(0, \sigma_1^2)$. Since the mean and variance are orthogonal in the sense of having diagonal expected Fisher's information matrix, Jeffreys equated $\sigma_1^2 = \sigma_2^2 = \sigma^2$. Because of this, Jeffreys suggested that the variances can be assigned the same (improper) noninformative prior $\pi^J(\sigma) = 1/\sigma$, since the indeterminate multiplicative constant for the prior would cancel in the Bayes factor. (See [20] for a formal justification.)

Since the unknown mean μ occurs in only H_2, it needs to be assigned a proper prior. Jeffreys came up with the following desiderata for such a prior that in retrospect appears as compelling: (i) It should be centered at zero (i.e., centered at the null hypothesis), (ii) have scale σ (i.e., have the information provided by one observation), (iii) be symmetric around zero, and (iv) have no moments. He then settled for the Cauchy prior Cauchy$(0, \sigma^2)$ as the simplest distribution that obeys the desiderata. In formula's, Jeffreys's conventional prior for this problem is

$$\pi_1^J(\sigma_1) = \frac{1}{\sigma_1}, \quad \pi_2^J(\mu, \sigma_2) = \frac{1}{\sigma_2} \cdot \frac{1}{\pi\sigma_2(1+\mu^2/\sigma_2^2)}. \tag{7.4}$$

This solution is justified as a Bayesian prior, as shown in [20].

7.3.2 Intrinsic Bayes Factor (IBF) Approach

For the q models $H_1, \ldots, H_q$, suppose that (ordinary, usually improper) noninformative priors $\pi_i^N(\theta_i)$, $i = 1, \ldots, q$, have been chosen, preferably as reference priors. The general strategy for defining IBFs starts with the definition of a proper and minimal deterministic training sample, $\mathbf{y}(l)$. Using a particular training sample scales the improper Bayes factor. But, in principle, a single training sample is arbitrary. So it is necessary to perform some sort of average among the corrections from different training samples.

A variety of different averages are possible; here consideration is given only to the Arithmetic IBF (AIBF) defined as

$$BF_{ji}^{AI} = BF_{ji}^N(\mathbf{y}) \cdot \frac{1}{L}\sum_{l=1}^{L} BF_{ij}^N(\mathbf{y}(l)). \tag{7.5}$$

It turns out by [4] that it is the AIBF, which is equivalent to the use of proper (conditional) priors.

7.3.3 Fractional Bayes Factor as an Average of Training Samples for Exchangeable Observations

The Fractional Bayes Factor (FBF) was introduced in [19]. The FBF uses a fraction, b, of each likelihood function, $f_i(\mathbf{y}|\theta_i)$, with the remaining $1-b$ fraction of the likelihood used for model discrimination. Using Bayes' rule, it follows that the FBF of model H_j to model H_i is then given by

$$BF_{ji}^{F(b)} = BF_{ji}^N(\mathbf{y}) \frac{\int f^b(\mathbf{y}|\theta_i)\pi_i^N(\theta_i)\, d\theta_i}{\int f^b(\mathbf{y}|\theta_j)\pi_j^N(\theta_j)\, d\theta_j} = BF_{ji}^N(\mathbf{y}) \frac{m_i^b(\mathbf{y})}{m_j^b(\mathbf{y})}. \tag{7.6}$$

The usual choice of b (as in the examples in [19] and the discussion in [3] of [19]) is $b = m/n$, where m is the minimal training sample size (i.e., the number of observations contained in a minimal training sample).

It has been pointed out in [10], that the FBF for exchangeable observations at least can be thought of as a Bayes factor with a correction obtained through the geometrical average of likelihoods over all training samples; that is, $m_j^b(\mathbf{y})$ can be obtained by integrating the geometric average of the product of the likelihoods over all training samples of equal size $r = b \times n$,

$$m_j^b(\mathbf{y}) = \int \left[\prod f(\mathbf{y}(l)|\theta_j)\right]^{\frac{1}{N(r)}} \pi^N(\theta_j)\, d\theta_j,$$

where the product is over all proper training samples of size r, and $N(r)$ is the total number of proper training samples of size $m \leq r < n$. This property is important to understand the conceptual connections between FBF's and IBF's.

7.3.4 The Intrinsic Prior Approach

Methods closely related to the IBF approach are the intrinsic prior and the Emprirical Expected Posterior Prior (EP) approach. The IBF approach can be thought of as the long sought device for the generation of good conventional priors for model selection in nested scenarios. See [12] and references therein. As part of the general evaluation strategy of the methods, Principle 1 proposes investigation of so-called *intrinsic priors* corresponding to a model selection method.

The following is a key approximation to a Bayes factor associated with priors π_j and π_i:

$$BF_{ji} = BF_{ji}^N \cdot \frac{\pi_j(\widehat{\theta}_j)\pi_i^N(\widehat{\theta}_i)}{\pi_j^N(\widehat{\theta}_j)\pi_i(\widehat{\theta}_i)}(1 + o(1)), \tag{7.7}$$

where $\widehat{\theta}_i$ and $\widehat{\theta}_j$ are the maximum likelihood estimates (m.l.e.s) under H_i and H_j. This approximation holds in considerably greater generality than does the Schwarz approximation and it is fundamental for Bayes factors theory and practice. On the other hand, methods like the FBF and IBF yield

$$BF_{ji} = BF_{ji}^N \cdot CF_{ij}, \tag{7.8}$$

where CF_{ij} is the it correction factor. To define intrinsic priors, equate (7.7) with (7.8), yielding

$$\frac{\pi_j(\widehat{\theta}_j)\pi_i^N(\widehat{\theta}_i)}{\pi_j^N(\widehat{\theta}_j)\pi_i(\widehat{\theta}_i)}(1 + o(1)) = CF_{ij}. \tag{7.9}$$

In the nested model scenario (H_i nested in H_j), for the AIBF and under mild assumptions, solutions are given by

$$\pi_i^I(\theta_i) = \pi_i^N(\theta_i), \quad \pi_j^I(\theta_j) = \pi_j^N(\theta_j) BF_j^*(\theta_j), \tag{7.10}$$

the last expression having a simpler expression for exchangeable observations:

$$BF_j^*(\theta_j) = \int f_j(\mathbf{y}(l)|\theta_j) \frac{m_i^N(\mathbf{y}(l))}{m_j^N(\mathbf{y}(l))}\, d\mathbf{y}(l).$$

The intrinsic prior π_j^I is proper under some generality. The following theorem was given in [4], assuming that the sampling model is absolutely continuous.

Theorem 1. *Suppose that either the null model is simple or the prior under the null model is proper. Then the intrinsic prior is proper. When the prior under the null is not proper, in [4], a justification of intrinsic priors is given.*

7.4 Informative Hypotheses

We now proceed to apply the objective Bayes factor methods to informative hypotheses. Suppose we have, say, four groups, with data $y_{i,j}, i = 1, \ldots, 4; j = 1, \ldots, n_i$, and the data are assumed to be

$$Y_{i,j} \sim \mathcal{N}(\mu_i, \sigma^2).$$

The null hypothesis

$$H_0 : \mu_1 = \mu_2 = \mu_3 = \mu_4, \sigma > 0$$

is compared against an informative hypothesis like

$$H_{1a} : \mu_3 < \{\mu_1 = \mu_4 = \mu\} < \mu_2, \sigma > 0.$$

This is not a standard comparison and has generated considerable interest under the name of informative hypotheses, important among other areas in psychology and is the subject matter of this book. We are going to follow two different strategies (and different methods within each), but all the methods are based on the fundamental Lemma 1. In the first strategy (denoted objective strategy and discussed in Section 7.5), we start just with standard noninformative priors for hypotheses H_k. In the second strategy (denoted conventional prior strategy and discussed in Section 7.6) we assume a prior which is conditionally proper on the parameters under test and noninformative in the remaining parameters. Denote by θ_k and σ the unknown parameters under the hypothesis H_k; then

$$\pi_k^N(\theta_k, \sigma) \propto 1/\sigma.$$

In the conventional prior approach, the above prior is assumed only for the null, and a conditional proper prior is assumed for the "parameters under test" – that is, the parameters that are bigger or smaller than the others (i.e., parameters which are subject to order constraints). Alternatively, reparametrizing H_{1a} calling $\theta = \mu$, $\theta_1 = \mu - \mu_3$, and $\theta_2 = \mu_2 - \mu$, the alternative hypothesis becomes $H_{1a} : \theta_1 > 0$ and $\theta_2 > 0$ and the parameters under test are $[\theta_1, \theta_2]$.

We need to compute, calling θ_a all the mean parameters under H_{1a}

$$BF^N_{H_{1a},H_0} = \frac{\int_{H_{1a}} f(y|\theta_a,\sigma)\pi^N(\theta_a,\sigma)\ d\theta_a\ d\sigma}{\int_{H_0} f(y|\theta_0,\sigma)\pi^N(\theta_0,\sigma)\ d\theta_0\ d\sigma}, \tag{7.11}$$

where θ_0 are the unknown parameters under H_0. Notice that using $\pi^N(\theta_a,\sigma)$ in (7.11) makes the comparison without a proper scaling, so we also have to compute a correction factor $CF_{0,1a}$, so that

$$BF_{H_{1a},H_0} = BF^N_{H_{1a},H_0} \times CF_{H_0,H_{1a}}. \tag{7.12}$$

The numerator in (7.11) is extremely difficult to evaluate, except perhaps in the very simple situations considered in depth in [3]. Equation (7.12) is even more difficult to be directly evaluated, due to the correction factor $CF_{H_0,H_{1a}}$. So we found the need to restate the problem, and we will see that such a restatement allows a massive simplification.

7.4.1 A Restatement of the Problem of Informative Hypotheses: Completing the Informative Hypothesis and Splitting the Bayes Factors

A useful restatement of the problem is achieved invoking a variation of the theme of the encompassing approach presented in Chapter 4, but instead of an overall encompassing model, we have specific encompassing models for each informative hypothesis. We call this novel approach Completing and Splitting informative hypotheses.

1. Completing: First of all, it is useful to denote by H^*_{1a} the hypothesis formed by replacing any inequality order constraints by inequalities – for example, regarding the illustration above, where $H_{1a}: \mu_3 < \{\mu_1 = \mu_4 = \mu\} < \mu_2$ then $H^*_{1a}: \mu_3 \neq \{\mu_1 = \mu_4 = \mu\} \neq \mu_2$.
2. Splitting: Denote by Θ_a the parameter space, which contains H^*_{1a}.

Lemma 1. *It turns out that*

$$BF^N_{H_{1a},H_0} = \frac{m^N_{1a}(\mathbf{y})}{m^N_0(\mathbf{y})} \times Pr(H_{1a}|\mathbf{y}, H^*_{1a}) \tag{7.13}$$

or, equivalently,

$$BF^N_{H_{1a},H_0} = BF^N_{H^*_{1a},H_0} \times Pr(H_{1a}|\mathbf{y}, H^*_{1a}),$$

where

$$Pr(H_{1a}|\mathbf{y}, H^*_{1a}) = \frac{\int_{H_{1a}} f(\mathbf{y}|\theta_a,\sigma)\pi^N(\theta_a,\sigma)\ d\theta_a\ d\sigma}{\int_{H^*_{1a}} f(\mathbf{y}|\theta_a,\sigma)\pi^N(\theta_a,\sigma)\ d\theta_a\ d\sigma},$$

$$m^N_{1a}(\mathbf{y}) = \int_{H^*_{1a}} f(\mathbf{y}|\theta_a,\sigma)\pi^N(\theta_a,\sigma)\ d\theta_a\ d\sigma,$$

and

$$m_0^N(\mathbf{y}) = \int_{H_0} f(\mathbf{y}|\theta_0, \sigma)\pi^N(\theta_0, \sigma)\, d\theta_0\, d\sigma.$$

The proof follows from the following observation: In (7.11) multiply and divide by (the marginal density of the completed hypotheses, which is assumed to exist)

$$m_{1a}^N(\mathbf{y}) = \int_{H_{1a}^*} f(\mathbf{y}|\theta_a, \sigma)\pi^N(\theta_a, \sigma)\, d\theta_a\, d\sigma.$$

The expression (7.13) is the completed and splitted version of (7.11) In the example above, H_{1a}^* becomes the completed hypothesis of H_{1a}, useful for splitting the Bayes factors. The equation (7.13) entails an enormous simplification of (7.11), (and potentially also of (7.12)), since the first factor is just the familiar $BF_{H_{1a}^*, H_0}^N$, calculated on unrestricted parameter spaces that can be dealt with the usual methods of objective Bayesian model selection (which is the reason to have introduced objective BFs methods in the previous section), and it is only the second term, which involves the awkward parameter region. But, a big *but*, the second term is, fortunately, a probability, and as such can be dealt effectively by MCMC methods like Gibbs Sampler, or more general samplers, in the following way: *Generate a sample from the posterior, and calculate the proportion of times that H_{1a} is obeyed.*

In the rest of the chapter, Lemma 1 will be used repeatedly. Equation (7.13) under normal linear models is readily available. The first ratio is known in closed form (e.g., see [4, 5]), and the second factor is a standard probability (of an awkward region) that can be dealt effectively by the Gibbs for normal linear models. It is important to notice that the computation of the probability in (7.13) is simple because, for each generated sample, it is easy to verify whether H_{1a} is obeyed or not and the probability is estimated as the proportion of samples on which the restrictions specified by H_{1a} are true.

An algorithm to compute the factors of Lemma 1 is given by the following two steps:

1. Compute the factor $BF_{H_{1a}^*, H_0}^N$, which is available in closed form in some generality that covers all normal linear models as shown for example in [4, 5].
2. The second factor $Pr(H_{1a}|\mathbf{y}, H_{1a}^*)$ is computed by any standard MCMC method (including Gibbs sampler in the context of normal linear models, for example using the BUGS software): Generate samples from the posterior density: $\pi(\theta_a, \sigma|\mathbf{y}, H_{1a}^*)$; name the samples $[\theta_a(1), \ldots, \theta_a(L)]$, and count how many of these samples fulfill the conditions that defines H_{1a}, like $[\theta_1(l) > 0$ and $\theta_2(l) > 0]$. Dividing by the total number of samples L gives an estimate of the posterior probability.

Comment in passing: Another way to deal (approximately) with (7.11) is through asymptotic expansions. This cannot be done directly via Laplace expansions, due to the nature of the parameter space of informative hypotheses,

G. Liu (PhD thesis, UPR, in progress). It is possible however, to use methods suited to this kind of comparisons such as in [11].

7.4.2 Extension to More Than One Informative Hypotheses

Suppose that it is also of interest to entertain the following informative hypothesis:

$$H_{1b} : \{\mu_1 = \mu_3 = \mu\} < \mu_4 < \mu_2.$$

We can extend the discourse of Lemma 1. First, we complete this informative hypothesis by

$$H^*_{1b} : \{\mu_1 = \mu_3 = \mu\} \neq \mu_4 \neq \mu_2.$$

Lemma 2. *For more than one informative hypothesis the (unscaled) Bayes factors can be calculated as*

$$BF^N_{H_{1a},H_{1b}} = \frac{m^N_{H^*_{1a}}}{m^N_{H^*_{1b}}} \times \frac{Pr(H_{1a}|\mathbf{y}, H^*_{1a})}{Pr(H_{1b}|\mathbf{y}, H^*_{1b})}.$$

The proof is by applying Lemma 1 to each of the hypotheses.

We now have Lemma 1 and Lemma 2 as computational devices, but we present in Sections 7.5 and 7.6 strategies to *scale* the Bayes factors (recall that the priors used so far in the informative hypotheses are improper and thus unscaled).

7.4.3 General Notation for Informative Hypotheses

We now propose a general notation, seemingly useful for general informative hypotheses when a large number of different hypotheses is entertained. Suppose that we have four means, without loss of generality. Rename the hypotheses as follows:

$$H_0 : \mu_1 = \mu_2 = \mu_3 = \mu_4, \; H_M : \text{Negation of } H_0,$$

$$H_{12,3,4} : \{\mu_1 = \mu_2\} < \mu_3 < \mu_4, \; H^*_{12,3,4} : \{\mu_1 = \mu_2\} \neq \mu_3 \neq \mu_4,$$

$$H_{1,23,4} : \mu_1 < \{\mu_2 = \mu_3\} < \mu_4, \; H^*_{1,23,4} : \mu_1 \neq \{\mu_2 = \mu_3\} \neq \mu_4.$$

This notation is general, useful for complex studies, and descriptive of the meaning of each hypothesis. However, in the present chapter (for simplicity and ease of exposition) in the illustration we will use the simpler notation: $H_{1a} = H_{3,14,2}$ and $H_{1b} = H_{13,4,2}$.

In the next two sections we will present several methods for scaling, dividing the methods in two cases: (i) Objective strategy (Section 7.5) and (ii) conventional prior strategy (Section 7.6).

7.5 Some Feasible Methods of Objective Bayes Factors for Informative Hypothesis

We begin with two procedures (this is nonexhaustive; other methods are feasible) that are objective and empirical. We selected these two methods for simplicity of the programming.

7.5.1 Fractional Bayes Factor

The fractional Bayes factor (FBF) is given by

$$BF^F_{H_{1a},H_0} = \frac{\int_{H_{1a}} f(\mathbf{y}|\theta_a)\pi^N(\theta_a)\, d\theta_a \int_{H_0} f^b(\mathbf{y}|\theta_0)\pi^N(\theta_0)\, d\theta_0}{\int_{H_0} f(\mathbf{y}|\theta_0)\pi^N(\theta_0)\, d\theta_0 \int_{H_{1a}} f^b(\mathbf{y}|\theta_a)\pi^N(\theta_a)\, d\theta_a}. \tag{7.14}$$

This expression can be converted into

$$BF^F_{H_{1a},H_0} = BF^F_{H^*_{1a},H_0} \times \frac{Pr^f(H_{1a}|\mathbf{y}, H^*_{1a})}{Pr^{f^b}(H_{1a}|\mathbf{y}, H^*_{1a})}. \tag{7.15}$$

The passage from (7.14) to (7.15) is obtained by multiplying and dividing by $\int_{H^*_{1a}} f(\mathbf{y}|\theta_a)\pi^N(\theta_a)\, d\theta_a$ and dividing and multiplying by $\int_{H^*_{1a}} f^b(\mathbf{y}|\theta_a)\pi^N(\theta_a)\, d\theta_a$ (assumed to exist).

If, as in Section 7.4.2, it is also of interest to entertain the informative hypothesis

$$H_{1b} : \{\mu_1 = \mu_3 = \mu\} < \mu_4 < \mu_2,$$

a possible solution is to replicate the previous analysis and complete and split as

$$H^*_{1b} : \{\mu_1 = \mu_3 = \mu\} \neq \mu_4 \neq \mu_2,$$

$$BF_{H_{1b},H_0} = BF_{H^*_{1b},H_0} \times \frac{Pr^f(H_{1b}|\mathbf{y}, H^*_{1b})}{Pr^{f^b}(H_{1b}|\mathbf{y}, H^*_{1b})}. \tag{7.16}$$

Thus, if we take the ratio between H_{1a} and H_{1b}, we get

$$BF_{H_{1a},H_{1b}} = BF^F_{H^*_{1a},H^*_{1b}} \times \frac{Pr^f(H_{1a}|\mathbf{y}, H^*_{1a}) \cdot Pr^{f^b}(H_{1b}|\mathbf{y}, H^*_{1b})}{Pr^f(H_{1b}|\mathbf{y}, H^*_{1b}) \cdot Pr^{f^b}(H_{1a}|\mathbf{y}, H^*_{1a})}. \tag{7.17}$$

The preceding marginals, or any other marginal that will be developed in the sequel, may be converted into probabilities, using the general formula presented in Section 7.2.1. In this chapter we assume equal prior probabilities for all the entertained hypothesis; that is, $P(H_j) = 1/q$, where q is the number of competing hypotheses. The formula in Section 7.2.1 is valid under any of the methods to be developed in the sequel, not only under the FBF.

7.5.2 Empirical Expected Posterior Prior (EP)

Now, take the observations $\mathbf{y}(l)$ a training sample from the empirical data. Again define the prior in two steps:

Step 1:

$$\pi_{H_{1a}^*}^{EP}(\theta_a) = \frac{1}{L}\sum \pi_{H_{1a}^*}^{EP}(\theta_a|\mathbf{y}(l)).$$

Step 2:

$$\pi_{H_{1a}}^{EP}(\theta_a) = \frac{\pi_{H_{1a}^*}^{EP}(\theta_a)}{Pr^{\pi^{EP}}(H_{1a}|H_{1a}^*)} I_{H_{1a}}(\theta_a).$$

By definition of the EP method and the prior specified in Step 2, we have

$$BF_{H_{1a},H_0}^{EP} = \frac{\int_{H_{1a}} f(\mathbf{y}|\theta_a)\pi^{EP}(\theta_a)\, d\theta_a}{\int_{H_0} f(\mathbf{y}|\theta_0)\pi^{EP}(\theta_0)\, d\theta_0} \times \frac{1}{Pr^{\pi^{EP}}(H_{1a}|H_{1a}^*)}.$$

We now multiply and divide by $\int_{H_{1a}^*} f(\mathbf{y}|\theta_a)\pi^{EP}(\theta_a)\, d\theta_a$, and we obtain the completed and splitted EP Bayes factor:

$$BF_{H_{1a},H_0} = BF_{H_{1a}^*,H_0}^{EP} \times \frac{Pr^{\pi^{EP}}(H_{1a}|\mathbf{y}, H_{1a}^*)}{Pr^{\pi^{EP}}(H_{1a}|H_{1a}^*)}. \tag{7.18}$$

Notice that the last ratio involves a fraction of posterior over a prior probability. Both are perfectly well defined and can be calculated through MCMC methods. The prior probability enters in the formula naturally, since it was introduced in Step 2. See [16] for a related equation.

7.6 Conventional Conditionally Proper Priors

We now continue with procedures that do not have an integrable prior on all parameters, but there is an integrable, and in fact proper (integrating one) conditional prior on the subset of the parameters involved in the hypothesis.

Denote by $\theta_c = [\theta_{c_1}, \ldots, \theta_{c_{k_1}}]$ the parameters under test, referred to in Section 7.4, and call θ_0 the remaining parameters, common with the null hypothesis, which are not subjected to the informative hypothesis (e.g., in H_{1a}, $\theta_a = [\theta_c, \theta_0]$). We make explicit the following assumption.

Assumption 1. The prior π is such that

$$\int_{H_1^*} \pi(\theta_c|\theta_0, \sigma)\, d\theta_c = 1,$$

for all $[\theta_0, \sigma]$ and

$$\int_{H_1} \pi(\theta_c|\theta_0, \sigma)\, d\theta_c$$

is the same for all $[\theta_0, \sigma]$.

Assumption 1 has three distinct components: (i) The conditional prior is integrable, (ii) it integrates one, and (iii) the integral for H_1 and H_1^* are independent of the values of θ_0 and σ. If (ii) is not fulfilled, it may be renormalized. We will be computing

$$\int_{H_1} \pi(\theta_c|\theta_0, \sigma)\, d\theta_c, \tag{7.19}$$

which will typically entail probabilities of the form

$$Pr(\theta_{c_1} \geq 0, \ldots, \theta_{c_{k_1}} \leq 0).$$

When θ_c is one dimensional (i.e. $k_1 = 1$), a balanced prior yields a priori probability of 1/2 (or close to a half), balancing in this way the probabilities of the hypothesis and the alternative. For multivariate θ_c, if the components θ_{c_i} are independent, then the probability in (7.19) will be equal to (or close to) to $1/2^{k_1}$, where k_1 is the number of inequality constraints in H_{1a}. Complete independence is seldom fulfilled, however, but reasonable conventional priors, should not give prior probabilities "too far" from $1/2^{k_1}$. We now see examples on which Assumption 1 is obeyed.

7.6.1 Conventional Zellner-Siow Prior

Start with the Zellner-Siow kind of conventional prior: Assume that the parameters under test (θ_c) have dimension k_1, and θ_0 are the common mean parameters on H_0 and H_{1a}^*. This latter hypothesis can be written as $\theta_c = [\theta_{c_1}, \ldots, \theta_{c_{k_1}}] = [0, \ldots, 0]$. Zellner and Siow's prior is

$$\pi^Z_{H_{1a}^*}(\theta_0, \theta_c|\sigma)\pi(\sigma) = \text{Cauchy}_{k_1}(\theta_c|0, (X^t X/n)^{-1}\sigma^2) \times 1/\sigma,$$

where X is the design matrix associated with θ_c and n is the sample size. Thus, the prior over the common parameters θ_0 is uniform, and the parameters on the null are assumed to be independent of the other parameters (in the sense of mutually orthogonal Fisher information matrix). This prior is centered at the null hypothesis (with location at zero) and with scale that is a simple (too simple perhaps, particularly for very unbalanced designs) version of the average (Fisher) information per observation. Now consider the restriction of that prior on H_{1a}^*

$$\pi^Z_{H_{1a}}(\theta_c|\sigma) = \text{Cauchy}_{k_1}(\theta_c|0, (X^t X/n)^{-1}\sigma^2) \times C \times I_{H_{1a}}(\theta_c),$$

where $C = 1/\int_{H_{1a}} \pi^Z_{H_{1a}^*}(\theta_c|\sigma)\, d\theta_c = 1/Pr^{\pi^Z_{H_{1a}^*}(\theta_c|\sigma)}(H_{1a})$ and $I_{H_{1a}}(\theta_c)$ is the indicator function of the alternative awkward hypothesis H_{1a} . By multiplying and dividing $BF^Z_{H_{1a},H_0}$ by $\int f(\mathbf{y}|\theta_a, \sigma)\pi^Z_{H_{1a}^*}(\theta_a, \sigma)\, d\theta_a d\sigma$, we obtain

$$BF^Z_{H_{1a},H_0} = BF^Z_{H^*_{1a},H_0} \times \frac{Pr^{\pi^Z_{1a}}(H_{1a}|\mathbf{y}, H^*_{1a})}{Pr^{\pi^Z_{1a}}(H_{1a}|\theta_0, \sigma, H^*_{1a})}. \tag{7.20}$$

Important point: The reason why we can calculate the denominator in (7.20) is because Assumption 1 is fulfilled here, as can be checked from Zellner and Siow's prior, as it is the case for the next priors in Sections 7.6.2 and 7.6.3.

We remark that the prior probability $Pr^{\pi^Z_{1a}}(H_{1a}|\theta_0, \sigma, H^*_{1a})$ should not be too far from $1/2^{k_1}$, where k_1 is the number of parameters under test. Otherwise, there is a strong prior correlation among parameters or the prior is unbalanced between the hypothesis and the alternative.

7.6.2 Intrinsic Prior

Let us call $\pi^I_{1a}(\theta_a)$ the intrinsic prior for H^*_{1a}, and recall that the intrinsic prior for H_0 is still the non-informative prior (see (7.10)): $\pi^I(\theta_0) = \pi^N(\theta_0)$. Under the alternatives, we are going to define the intrinsic prior under informative hypothesis as

$$\pi^I_{H_{1a}}(\theta_a) = \pi^I_{1a}(\theta_a) \times \frac{I_{H_{1a}}(\theta_a)}{Pr^{\pi^I}(H_{1a}|\theta_0, \sigma, H^*_{1a})}.$$

Equation (7.10), which defines the intrinsic prior under H_{1a}, can be rewritten as in [7]:

$$\pi^I_1(\theta_a, \sigma) = \pi^N_1(\theta_a) \times \int f(\mathbf{y}(l)|\theta_a, \sigma) m_0(\mathbf{y}(l))/m_1(\mathbf{y}(l))\; d\mathbf{y}(l). \tag{7.21}$$

Now, by completing and splitting the equation we find

$$BF^I_{H_{1a},H_0} = \frac{\int f(\mathbf{y}|\theta_a, \sigma)\pi^I(\theta_a, \sigma) d\theta d\sigma}{\int f(\mathbf{y}|\theta_0, \sigma)\pi^I(\theta_0, \sigma)} \times \frac{Pr^{\pi^I}(H_{1a}|\mathbf{y}, H^*_{1a})}{Pr^{\pi^I}(H_{1a}|\theta_0, \sigma, H^*_{1a})}. \tag{7.22}$$

For linear models, exact equations for the intrinsic priors are available; see [4, 5, 6].

7.6.3 Intrinsic EP Prior

The intrinsic prior which is also the expected posterior EP prior, generating $\mathbf{y}(l)$ from the simplest of models (i.e., from $m_0(\mathbf{y}(l))$ in H_0) would be in this case defined in two steps as

Step 1:

$$\pi^{H_1,I}(\theta_a) = \int \pi^{H_1,N}(\theta_a|\mathbf{y}(l)) m_0(\mathbf{y}(l))\; d\mathbf{y}(l), \tag{7.23}$$

which turns out to be mathematically equivalent with the intrinsic prior identity in (7.21). To see this, use Bayes' rule in (7.10) to write $m^N_1(\mathbf{y}(l)) =$

$\frac{f(\mathbf{y}(l)|\theta_a)\pi^N(\theta_a)}{\pi(\theta_a|\mathbf{y}(l))}$, and then simplification leads to the result. Equation (7.23) is approximated by $\frac{1}{L}\sum \pi^{H_1,N}(\theta_a|\mathbf{y}^*(l))$, where $\mathbf{y}^*(l)$ are random training samples, generated from $m_0(\mathbf{y}^*(l))$.

Step 2:

$$\pi^{H_{1a},I}(\theta_a) = \frac{\pi^{H^*_{1a},I}(\theta_a)}{Pr^{\pi^{H_1,I}(H_{1a}|\theta_0,\sigma)}} \times I_{H_{1a}}(\theta_a). \tag{7.24}$$

This definition through the encompassing model leads to a substantial simplification, as compared with a direct definition in H_{1a}.

The resulting Bayes factor is then

$$BF^I_{H_{1a},H_0} = \frac{\int_{H^*_{1a}} f(\mathbf{y}|\theta_a)\pi^I(\theta_a)\; d\theta_a}{\int f(\mathbf{y}|\theta_0)\pi^N(\theta_0)\; d\theta_0} \times \frac{Pr^{\pi^I}(H_{1a}|\mathbf{y}, H^*_{1a})}{Pr^{\pi^I}(H_{1a}|\theta_0, \sigma, H^*_{1a})} \tag{7.25}$$

or, equivalently,

$$BF^I_{H_{1a},H_0} = BF^I_{H^*_{1a},H_0} \times \frac{Pr^{\pi^I}(H_{1a}|\mathbf{y}, H^*_{1a})}{Pr^{\pi^I}(H_{1a}|\theta_0, \sigma, H^*_{1a})}.$$

Note that the exact analytical solution for the first factor in (7.25) exists for linear models, as presented in [4, 5, 6] in terms of the Kummer function $M(a, b, z)$ (for which efficient algorithms are available) and that is defined as (see [1])

$$M(a, b, z) = \frac{\Gamma(b)}{\Gamma(a)} \sum_{j=0}^{\infty} \frac{\Gamma(a+j)}{\Gamma(b+j)} \cdot \frac{z^j}{j!}.$$

For the point null hypotheses H^*_{1a}, the intrinsic prior is defined as

$$\pi^I(\theta_a|\sigma, \theta_0) = C\frac{2^{k_1/2}}{(2k_1)^{1/2}} \cdot \frac{\exp(-\lambda/2)}{\sigma 2^k_1/2\Gamma((p+2)/2)} M(1/2, (k+2)/2, \lambda/2),$$

where $C = \pi^{-k_1/2}\frac{\Gamma((k_1+1)/2)}{\Gamma((1/2))}$ and the noncentrality parameter λ, for a training sample of size $2k$, is based on a convenient choice.

For the informative hypotheses H_{1a}, the equation becomes

$$\pi^I_{H_{1a}}(\theta_a|\sigma, \theta_0) \propto \pi^I(\theta_a|\sigma, \theta_0) \cdot I_{H_{1a}}(\theta_a). \tag{7.26}$$

This prior is called the *restricted intrinsic prior*. This prior can now be used to get the (scaled) Bayes factors and posterior model probabilities.

7.7 Numerical Results

In this section, and as illustrations, we present the numerical results using one method for each of the two classes of methodology introduced in Sections 7.5

Table 7.1. Residual sums of squares (RSS) of completed hypotheses

Models	RSS
H_0	2196.47
H_2	237.63
H_{1a}^*	260.48
H_{1b}^*	253.84

and 7.6. We use the fractional method of Section 7.5.1 and the Zellner and Siow conventional priors, described in Section 7.6.1. We selected these two methods for simplicity, because we were able to implement them using the package BUGS, so no extensive programming was involved. Future useful work would be to implement the other methods introduced here, for an exhaustive comparison. We are going to present numerical results for selected hypotheses, presumably the most important ones.

7.7.1 Illustration: Dissociative Identity Disorder

The example on interidentity amnesia in dissociative identity disorder is taken from [13]. The hypotheses are

$$H_0 : \mu_1 = \mu_2 = \mu_3 = \mu_4$$

its negation

$$H_2 : \text{ not } H_0,$$

and

$$H_{1a} : \mu_{con} > \{\mu_{amn} = \mu_{pat}\} > \mu_{sim}$$

that is,

$$\mu_3 < \{\mu_4 = \mu_1 = \mu\} < \mu_2$$

or, equivalently,

$$H_{1a} : \mu_3 < \mu < \mu_2,$$

and, finally,

$$H_{1b} : \mu_{con} > \mu_{amn} > \{\mu_{pat} = \mu_{sim}\},$$

$$\{\mu_1 = \mu_3 = \mu\} < \mu_4 < \mu_2,$$

or

$$H_{1b} : \mu < \mu_4 < \mu_2.$$

Before going into the methods, in Table 7.1 the residual sum of squares of the null and completed hypotheses are presented. For the completed hypotheses, we see that H_{1b}^* is doing somewhat better than H_{1a}^*.

Table 7.2. Posterior and prior probabilities implied by the fractional Bayes factor for H_{1a} given H_{1a}^* and for H_{1b} given H_{1b}^*

Model	Posterior ($b = 1$)	Prior ($b = \frac{6}{94}$)
H_{1a}	1.0	.981
H_{1b}	1.0	.984

Table 7.3. Posterior probabilities given by the FBF for H_0, H_{1a}, H_{1b}

Models	FBF	PMP
H_0 vs H_2	7.30e-41	$1.31\text{e} - 40$
H_0 vs H_{1a}	5.39e-40	.24
H_0 vs H_{1b}	1.73e-40	.76

7.7.2 Fractional BF

We employ here the theory of Section 7.5.1. We use the "zeros trick" in BUGS (see BUGS help) in order to define the likelihood raised to the power $b = 6/94$. Justification of this choice for b is in Berger and Mortera's discussion of [19].

According to (7.15) (for the ratio of probabilities) we will be monitoring, via MCMC (e.g., using BUGS), the proportion of times on which either H_{1a} or H_{1b} is obeyed for samples drawn from the posterior distributions with the likelihoods raised to the power $b = 6/94$ and $b = 1$, respectively, that we call prior and posterior probabilities, the former rather metaphorically.

For the first factor, the ratio of marginal densities in (7.15), we have the equation

$$BF_{ji}^{F(b)} = \frac{\Gamma[(n-k_j)/2]\Gamma[(m-k_i)/2)]}{\Gamma[(n-k_i)/2)]\Gamma[(m-k_j)/2]} \times \left[\frac{R_i}{R_j}\right]^{(n-m)/2},$$

where k_j is the number of adjustable parameters under H_j, $m = max[k_j] + 1$, and R_j is the residual sum of squares under H_j.

Comments: (i) The prior probabilities of the hypotheses in Table 7.2, using $b = 6/94$, are far too high. The reason for this, is that the fractional "prior," however diluted by the power $b << 1$, is still centered at posterior modal values. This is not wholly satisfactory as a prior, since we would expect a prior centered at the null hypothesis. (ii) We used (7.3) in Section 7.2.1 for PMPs and (7.18) in order to convert the Bayes factors into probabilities. We did not compute $P(H_2|\mathbf{y})$, since in a sense this hypothesis "intersects" with H_{1a} and H_{1b}. However, if desired, using the Bayes factors of Table 7.2, and now giving each of the four models prior probabilities of $1/4$, then $P(H_2|\mathbf{y})$ is obtained from (7.18). Here we only considered three models and gave them equally $1/3$ of prior probability. (iii) As expected, and shown in Table 7.3, H_{1b} is more probable than H_{1a} but not overwhelmingly so, in a proportion of about $3/1$. (iv) The fractional is perhaps the simplest to calculate from the factors considered here.

7.7.3 Zellner-Siow Conventional Prior

We follow here the theory of Section 7.6.1. It should be noted that a parameter transformation is needed here, in order to assess the conventional prior. In this illustration some ingenuity is needed, and we give the details of the implementation of this conventional prior method.

As in the introduction, we have the ANOVA model

$$\mathbf{Y} = X\theta + \epsilon,$$

on which the first column of X should be a column of ones. Denote $\theta = [\theta_0, \alpha_1, \alpha_2]$. The hypothesis $H^*_{1a} : \mu_3 \neq \mu \neq \mu_2$ should become $\alpha_1 \neq 0; \alpha_2 \neq 0$. In the conventional prior setup, a proper conditional prior is given to the parameters under test (the α's), but an improper uniform prior is given to the common parameter θ_0. For this to be justifiable as a Bayes factor, then the first column of the matrix X ought to be orthogonal to the other columns (i.e., $\mathbf{x}_1^t \cdot \mathbf{x}_i = 0, i = 2, 3$). We denote by n_i the number of observations in the group i. So we have $n_3 = 25$ cases with mean μ_3, $n_2 = 25$ cases with mean μ_2, and $n = 94$ is the total number of cases, since $n_1 = 19$ and $n_4 = 25$. For H_{1a}, in the column of observations we place first the observations of the first and fourth group (assumed equal), second the observations of the third group (presumed to be lowest) and, finally, the observations belonging to the second group (presumed to be highest). Recall that the first and the next two columns of the design matrix ought to be orthogonal. So the design matrix X_{1a} for H_{1a} is set to be

$$X_{1a} = \begin{bmatrix} 1 & -n_3/n & -n_2/n \\ \vdots & \vdots & \vdots \\ 1 & -n_3/n & -n_2/n \\ 1 & (1 - n_3/n) & -n_2/n \\ \vdots & \vdots & \vdots \\ 1 & (1 - n_3/n) & -n_2/n \\ 1 & -n_3/n & (1 - n_2/n) \\ \vdots & \vdots & \vdots \\ 1 & -n_3/n & (1 - n_2/n) \end{bmatrix}.$$

Notice that $\mathbf{x}_1^t \cdot [\mathbf{x}_2, \mathbf{x}_3] = 0$. It turns out, as can be checked after simplifications, that

$$H_{1a} : \mu_3 < \mu < \mu_2 \text{ is equivalent to: } \alpha_1 < 0, \alpha_2 > 0.$$

Numerically (which will be needed for the bivariate Cauchy to be used) it is obtained that

$$[X_{1a}^t X_{1a}]^{-1} = \begin{bmatrix} 0.011 & 0 & 0 \\ 0 & 0.063 & 0.023 \\ 0 & 0.023 & 0.063 \end{bmatrix},$$

and $[X_{1a}^{*t}X_{1a}^{*}]^{-1}$ is the two by two lower right symmetric submatrix, which is the only part used in the prior.

Turning now to the other hypothesis,

$$H_{1b} : \{\mu_1 = \mu_3 = \mu\} < \mu_4 < \mu_2,$$

we place, in the vector $\mathbf{y}$ of observations, first the amalgamated first and third group, second the fourth group (assumed in the middle), and, finally, the second group (presumed highest). For H_{1b} we set the following design matrix:

$$X_{1b} = \begin{bmatrix} 1 & -(n_2+n_4)/n & -n_2/n \\ \vdots & \vdots & \vdots \\ 1 & -(n_2+n_4)/n & -n_2/n \\ 1 & (1-(n_2+n_4)/n) & -n_2/n \\ \vdots & \vdots & \vdots \\ 1 & (1-(n_2+n_4)/n) & -n_2/n \\ 1 & (1-(n_2+n_4)/n) & (1-n_2/n) \\ \vdots & \vdots & \vdots \\ 1 & (1-(n_2+n_4)/n) & (1-n_2/n) \end{bmatrix}.$$

Notice that $\mathbf{x}_1^t \cdot [\mathbf{x}_2, \mathbf{x}_3] = 0$. It turns out, as can be checked after simplifications, that

$$H_{1b} : \{\mu_1 = \mu_3 = \mu\} < \mu_4 < \mu_2 \text{ is equivalent to: } \alpha_1 > 0, \alpha_2 > 0.$$

Numerically (which will be needed for the bivariate Cauchy to be used) it obtains that

$$[X_{1b}^t X_{1b}]^{-1} = \begin{bmatrix} 0.011 & 0 & 0 \\ 0 & 0.063 & -0.04 \\ 0 & -0.04 & 0.08 \end{bmatrix},$$

and the necessary matrix for the prior $[X_{1b}^{*t}X_{1b}^{*}]^{-1}$ is the two by two lower right symmetric submatrix.

Now we turn to the prior assessments: First, the hypotheses are

$$H_0 : \alpha_1 = \alpha_2 = 0; \quad H_{1a}^* : \alpha_1 \text{ and } \alpha_2 \neq 0, H_{1a} : \alpha_1 < 0, \alpha_2 > 0.$$

Clearly, the design matrix under the null hypothesis is just a vector of ones. It is customary here to assume references priors, under H_0 and conditionally proper under H_{1a}^*:

$$\pi_0^N(\theta_0, \sigma) = 1/\sigma$$

and

$$\pi_1^N(\theta_0, \alpha_1, \alpha_2, \sigma) = \text{Cauchy}[(\alpha_1, \alpha_2)|(0,0), (X_{1a}^{*t}X_{1a}^{*}/n)^{-1}\sigma^2] \cdot 1/\sigma.$$

Next, let us check that Zellner and Siow's prior obeys Assumption 1 of Section 7.6. A priori the probability of the entertained hypothesis should not depend upon the conditioning parameters by symmetry of the Cauchy distribution centered in zero. The Zellner and Siow prior for model selection is, in our case of informative hypotheses, equal to

$$\pi(\alpha|\sigma) = c\frac{\sqrt{\det[X^{*t}X^{*}/(n\sigma^2)]}}{(1+\alpha^t X^{*t}X^{*}\alpha/(n\sigma^2))^{3/2}}, \tag{7.27}$$

where $c = \Gamma(3/2)/\pi^{3/2}$ and $\alpha^t = (\alpha_1, \alpha_2)$.

The distribution (7.27) is a bidimensional Cauchy with location zero and scale (quasivariance) matrix equal to $(X^{*t}X^{*}/(n\sigma^2))^{-1}$. We need to compute the a priori probability of, say, H_{1b}:

$$Pr(\alpha_1 > 0, \alpha_2 > 0) = \int_0^\infty \int_0^\infty \pi(\alpha|\sigma)\ d\alpha_1\ d\alpha_2. \tag{7.28}$$

This probability apparently depends upon σ which is unknown. Fortunately, however, the following transformations show that the probability is independent of σ. Make the transformation

$$\lambda_1 = \alpha_1/\sqrt{n\sigma^2},\ \text{and}\ \lambda_2 = \alpha_2/\sqrt{n\sigma^2}. \tag{7.29}$$

Note also that $\sqrt{\det[X^{*t}X^{*}/(n\sigma^2)]} = \sqrt{\det[X^{*}X^{*}]}/(n\sigma^2)$. Then it follows, applying the change of variables formula's that

$$Pr(\alpha_1 > 0, \alpha_2 > 0) = Pr(\lambda_1 > 0, \lambda_2 > 0) =$$

$$\int_0^\infty \int_0^\infty c\frac{\sqrt{\det[X^tX]}}{(1+\lambda^t X^t X\lambda)^{3/2}}\ d\lambda_1\ d\lambda_2. \tag{7.30}$$

This proves that our version of Zellner and Siow prior obeys Assumption 1.

The double integral can be solved in at least two ways: first, directly by numerical quadrature or, second, by generating several variables $[\lambda_1(l), \lambda_2(l)]$ with the distribution above and counting what fraction of them are simultaneously positive. For the second we need an efficient bivariate Cauchy generator. It has been pointed out to us that in fact there is a closed-form solution to the integral, at least under certain assumptions (Dr. John Cook, personal communication). Since it is simple and stable, we performed the computations by simulation. In the numerical illustrations presented below, we computed the prior probabilities by simulation of 100,000 bivariate Cauchy random variables.

From the a posteriori densities and simulating via BUGS (modeling a Cauchy density as a scale mixture of a Normal and a Gamma, or using the zeros trick) we get the results for H_{1a} and H_{1b} in Table 7.4.

For the Bayes factors in (7.20), there is the following approximation in [21]:

Table 7.4. Posterior and prior probabilities implied by the Zellner and Siow prior for H_{1a} given H_{1a}^* and for H_{1b} given H_{1b}^*

Model	Posterior	Prior
H_{1a}	1.0	.190
H_{1b}	1.0	.153

Table 7.5. Posterior probabilities given by Zellner and Siow prior for H_0, H_{1a}, H_{1b}

Models	BF^Z	PMP
H_0 vs H_2	2.83e-40	1.72e–39
H_0 vs H_{1a}	6.14e-39	.28
H_0 vs H_{1b}	2.38e-39	.72

$$BF_{12}^Z = \pi^{1/2}/\Gamma((k_1+1)/2)((n-k_1)/2)^{k_1/2}(R_1/R_0)^{(n-k_1-1)/2}.$$

In Table 7.5 we display the Bayes factors and posterior model probabilities for Zellner and Siow's prior.

Comments: (i) Although we may calculate the posterior probability of H_2 using the Bayes factors above, we do not, since H_2 does not appear to be the most interesting hypothesis and also because of interpretation: Both $H_{1a}^\star$ and $H_{1b}^\star$ are well embedded in H_2. But $P(H_2|\mathbf{y})$ can be easily calculated if desired, as we pointed out in the previous method. (ii) The prior probabilities of H_{1a} and H_{1b} are now theoretically more pleasant to the mind and natural. However, the difference with the posterior probabilities from the FBF, are rather small. This is reassuring and comes from the fact that the distortion in the prior probabilities in the fractional occurs in both the models, which have no negligible probability, therefore "canceling out." If in an example, say, the probability of H_0 would have been higher, the difference between the methods would have been more pronounced, giving the Zellner-Siow method the edge here.

7.8 Conclusions

1. The informative hypothesis problem becomes feasible when it is split in two problems, first by completing the informative hypothesis and second by splitting into two factors: (i) a ratio of objective Bayes factor of appropriate encompassing hypotheses that we call completions of the informative hypotheses and (ii) a ratio of prior to posterior probabilities of the informative hypothesis.
2. The first factor can be dealt with the usual methods of objective Bayes factors. We have presented several techniques, so that researchers can implement them as they judge more appropriate.
3. The second factor can be expressed, with some ingenuity, as an MCMC problem of estimation.

Acknowledgements. The authors were supported by National Science Foundation grants DMS-0604896 and DUE-0630927. The authors are grateful to Dr. M.E. Pérez and Dr. John Cook for useful discussions and to the Editors for several detailed suggestions that improved the manuscript. The first author is also grateful for the support of Utrecht University and Viajes Caribe Inc.

References

[1] Abramowitz, M., Stegun, I.: Handbook of Mathematical Functions. Applied Mathematics Series, Vol. 55. Washington, DC, National Bureau of Standards (1970)

[2] Akaike, H.: Information theory and the extension of the maximum likelihood principle. In Second International Symposium on Information Theory. Petrov, B.N., Csaki, F. (eds). Budapest, Akademia Kiado (1973)

[3] Berger, J., Mortera, J.: Default Bayes factors for one-sided hypothesis testing. Journal of the American Statistical Association, **31**, 542–554 (1999)

[4] Berger, J., Pericchi, L.: The intrinsic Bayes factor for model selection and prediction. Journal of the American Statistical Association, **91**, 109–122 (1996)

[5] Berger, J., Pericchi, L.: The intrinsic Bayes factor for linear models. In Bernardo, J.M. et al. (eds) Bayesian Statistics 5. London, Oxford University Press (1996)

[6] Berger, J., Pericchi, L.: On the justification of default and intrinsic Bayes factors. In Lee, J.C. et al. (eds) Modeling and Prediction. Berlin, Springer (1997)

[7] Berger, J., Pericchi, R.: Objective Bayesian methods for model selection: Introduction and comparison. In: Lahiri, P. (ed) Model Selection. Beachwood, OH, Institute of Mathematical Statistics Lecture Notes Monograph Series **38**, 135–207 (2001)

[8] Berger, J., Pericchi, L.: Training samples in objective Bayesian model selection. Annals of Statistics, **32**, 841–869 (2004)

[9] Berger, J., Pericchi, L., Varshavsky, J.: Bayes factors and marginal distributions in invariant situations. Sankhya, Series A, **60**, 135–321 (1998)

[10] De Santis, F., Spezzaferri, F.: Alternative Bayes factors for model selection. Canadian Journal of Statistics, **25**, 503–515 (1997)

[11] Dudley, R.M., Haughton, D.: Information criteria for multiple data sets and restricted parameters. Statistica Sinica, **7**, 265–284 (1997)

[12] Giron, F.J., Moreno E., Casella, G.: Objective Bayesian analysis of multiple changepoints models (with discussion). Bayesian Statistics 9. Oxford, Oxford University Press (2006)

[13] Huntjens, R.J.C., Peters, M.L., Woertman, L., Bovenschen, L.M., Martin, R.C., Postma, A.: Inter-identity amnesia in dissociative identity disorder: A simulated memory impairment? Psychological Medicine, **36**, 857–863 (2006)

[14] Jeffreys, H.: Theory of Probability (3rd. ed.). Oxford, Oxford University Press (1961)

[15] Kato, B.S., Hoijtink, H.: A Bayesian approach to inequality constrained linear mixed models: Estimation and model selection. Statistical Modelling, **6**, 231–249 (2006)

[16] Klugkist, I., Hoijtink, H.: The Bayes factor for inequality and about equality constrained models. Computational Statistics and Data Analysis, **51**, 6367–6379 (2007)

[17] Laudy, O., Hoijtink, H.: Bayesian methods for the analysis of inequality constrained contingency tables. Statistical Methods in Medical Research, **16**, 123–138 (2007)

[18] Lempers, F.B.: Posterior Probabilities of Alternative Linear Models. Rotterdam, University of Rotterdam Press (1971)

[19] O'Hagan, A.: Fractional Bayes factors for model comparisons. Journal of the Royal Statistical Society, Series B, **57**, 115–149 (1995)

[20] Pericchi, L.R.: Model selection and hypothesis testing based on objective probabilities and Bayes factors. In Dey, D.P., Rao, C.R. (eds) Bayesian Thinking, Modeling and Computation. Handbook of Statistics, Vol. 25. Amsterdam, Elsevier (2005)

[21] Zellner, A., Siow, A.: Posterior odds for selected regression hypothesis. In Bernardo, J.M. et al. (eds) Bayesian Statistics 1. Valencia, Valencia University Press (1980)

8

The Bayes Factor Versus Other Model Selection Criteria for the Selection of Constrained Models

Ming-Hui Chen[1] and Sungduk Kim[2]

[1] Department of Statistics, University of Connecticut, 215 Glenbrook Road, U-4120, Storrs, CT 06269, USA mhchen@stat.uconn.edu
[2] Division of Epidemiology, Statistics and Prevention Research, National Institute of Child Health and Human Development, NIH, Rockville, MD 20852, USA kims2@mail.nih.gov

8.1 Introduction

Model assessment and model comparison are a crucial part of statistical analysis. Due to recent computational advances, sophisticated techniques for Bayesian model assessment are becoming increasingly popular. There is a rich literature on Bayesian methods for model assessment and model comparison, including [1, 3, 6, 9, 10, 13, 14, 16, 17, 18, 24, 26, 28, 30, 32, 33, 34, 36]. The scope of Bayesian model assessment can be investigated via model diagnostics, goodness of fit measures, or posterior model probabilities (or Bayes factors). A comprehensive account of model diagnostics and related methods for model assessment is given in [15].

Many of the proposed Bayesian methods for model comparison usually rely on posterior model probabilities or Bayes factors. It is well known that Bayes factors and posterior model probabilities are generally sensitive to the choices of prior parameters, and thus one cannot simply select vague proper priors to get around the elicitation issue. Alternatively, criterion-based methods can be attractive in the sense that they do not require proper prior distributions in general and thus have an advantage over posterior model probabilities. Several recent articles advocating the use of Bayesian criteria for model assessment include [2, 3, 13, 14, 24, 26, 30, 36].

For comparing the inequality constrained models, the posterior model probabilities are considered in [21], and the Bayes factor approach using encompassing priors is discussed in detail in [29]. However, the literature on criterion-based methods other than the Bayes factor or posterior model probabilities for inequality constrained models is still sparse. In the subsequent sections, several criterion-based methods proposed in the recent literature are considered for the selection of constrained ANOVA models. A class of priors for constrained analysis of variance (ANOVA) models based on the conjugate prior of [5] is constructed and the properties of these priors are examined. In addition, computational issues for various Bayesian criteria will be addressed for testing different hypotheses.

8.2 The Model and Notation

Suppose that y_{ij} equals the score of the ith person in the jth group on the dependent variable and μ_j denotes the mean of the persons in group j with respect to y_{ij} for $i = 1, 2, \ldots, n_j$, where n_j is the size of group j and $j = 1, 2, \ldots, J$. Write $\boldsymbol{y}_j = (y_{1j}, y_{2j}, \ldots, y_{n_j j})'$, $\boldsymbol{y} = (\boldsymbol{y}_1', \boldsymbol{y}_2', \ldots, \boldsymbol{y}_J')'$, and $\boldsymbol{\mu} = (\mu_1, \mu_2, \ldots, \mu_J)'$. We assume that y_{ij} independently follows from a normal distribution with mean μ_j and variance σ^2; that is,

$$y_{ij} \sim \mathcal{N}(\mu_j, \sigma^2) \tag{8.1}$$

for $i = 1, 2, \ldots, n_j$ and $j = 1, 2, \ldots, J$. Let $n = \sum_{j=1}^J n_j$ denote the total sample size, let $D = (n, \boldsymbol{y})$ denote the observed data, and let $\boldsymbol{\theta} = (\boldsymbol{\mu}, \sigma^2)$. Then the likelihood function is given by

$$L(\boldsymbol{\theta}|D) = L(\boldsymbol{\mu}, \sigma^2|D) = (2\pi\sigma^2)^{-\frac{n}{2}} \exp\left\{ -\frac{1}{2\sigma^2} \sum_{j=1}^{J} \sum_{i=1}^{n_j} (y_{ij} - \mu_j)^2 \right\}. \tag{8.2}$$

Note that (8.2) is the likelihood function corresponding to the standard one-way ANOVA model.

8.3 The Prior and Posterior Distributions

For the ANOVA model given by (8.2), we let Θ denote the constrained parameter space with the form

$$\Theta = \{\boldsymbol{\theta} = (\boldsymbol{\mu}, \sigma^2) : \ \boldsymbol{\mu} \in \Theta_{\boldsymbol{\mu}} \text{ and } \sigma^2 > 0\}, \tag{8.3}$$

where $\Theta_{\boldsymbol{\mu}}$ is the constrained parameter space for $\boldsymbol{\mu}$. The various specific forms of $\Theta_{\boldsymbol{\mu}}$ will be given in the later sections. Note that for the unconstrained parameter space, we have $\Theta_{\boldsymbol{\mu}} = R^J$, which implies $\Theta = R^J \times (0, \infty)$.

In the context of Bayesian model selection of inequality constrained models, a prior distribution for $\boldsymbol{\theta} = (\boldsymbol{\mu}, \sigma^2)$ needs to be specified for each model. To this end, we consider a conjugate prior of [5]. Following [5], for the ANOVA model given in (8.2), we first specify a conditional prior for $\boldsymbol{\mu}$ given σ, which takes the form

$$\pi(\boldsymbol{\mu}|\sigma^2, \boldsymbol{y}_0, \boldsymbol{a}_0) \propto \left[\prod_{j=1}^{J} \{L(\boldsymbol{y}_{0j}|\mu_j, \sigma^2)\}^{a_{0j}} \right] I_{\Theta_{\boldsymbol{\mu}}}(\boldsymbol{\mu}), \tag{8.4}$$

where the indicator function $I_{\Theta_{\boldsymbol{\mu}}}(\boldsymbol{\mu}) = 1$ if $\boldsymbol{\mu} \in \Theta_{\boldsymbol{\mu}}$ and 0 otherwise,

$$L(\boldsymbol{y}_{0j}|\mu_j, \sigma^2) = (2\pi\sigma^2)^{-\frac{n_j}{2}} \exp\Big\{ -\frac{1}{2\sigma^2}\sum_{i=1}^{n_j}(y_{0ij} - \mu_j)^2 \Big\},$$

$\boldsymbol{y}_{0j} = (y_{01j}, y_{02j}, \ldots, y_{0n_jj})'$ is an $n_j \times 1$ vector of prior predictive values of the response variables for the subjects in the jth group, $\boldsymbol{y}_0 = (\boldsymbol{y}_{01}', \boldsymbol{y}_{02}', \ldots, \boldsymbol{y}_{0J}')'$, $a_{0j} > 0$ is a scalar prior parameter, and $\boldsymbol{a}_0 = (a_{01}, a_{02}, \ldots, a_{0J})'$. From (8.4), we can rewrite the conditional prior distribution for $\boldsymbol{\mu}$ given σ^2 as

$$\pi(\boldsymbol{\mu}|\sigma^2, \boldsymbol{y}_0, \boldsymbol{a}_0) \propto \Big\{ \prod_{j=1}^{J} \phi\Big(\mu_j|\bar{y}_{0j}, \frac{\sigma^2}{a_{0j}n_j}\Big)\Big\} I_{\Theta_{\boldsymbol{\mu}}}(\boldsymbol{\mu}), \tag{8.5}$$

where $\bar{y}_{0j} = \frac{1}{n_j}\sum_{i=1}^{n_j} y_{0ij}$, and $\phi\Big(\mu_j|\bar{y}_{0j}, \frac{\sigma^2}{a_{0j}n_j}\Big)$ is the density function of the normal distribution $\mathcal{N}\Big(\bar{y}_{0j}, \frac{\sigma^2}{a_{0j}n_j}\Big)$ evaluated at μ_j for $j = 1, 2, \ldots, J$. The normalized conditional prior distribution for $\boldsymbol{\mu}$ given σ^2 is given by

$$\pi(\boldsymbol{\mu}|\sigma^2, \boldsymbol{y}_0, \boldsymbol{a}_0) = \frac{1}{C_0(\sigma^2, \boldsymbol{y}_0, \boldsymbol{a}_0)}\Big\{ \prod_{j=1}^{J} \phi\Big(\mu_j|\bar{y}_{0j}, \frac{\sigma^2}{a_{0j}n_j}\Big)\Big\} I_{\Theta_{\boldsymbol{\mu}}}(\boldsymbol{\mu}), \tag{8.6}$$

where

$$C_0(\sigma^2, \boldsymbol{y}_0, \boldsymbol{a}_0) = \int_{\Theta_{\boldsymbol{\mu}}} \Big\{ \prod_{j=1}^{J} \phi\Big(\mu_j|\bar{y}_{0j}, \frac{\sigma^2}{a_{0j}n_j}\Big)\Big\}\, d\boldsymbol{\mu}. \tag{8.7}$$

To complete the prior specification, we assume an inverse gamma distribution for σ^2, which is given by

$$\pi(\sigma^2|\alpha_0, \beta_0) = \frac{\beta_0^{\alpha_0}}{\Gamma(\alpha_0)} \times (\sigma^2)^{-(\alpha_0+1)} \exp\big(-\frac{\beta_0}{\sigma^2}\big), \tag{8.8}$$

where α_0 and β_0 are prespecified hyperparameters. We write

$$\sigma^2|\alpha_0, \beta_0 \sim \mathcal{IG}(\alpha_0, \beta_0).$$

Combining (8.6) and (8.8) gives the joint prior of $\boldsymbol{\theta} = (\boldsymbol{\mu}, \sigma^2)$ as follows:

$$\pi(\boldsymbol{\theta}|\boldsymbol{y}_0, \boldsymbol{a}_0, \alpha_0, \beta_0) = \pi(\boldsymbol{\mu}|\sigma^2, \boldsymbol{y}_0, \boldsymbol{a}_0)\pi(\sigma^2|\alpha_0, \beta_0). \tag{8.9}$$

Next, we discuss the properties of the prior given in (8.6). As discussed in [5], y_{0ij} can be viewed as a prior prediction for the marginal mean of y_{ij}. Thus, in eliciting $\boldsymbol{y}_0$, we may focus only on a prediction (or guess) for $E(y_{ij})$, which narrows the possibilities for choosing y_{0ij}. When the prior information about y_{0ij} is not available, the specification of all y_{0ij} equal has an appealing interpretation. A prior specification with $y_{0ij} = 0$ for $i = 1, 2, \ldots, n_j$ and

$j = 1, 2, \ldots, J$ implies a prior in which the prior modes of μ_j's are 0 under the unconstrained ANOVA model. This is intuitively appealing since in this case the prior prediction on y_{0ij} does not depend on the ith subject's specific information. In addition, $\boldsymbol{y}_0 = \mathbf{0}$ greatly eases the posterior computation under the inequality constrained model. The parameter a_{0j} in (8.6) can be generally viewed as a precision parameter that quantifies the strength of our prior belief in $\boldsymbol{y}_{0j}$. Note that when $\Theta_{\boldsymbol{\mu}} = R^J$, $y_{0ij} = 0$, and $a_{0j} n_j = 1/\kappa_0$, (8.6) and (8.8) essentially reduce to a standard conjugate prior for $(\boldsymbol{\mu}, \sigma^2)$, that is, we specify

$$\mu_j | \sigma^2, \kappa_0 \sim \mathcal{N}(0, \kappa_0 \sigma^2)$$

independently for $j = 1, 2, \ldots, J$ and $\sigma^2 | \alpha_0, \beta_0 \sim \mathcal{IG}(\alpha_0, \beta_0)$. Another special case of (8.4) is an improper uniform prior for $\boldsymbol{\mu}$ over the constrained parameter space $\Theta_{\boldsymbol{\mu}}$, which is obtained by taking $a_{0j} \to 0$. In this case, no prior normalizing constant is available. In the context of Bayesian model selection, (8.6) and (8.8) specify the priors for all constrained or unconstrained models in the model space in an automatic and systematic fashion in the sense that the forms of the priors are automatically determined by the models with or without constraints and the hyperparameters $(\boldsymbol{a}_0, \alpha_0, \beta_0)$ need to be specified only once for all models under considerations.

Due to the nature of the constrained parameter problem, the prior normalizing constant $C_0(\sigma^2, \boldsymbol{y}_0, \boldsymbol{a}_0)$ is often analytically intractable. However, for many special cases, $C_0(\sigma^2, \boldsymbol{y}_0, \boldsymbol{a}_0)$ is either analytically available or free from σ^2. For example, when $\Theta_{\boldsymbol{\mu}} = R^J$, the prior normalizing constant $C_0(\sigma^2, \boldsymbol{y}_0, \boldsymbol{a}_0) = 1$. When $\boldsymbol{y}_0 = \mathbf{0}$, $C_0(\sigma^2, \boldsymbol{y}_0 = \mathbf{0}, \boldsymbol{a}_0)$ in (8.7) can be free from σ^2 under certain unbounded inequality constraints. As an illustration, we consider the monotone constraints, $\mu_1 < \mu_2 < \cdots < \mu_J$. By taking the one-to-one transformation $\widetilde{\mu}_j = \mu_j / \sigma$ for $j = 1, 2, \ldots, J$, the constrained parameter space $\Theta_{\boldsymbol{\mu}}$ is transformed to $\Theta_{\widetilde{\boldsymbol{\mu}}} = \{\widetilde{\boldsymbol{\mu}} : \widetilde{\mu}_1 < \widetilde{\mu}_2 < \cdots < \widetilde{\mu}_J\}$, where $\widetilde{\boldsymbol{\mu}} = (\widetilde{\mu}_1, \widetilde{\mu}_2, \ldots, \widetilde{\mu}_J)'$. As the result of this transformation, we obtain

$$\begin{aligned} C_0(\sigma^2, \boldsymbol{y}_0, \boldsymbol{a}_0) &= \int_{\Theta_{\boldsymbol{\mu}}} \left\{ \prod_{j=1}^{J} \phi\Big(\mu_j | 0, \frac{\sigma^2}{a_{0j} n_j}\Big) \right\} d\boldsymbol{\mu} \\ &= \int_{\Theta_{\widetilde{\boldsymbol{\mu}}}} \left\{ \prod_{j=1}^{J} \phi\Big(\widetilde{\mu}_j | 0, \frac{1}{a_{0j} n_j}\Big) \right\} d\widetilde{\boldsymbol{\mu}} \equiv C_0(\boldsymbol{a}_0), \end{aligned}$$

where $C_0(\boldsymbol{a}_0)$ is free from σ^2, as $\Theta_{\widetilde{\boldsymbol{\mu}}}$ is essentially the same as $\Theta_{\boldsymbol{\mu}}$. When $C_0(\sigma^2, \boldsymbol{y}_0, \boldsymbol{a}_0)$ is an analytically intractable function of σ^2, the posterior computation is difficult to carry out. For example, when $\boldsymbol{y}_0 \neq \mathbf{0}$ and $\boldsymbol{\mu}$ is subject to the monotone constraints, $C_0(\sigma^2, \boldsymbol{y}_0, \boldsymbol{a}_0)$ is an analytically intractable function of σ^2. Also, if the constraints, $|\mu_j - \mu_{j'}| \leq K_0$ for certain $j < j'$, where $K_0 > 0$ is a fixed constant, are of interest, $C_0(\sigma^2, \boldsymbol{y}_0, \boldsymbol{a}_0)$ does depend on σ^2 as well. However, for these cases, the Monte Carlo method developed in [7] can be applied for sampling from the posterior distribution and computing various posterior summaries.

To ease the posterior computation, we assume that the constrained parameter space $\Theta_{\boldsymbol{\mu}}$ satisfies the following invariant condition:

$$C_0(\tau\sigma^2, \boldsymbol{y}_0 = \boldsymbol{0}, \boldsymbol{a}_0) = C_0(\sigma^2, \boldsymbol{y}_0 = \boldsymbol{0}, \boldsymbol{a}_0) \text{ for all } \tau > 0. \tag{8.10}$$

It is easy to see that the invariant condition (8.10) implies that $C_0(\sigma^2, \boldsymbol{y}_0 = \boldsymbol{0}, \boldsymbol{a}_0)$ is free from σ^2.

We note that if condition (8.10) does not hold, the prior normalizing constant $C_0(\sigma^2, \boldsymbol{y}_0 = \boldsymbol{0}, \boldsymbol{a}_0)$ is not free from σ^2. In this case, sampling (μ, σ^2) from the joint prior distribution given in (8.9) and the respective posterior distribution becomes difficult.

Based on the above discussions, we take $\boldsymbol{y}_0 = \boldsymbol{0}$ and $a_{01} = a_{02} = \cdots = a_{0j} = a_0$ in the rest of this chapter. With these special choices of $\boldsymbol{y}_0$ and $\boldsymbol{a}_0$, we rewrite $\pi(\boldsymbol{\mu}|\sigma^2, \boldsymbol{y}_0, \boldsymbol{a}_0)$ with $C_0(\sigma^2, \boldsymbol{y}_0, \boldsymbol{a}_0)$ in (8.6) as $\pi(\boldsymbol{\mu}|\sigma^2, a_0)$ with $C_0(\sigma^2, a_0)$. We note that when $\boldsymbol{y}_0 = \boldsymbol{0}$ and $a_{01} = a_{02} = \cdots = a_{0j} = a_0$, the invariant condition (8.10) also implies that $C_0(\sigma^2, a_0)$ is free from a_0 as the prior normalizing constant depends on σ^2 and a_0 only through σ^2/a_0. Using the joint prior given in (8.9), the posterior distribution is thus given by

$$\begin{aligned}\pi(\boldsymbol{\theta}|D) &\propto L(\boldsymbol{\theta}|D)\pi(\boldsymbol{\theta}|a_0, \alpha_0, \beta_0)\\ &= L(\boldsymbol{\mu}, \sigma^2|D)\pi(\boldsymbol{\mu}|\sigma^2, a_0)\pi(\sigma^2|\alpha_0, \beta_0),\end{aligned} \tag{8.11}$$

where $L(\boldsymbol{\theta}|D)$ is defined in (8.2).

Under the unconstrained ANOVA model (i.e., $\Theta_{\boldsymbol{\mu}} = R^J$), the posterior distributions of $\boldsymbol{\mu}$ and σ^2 are available in closed form as the joint prior given in (8.9) is conjugate. In this special case, after some algebra, we obtain that the conditional posterior distribution of μ_j given σ^2 and D is

$$\mu_j \mid \sigma^2, D \sim \mathcal{N}\Big(\widehat{\mu}_j, \widehat{\sigma}^2_{\mu_j}\Big), \tag{8.12}$$

where

$$\widehat{\mu}_j = \frac{1}{1+a_0}\bar{y}_j \text{ and } \widehat{\sigma}^2_{\mu_j} = \frac{\sigma^2}{(1+a_0)n_j},$$

for $j = 1, 2, \ldots, J$, and the marginal posterior distribution of σ^2 is

$$\sigma^2 \mid D \sim \mathcal{IG}(A, B), \tag{8.13}$$

where

$$A = \alpha_0 + \frac{n}{2} \text{ and } B = \beta_0 + \frac{1}{2}\sum_{j=1}^{J}\Big\{\sum_{i=1}^{n_j}(y_{ij} - \bar{y}_j)^2 + \frac{a_0 n_j}{1+a_0}(\bar{y}_j)^2\Big\}. \tag{8.14}$$

8.4 Bayesian Model Selection Criteria

In this section, we consider four Bayesian model selection criteria, namely the L measure (see [3, 13, 24]), Deviance Information Criterion (DIC) (see [36]), Conditional Predictive Ordinate (CPO) statistic (see [11, 12, 15]), and marginal likelihood or Bayes factor (see [28]) for the inequality constrained models.

8.4.1 The L Measure

The L measure is constructed from the posterior predictive distribution of the data. We note that we use L as L measure and Likelihood and it should be clear from the context which is meant. Let $\boldsymbol{z} = (\boldsymbol{z}'_1, \boldsymbol{z}'_2, \ldots, \boldsymbol{z}'_J)'$ denote future values of an imagined replicate experiment, where $\boldsymbol{z}_j = (z_{1j}, z_{2j}, \ldots, z_{n_j j})'$ for $j = 1, 2, \ldots, J$; that is, $\boldsymbol{z}$ is a future response vector with the same sampling density as $\boldsymbol{y}|\boldsymbol{\theta}$. Then, the L measure is defined as

$$
\begin{aligned}
L(\nu) &= \sum_{j=1}^{J}\sum_{i=1}^{n_j} \mathrm{Var}(z_{ij}|D) + \nu \sum_{j=1}^{J}\sum_{i=1}^{n_j} \{E(z_{ij}|D) - y_{ij}\}^2 \\
&= \sum_{j=1}^{J}\sum_{i=1}^{n_j} \Big[E\{\mathrm{Var}(z_{ij}|\boldsymbol{\theta})|D\} + \mathrm{Var}\{E(z_{ij}|\boldsymbol{\theta})|D\}\Big] \\
&\quad + \nu \sum_{j=1}^{J}\sum_{i=1}^{n_j} \Big[E\{E(z_{ij}|\boldsymbol{\theta})|D\} - y_{ij}\Big]^2, \qquad (8.15)
\end{aligned}
$$

where $0 < \nu < 1$ and all expectations and variances are taken with respect to the posterior distribution $\pi(\boldsymbol{\theta}|D)$. The smaller the L measure, the better the model fits the data. The quantity ν plays a major role in (8.15). This quantity is specified as $\nu = 1$ in [26], and this special choice of ν gives equal weight to the squared bias and variance components. However, there is no theoretical justification for such a weight and, indeed, using $\nu = 1$ may not be desirable in certain situations. Allowing ν to vary between zero and one gives the user a great deal of flexibility in the trade-off between bias and variance. For the linear model, it is theoretically shown in [24] that certain values of ν yield highly desirable properties of the L measure. It is also demonstrated in [24] that the choice of ν has much potential influence on the properties of the L measure. Based on the theoretical exploration of [24], $\nu = \frac{1}{2}$ is a desirable and justifiable choice for model selection.

Under the unconstrained ANOVA model, the closed-form expression of $L(\nu)$ can be obtained. Using (8.12) and (8.13), we have

$$
E\{E(z_{ij}|\boldsymbol{\theta})|D\} = E(\mu_j|D) = E\{E(\mu_j|\sigma, D)|D\} = \widehat{\mu}_j = \frac{1}{1+a_0}\bar{y}_j, \qquad (8.16)
$$

$$\begin{aligned}
&\mathrm{Var}\{E(z_{ij}|\boldsymbol{\theta})|D\} = \mathrm{Var}(\mu_j|D) \\
&= E\{\mathrm{Var}(\mu_j|\sigma^2, D)|D\} + \mathrm{Var}\{E(\mu_j|\sigma^2, D)|D\} \\
&= E\Big\{\frac{\sigma^2}{(1+a_0)n_j}\Big|D\Big\} = \frac{B}{(A-1)(1+a_0)n_j},
\end{aligned} \tag{8.17}$$

where A and B are defined in (8.14) and

$$E\{\mathrm{Var}(z_{ij}|\boldsymbol{\theta})|D\} = E(\sigma^2|D) = \frac{B}{A-1}. \tag{8.18}$$

Plugging (8.16), (8.17), and (8.18) into (8.15) gives the explicit expression of the L measure $L(\nu)$.

For the inequality constrained ANOVA model, the analytical evaluation of $L(\nu)$ does not appear possible. However, it is fairly easy to compute $L(\nu)$ via a Monte Carlo method. First, we briefly discuss how to sample $(\boldsymbol{\mu}, \sigma^2)$ from the posterior distribution in (8.11) via Gibbs sampling. For most inequality constraints on $\boldsymbol{\mu}$ so that $\Theta_{\boldsymbol{\mu}}$ satisfies (8.10), the conditional posterior distribution of μ_j given $\mu_{j'}$, $j' \neq j$, σ^2, and D is a truncated normal distribution. Therefore, we can use the algorithm given in [19] to generate μ_j. With the conjugate prior in (8.9), the conditional posterior distribution of σ^2 is an inverse gamma distribution. Thus, the Gibbs sampling algorithm is easy to implement.

Next, we discuss how to compute the L measure. Suppose that $\{\boldsymbol{\theta}_k = (\boldsymbol{\mu}_k = (\mu_{1k}, \mu_{2k}, \ldots, \mu_{Jk})', \sigma_k^2),\ k = 1, 2, \ldots, K\}$ is a Gibbs sample of size K from the posterior distribution in (8.11). Then, a Monte Carlo estimate of $E(z_{ij}|D) = E\{E(z_{ij}|\boldsymbol{\theta})|D\} = E(\mu_j|D)$ is given by

$$\widehat{\mu}_j = \frac{1}{K}\sum_{k=1}^{K} \mu_{jk},\ j = 1, 2, \ldots, J. \tag{8.19}$$

In (8.15), we observe

$$\begin{aligned}
&\mathrm{Var}(z_{ij}|D) = E(z_{ij}^2|D) - \{E(z_{ij}|D)\}^2 \\
&= E\{E(z_{ij}^2|\boldsymbol{\theta})|D\} - \{E(z_{ij}|D)\}^2 = E(\sigma^2 + \mu_j^2|D) - \{E(z_{ij}|D)\}^2.
\end{aligned}$$

Thus, a Monte Carlo estimate of $\mathrm{Var}(z_{ij}|D)$ is given by

$$\widehat{\mathrm{Var}}(z_{ij}|D) = \frac{1}{K}\sum_{k=1}^{K}(\sigma_k^2 + \mu_{jk}^2) - \widehat{\mu}_j^2. \tag{8.20}$$

Plugging (8.19) and (8.20) into the first equation in (8.15) gives a Monte Carlo estimate of the L measure $L(\nu)$.

8.4.2 The DIC Measure

The Deviance Information Criteria (DIC) measure proposed in [36] is defined as

$$\text{DIC} = d(\overline{\boldsymbol{\theta}}) + 2p_d, \tag{8.21}$$

where $d(\boldsymbol{\theta})$ is a deviance function and $\overline{\boldsymbol{\theta}}$ is the posterior estimate of $\boldsymbol{\theta}$ depending on the observed data D. For example, $\overline{\boldsymbol{\theta}}$ can be posterior mean, posterior median, and posterior mode. In (8.21), p_d is the effective number of model parameters, which is calculated as

$$p_d = \overline{d(\boldsymbol{\theta})} - d(\overline{\boldsymbol{\theta}}), \tag{8.22}$$

where $\overline{d(\boldsymbol{\theta})} = E\{d(\boldsymbol{\theta})|D\}$ and the expectation is taken with respect to the posterior distribution $\pi(\boldsymbol{\theta}|D)$.

For the constrained ANOVA model given in (8.2), we set $\boldsymbol{\theta} = (\boldsymbol{\mu}, \sigma^2)$ and take the deviance function to be of the form

$$d(\boldsymbol{\theta}) = -2\log L(\boldsymbol{\theta}|D) = n\log(2\pi\sigma^2) + \frac{1}{\sigma^2}\sum_{j=1}^{J}\sum_{i=1}^{n_j}(y_{ij} - \mu_j)^2. \tag{8.23}$$

The DIC defined in (8.21) is a Bayesian measure of predictive model performance, decomposed into a measure of fit ($d(\overline{\boldsymbol{\theta}})$) and a measure of model complexity (p_d). The smaller the DIC value, the better the model will predict new observations generated in the same way as the data.

Similarly to the L measure, a Monte Carlo estimate of the DIC measure can be obtained using a Gibbs sample, $\{\boldsymbol{\theta}_k = (\boldsymbol{\mu}_k = (\mu_{1k}, \mu_{2k}, \ldots, \mu_{Jk})', \sigma_k^2),\ k = 1, 2, \ldots, K\}$, from the posterior distribution in (8.11). First, it is straightforward to obtain Monte Carlo estimates of the posterior means or medians of μ_j and σ^2 using the Gibbs sample. Second, a Monte Carlo estimate of $\overline{d(\boldsymbol{\theta})}$ is simply the sample mean of $\{d(\boldsymbol{\theta}_k),\ k = 1, 2, \ldots, K\}$. Thus, the DIC measure is easy to compute.

8.4.3 The Conditional Predictive Ordinate

Under the constrained ANOVA model given in (8.2), for the ith observation in the jth group let $f(y_{ij}|\boldsymbol{\theta})$ denote the density of y_{ij}. Then, the CPO statistic is the posterior predictive density of y_{ij} based on the data $D^{(-ij)}$ with the ith observation in the jth group deleted. Mathematically, the CPO statistic is defined as

$$\begin{aligned}\text{CPO}_{ij} &= f(y_{ij}|D^{(-ij)}) = \int f(y_{ij}|\boldsymbol{\theta})\pi(\boldsymbol{\theta}|D^{(-ij)})\ d\boldsymbol{\theta} \\ &= \int \frac{1}{(2\pi\sigma^2)^{\frac{1}{2}}}\exp\Big\{-\frac{1}{2\sigma^2}(y_{ij} - \mu_j)^2\Big\}\pi(\boldsymbol{\theta}|D^{(-ij)})\ d\boldsymbol{\theta}, \quad (8.24)\end{aligned}$$

where $\pi(\boldsymbol{\theta}|D^{(-ij)})$ is the posterior distribution based on the data $D^{(-ij)}$. Note that here we define $\pi(\boldsymbol{\theta}|D^{(-ij)})$ as

$$\pi(\boldsymbol{\theta}|D^{(-ij)}) \propto \Big[\prod\prod_{(i',j')\neq(i,j)} f(y_{i'j'}|\boldsymbol{\theta})\Big]\pi(\boldsymbol{\theta}|\boldsymbol{y}_0 = \mathbf{0}, a_0, \alpha_0, \beta_0),$$

where the prior $\pi(\boldsymbol{\theta}|\boldsymbol{y}_0 = \mathbf{0}, a_0, \alpha_0, \beta_0)$ is defined in (8.9). After some algebra, we can show that

$$\mathrm{CPO}_{ij} = \left[\int (2\pi\sigma^2)^{\frac{1}{2}} \exp\left\{\frac{1}{2\sigma^2}\left(y_{ij} - \mu_j\right)^2\right\} \pi(\boldsymbol{\theta}|D) d\boldsymbol{\theta}\right]^{-1}, \tag{8.25}$$

where $\pi(\boldsymbol{\theta}|D)$ is the posterior distribution based on the entire observed data D.

As suggested in [25], a natural summary statistic of the CPO_{ij}'s is the logarithm of the pseudomarginal likelihood (LPML) defined as

$$\mathrm{LPML} = \sum_{j=1}^{J}\sum_{i=1}^{n_j} \log(\mathrm{CPO}_{ij}). \tag{8.26}$$

We use LPML as a criterion-based measure for model selection. The larger the LPML measure, the better the model fits the data.

As for the L measure and the DIC measure, we again use the Gibbs sample $\{\boldsymbol{\theta}_k = (\boldsymbol{\mu}_k = (\mu_{1k}, \mu_{2k}, \ldots, \mu_{Jk})', \sigma_k^2),\ k = 1, 2, \ldots, K\}$ from the posterior distribution in (8.11) to compute the CPO statistic. A Monte Carlo estimate of CPO_{ij} in (8.25) is given by

$$\widehat{\mathrm{CPO}}_{ij} = \left[\frac{1}{K}\sum_{k=1}^{K}(2\pi\sigma_k^2)^{\frac{1}{2}} \exp\left\{\frac{1}{2\sigma_k^2}\left(y_{ij} - \mu_{jk}\right)^2\right\}\right]^{-1},$$

for $i = 1, 2, \ldots, n_j$ and $j = 1, 2, \ldots, J$.

8.4.4 The Marginal Likelihood

For the constrained ANOVA model in (8.2) with the prior in (8.9), the marginal likelihood is given by

$$\begin{aligned} m(D) = \int_0^\infty \int_{\Theta_{\boldsymbol{\mu}}} \Bigg[&(2\pi\sigma^2)^{-\frac{n}{2}} \exp\left\{-\frac{1}{2\sigma^2}\sum_{j=1}^{J}\sum_{i=1}^{n_j}(y_{ij} - \mu_j)^2\right\} \\ &\times \frac{1}{C_0(\sigma^2, a_0)}\left\{\prod_{j=1}^{J}\phi\Big(\mu_j|0, \frac{\sigma^2}{a_0 n_j}\Big)\right\} \\ &\times \frac{\beta_0^{\alpha_0}}{\Gamma(\alpha_0)} \times (\sigma^2)^{-(\alpha_0+1)} \exp\Big(-\frac{\beta_0}{\sigma^2}\Big)\Bigg] d\boldsymbol{\mu}\, d\sigma^2, \end{aligned} \tag{8.27}$$

where $C_0(\sigma^2, a_0) = C(\sigma^2, \boldsymbol{y}_0 = \mathbf{0}, \boldsymbol{a}_0)$ with $a_{01} = a_{02} = \cdots = a_{0J} = a_0$ and $C_0(\sigma^2, \boldsymbol{y}_0, \boldsymbol{a}_0)$ is defined in (8.7).

When $\Theta_{\boldsymbol{\mu}} = R^J$, $m(D)$ can be evaluated analytically. In this case, the marginal likelihood, denoted by $m_U(D)$, is given by

$$m_U(D) = (2\pi)^{-\frac{n}{2}} \frac{\beta_0^{\alpha_0}\Gamma(\alpha_0 + \frac{n}{2})}{\Gamma(\alpha_0)} \Big(\frac{a_0}{1+a_0}\Big)^{\frac{J}{2}}$$

$$\times \Bigg[\beta_0 + \frac{1}{2}\sum_{j=1}^{J}\Big\{\sum_{i=1}^{n_j}(y_{ij} - \bar{y}_j)^2 + \frac{a_0 n_j}{1+a_0}(\bar{y}_j)^2\Big\}\Bigg]^{-(\alpha_0+\frac{n}{2})} . \tag{8.28}$$

When $\Theta_{\boldsymbol{\mu}} \subset R^J$ but $\Theta_{\boldsymbol{\mu}} \neq R^J$, we assume that $\Theta_{\boldsymbol{\mu}}$ satisfies the invariant condition given in (8.10). As discussed in Section 8.3, this assumption ensures that $C_0(\sigma^2, a_0)$ in (8.27) is free from both σ^2 and a_0. To compute $m(D)$ in (8.27), we write

$$\pi^*(\boldsymbol{\mu}, \sigma^2|D) = (2\pi\sigma^2)^{-\frac{n}{2}} \exp\Bigg\{-\frac{1}{2\sigma^2}\sum_{j=1}^{J}\sum_{i=1}^{n_j}(y_{ij} - \mu_j)^2\Bigg\} \prod_{j=1}^{J}\phi\Big(\mu_j|0, \frac{\sigma^2}{a_0 n_j}\Big)$$

$$\times \frac{\beta_0^{\alpha_0}}{\Gamma(\alpha_0)}(\sigma^2)^{-(\alpha_0+1)} \exp\Big(-\frac{\beta_0}{\sigma^2}\Big). \tag{8.29}$$

Let $\Theta = \Theta_{\boldsymbol{\mu}} \times R^+$, where $R^+ = (0, \infty)$. Suppose $\boldsymbol{\theta}^* = (\boldsymbol{\mu}^*, \sigma^{*2}) \in \Theta$, where $\boldsymbol{\mu}^*$ and σ^{*2} are arbitrary (but in accordance with the constraints) fixed values of $\boldsymbol{\mu}$ and σ^2. Using the identity established in [8], we have

$$m(D) = \frac{\pi^*(\boldsymbol{\mu}^*, \sigma^{*2}|D)}{C_0(\sigma^{*2}, a_0)\pi(\boldsymbol{\mu}^*, \sigma^{*2}|D)}, \tag{8.30}$$

where $\pi(\boldsymbol{\mu}, \sigma^2|D)$ is the posterior distribution with a support of Θ.

Let $\pi_U(\boldsymbol{\mu}, \sigma^2|D)$ denote the posterior distribution under the unconstrained ANOVA model. Note that

$$\begin{aligned}
\pi(\boldsymbol{\mu}^*, \sigma^{*2}|D) &= \frac{\pi^*(\boldsymbol{\mu}^*, \sigma^{*2}|D)/C_0(\sigma^{*2}, a_0)}{\int_\Theta \{\pi^*(\boldsymbol{\mu}, \sigma^2|D)/C_0(\sigma^2, a_0)\}\, d\boldsymbol{\mu}\, d\sigma^2} \\
&= \frac{\pi^*(\boldsymbol{\mu}^*, \sigma^{*2}|D)}{\int_\Theta \pi^*(\boldsymbol{\mu}, \sigma^2|D)\, d\boldsymbol{\mu}\, d\sigma^2} \text{ (as } C_0(\sigma^{*2}, a_0) = C_0(\sigma^2, a_0)) \\
&= \frac{\pi^*(\boldsymbol{\mu}^*, \sigma^{*2}|D)}{m_U(D)\int_{R^J \times R^+} I_\Theta(\boldsymbol{\mu}, \sigma^2)\pi_U(\boldsymbol{\mu}, \sigma^2|D)\, d\boldsymbol{\mu}\, d\sigma^2} \\
&= \frac{\pi^*(\boldsymbol{\mu}^*, \sigma^{*2}|D)}{m_U(D)E_U\{I_\Theta(\boldsymbol{\mu}, \sigma^2)|D\}},
\end{aligned} \tag{8.31}$$

where $m_U(D)$ is given in (8.28) and the expectation E_U is taken with respect to $\pi_U(\boldsymbol{\mu}, \sigma^2|D)$. Plugging (8.31) into (8.30) leads to

$$m(D) = \frac{m_U(D)E_U\{I_\Theta(\boldsymbol{\mu}, \sigma^2)|D\}}{C_0(\sigma^{*2}, a_0)}. \tag{8.32}$$

Taking the natural logarithm of both sides of (8.32) gives

$$\log m(D) = \log m_U(D) + \log E_U\{I_\Theta(\boldsymbol{\mu}, \sigma^2)|D\} - \log C_0(\sigma^{*2}, a_0). \quad (8.33)$$

Note that $m(D)/m_U(D)$ is the Bayes factor for comparing the inequality constrained model to the unconstrained model. Additionally note that the identity in (8.32) is also discussed in [29].

The identity in (8.33) allows us to develop a Monte Carlo method for computing the marginal likelihood $m(D)$ of the constrained model. To this end, we let $\{(\boldsymbol{\mu}_k, \sigma_k^2),\ k = 1, 2, \ldots, K\}$ denote a random sample from the posterior distribution $\pi_U(\boldsymbol{\mu}, \sigma^2|D)$. Then, an Monte Carlo estimate of $E_U\{I_\Theta(\boldsymbol{\mu}, \sigma^2)|D\}$ is given by

$$\widehat{E}_U\{I_\Theta(\boldsymbol{\mu}, \sigma^2)|D\} = \frac{1}{K}\sum_{k=1}^{K} I_\Theta(\boldsymbol{\mu}_k, \sigma_k^2), \quad (8.34)$$

which is the proportion of the samples $(\boldsymbol{\mu}_k, \sigma_k^2)$'s that fall in Θ. Similarly, let $\{\boldsymbol{\mu}_{0k},\ k = 1, 2, \ldots, K_0\}$ denote a random sample from the unconstrained prior distribution $\prod_{j=1}^{J} \phi\left(\mu_j|0, \frac{\sigma^{*2}}{a_0 n_j}\right)$. Then, a Monte Carlo estimate of $C_0(\sigma^{*2}, a_0)$ is given by

$$\widehat{C}_0(\sigma^{*2}, a_0) = \frac{1}{K_0}\sum_{k=1}^{K_0} I_{\Theta_{\boldsymbol{\mu}}}(\boldsymbol{\mu}_{0k}). \quad (8.35)$$

Plugging (8.34) and (8.35) into either (8.32) or (8.33) gives a Monte Carlo estimate of $m(D)$ or $\log m(D)$. Note that the Monte Carlo estimates given in (8.34) and (8.35) are also discussed in [29]. However, for unfavorable inequality constraints, the Monte Carlo estimates may not be efficient.

Another Monte Carlo method can be developed by using the Gibbs stopper estimator (see [38]) of $\pi(\boldsymbol{\theta}^*|D) = \pi(\boldsymbol{\mu}^*, \sigma^{*2}|D)$ given in (8.30). Note that when the support Θ is a constrained parameter space in (8.30), $\pi(\boldsymbol{\theta}|D) = \pi(\boldsymbol{\mu}, \sigma^2|D)$ is the posterior distribution in which the constraints are accounted for. The Gibbs sampling kernel moving from $\boldsymbol{\theta}$ to $\boldsymbol{\theta}^*$ takes the form

$$\begin{aligned} q(\boldsymbol{\theta}, \boldsymbol{\theta}^*) &= \pi(\mu_1^*|\mu_2, \ldots, \mu_J, \sigma^2, D)\pi(\mu_2^*|\mu_1^*, \mu_3, \ldots, \mu_J, \sigma^2, D) \\ &\quad \times \cdots \times \pi(\mu_J^*|\mu_1^*, \ldots, \mu_{J-1}^*, \sigma^2, D)\pi(\sigma^{*2}|\boldsymbol{\mu}^*, D), \end{aligned} \quad (8.36)$$

where $\pi(\mu_j|\mu_1, \ldots, \mu_{j-1}, \mu_{j+1}, \ldots, \mu_J, \sigma^2, D)$ is the conditional posterior density of μ_j given $\mu_1, \ldots, \mu_{j-1}, \mu_{j+1}, \ldots, \mu_J, \sigma^2$ and D for $j = 1, 2, \ldots, J$, and $\pi(\sigma^2|\boldsymbol{\mu}, D)$ denotes the conditional posterior density of σ^2 given $\boldsymbol{\mu}$ and D. Then, we have

$$\int_\Theta q(\boldsymbol{\theta}, \boldsymbol{\theta}^*)\pi(\boldsymbol{\theta}|D)\ d\boldsymbol{\theta} = \pi(\boldsymbol{\theta}^*|D).$$

Suppose that $\{\boldsymbol{\theta}_k = (\boldsymbol{\mu}_k, \sigma_k^2),\ k = 1, 2, \ldots, K\}$ is a Gibbs sample of size K from the posterior distribution $\pi(\boldsymbol{\theta}^*|D)$. Then, the Gibbs stopper estimator of [38] is given by

$$\widehat{\pi}(\boldsymbol{\theta}^*|D) = \frac{1}{K}\sum_{k=1}^{K} q(\boldsymbol{\theta}_k, \boldsymbol{\theta}^*). \quad (8.37)$$

Under the inequality constraints, $\pi(\mu_j|\mu_1,\ldots,\mu_{j-1},\mu_{j+1},\ldots,\mu_J,\sigma^2,D)$ is the density corresponding to a truncated normal distribution for $j=1,2,\ldots,J$ and $\pi(\sigma^2|\boldsymbol{\mu},D)$ is the density of an inverse gamma distribution. Therefore, the closed-form expressions of these conditional posterior distributions are available. The Gibbs stopper estimator can also be used for computing the prior normalizing constant $C_0(\sigma^{*2},a_0)$ in (8.30). To see this, let

$$\pi(\boldsymbol{\mu}|\sigma^{*2},a_0)\propto\prod_{j=1}^{J}\phi\Big(\mu_j|0,\frac{\sigma^{*2}}{a_0n_j}\Big)I_{\Theta_{\boldsymbol{\mu}}}(\boldsymbol{\mu}).$$

Then, we have

$$C_0(\sigma^{*2},a_0)=\frac{\prod_{j=1}^{J}\phi\Big(\mu_{0j}^*|0,\frac{\sigma^{*2}}{a_0n_j}\Big)}{\pi(\boldsymbol{\mu}_0^*|\sigma^{*2},a_0)},$$

where $\boldsymbol{\mu}_0^*=(\mu_{01}^*,\ldots,\mu_{0J}^*)'$ is a fixed value in $\Theta_{\boldsymbol{\mu}}$, which may be different from $\boldsymbol{\mu}^*$ used in (8.30). The Gibbs sampling kernel used for sampling $\boldsymbol{\mu}$ from the prior distribution $\pi(\boldsymbol{\mu}|\sigma^{*2},a_0)$ is given as follows:

$$q_0(\boldsymbol{\mu},\boldsymbol{\mu}_0^*)=\pi(\mu_{01}^*|\mu_2,\ldots,\mu_J,\sigma^{*2},a_0)\cdots\pi(\mu_{0J}^*|\mu_{01}^*,\ldots,\mu_{0,J-1}^*,\sigma^{*2},a_0).$$

Since

$$\int_{\Theta_{\boldsymbol{\mu}}}q_0(\boldsymbol{\mu},\boldsymbol{\mu}_0^*)\pi(\boldsymbol{\mu}|\sigma^{*2},a_0)\,d\boldsymbol{\mu}=\pi(\boldsymbol{\mu}^*|\sigma^{*2},a_0),$$

a Monte Carlo estimator of $C_0(\sigma^{*2},a_0)$ may be computed as follows:

$$\widehat{C}_0(\sigma^{*2},a_0)=\frac{1}{K_0}\sum_{k=1}^{K_0}q_0(\boldsymbol{\mu}_{0k},\boldsymbol{\mu}_0^*),\qquad(8.38)$$

where $\{\boldsymbol{\mu}_{0k},\ k=1,2,\ldots,K_0\}$ is a Gibbs sample from the prior distribution $\pi(\boldsymbol{\mu}|\sigma^{*2},a_0)$. Using (8.37) and (8.38), a Monte Carlo estimator of $m(D)$ on the log scale is given by

$$\begin{aligned}\log\widehat{m}(D)=&-\frac{n}{2}\log(2\pi\sigma^{*2})-\frac{1}{2\sigma^{*2}}\sum_{j=1}^{J}\sum_{i=1}^{n_j}(y_{ij}-\mu_j^*)^2\\&+\alpha_0\log\beta_0-\log\Gamma(\alpha_0)-(\alpha_0+1)\log\sigma^{*2}-\frac{\beta_0}{\sigma^{*2}}\\&+\log\Big\{\frac{1}{K_0}\sum_{k=1}^{K_0}q_0(\boldsymbol{\mu}_{0k},\boldsymbol{\mu}_0^*)\Big\}-\log\Big\{\frac{1}{K}\sum_{k=1}^{K}q(\boldsymbol{\theta}_k,\boldsymbol{\theta}^*)\Big\}.\end{aligned}\qquad(8.39)$$

The Gibbs stopper estimator can be potentially useful when $E_U\{I_\Theta(\boldsymbol{\mu},\sigma^2)|D\}$ or $C_0(\sigma^{*2},a_0)$ in (8.32) is small. When $E_U\{I_\Theta(\boldsymbol{\mu},\sigma^2)|D\}$ and $C_0(\sigma^{*2},a_0)$ in (8.32) are reasonably large, the Monte Carlo estimators given in (8.34) and (8.35) are quite accurate. Thus, the Gibbs stopper estimators are not needed in this case.

Table 8.1. List of hypotheses

H_{01}:	$\mu_{con} = \mu_{amn} = \mu_{pat} = \mu_{sim}$
H_{1a1}:	$\mu_{con} > \{\mu_{amn} = \mu_{pat}\} > \mu_{sim}$
H_{1b1}:	$\mu_{con} > \mu_{amn} > \{\mu_{pat} = \mu_{sim}\}$
H_{1a2}:	$\mu_{con} > \{\mu_{amn}, \mu_{pat}\} > \mu_{sim}$
H_{1b2}:	$\mu_{con} > \mu_{amn} > \{\mu_{pat}, \mu_{sim}\}$
H_2:	$\mu_{con}, \mu_{amn}, \mu_{pat}, \mu_{sim}$

8.5 Examples

8.5.1 The Dissociative Identity Disorder Data

The illustration is based on the DID data [22] introduced in Chapter 2. Recall that DID-patients were compared with three different types of control groups on their ability to recall information that they obtained in a prior phase of the experiment. The groups are (1) DID-patients, (2) Normal controls, (3) Simulators, and (4) True amnesiacs. The response variable (y_{ij}) is the recognition score. Let $\mu_{con} = \mu_{control}$, $\mu_{amn} = \mu_{trueamnesiacs}$, $\mu_{pat} = \mu_{DID-patients}$, and $\mu_{sim} = \mu_{DIDsimulators}$. Let $\boldsymbol{\theta} = (\boldsymbol{\mu}, \sigma)$. The hypotheses considered in this example are given in Table 8.1. Instead of performing conventional hypothesis testing, we adopt the notion that each hypothesis defines a model. Thus, we compare the six models defined by the six hypotheses listed in Table 8.1.

To apply the general notation introduced in the earlier sections, the model corresponding to H_2 is considered as the unconstrained model for the inequality constrained models corresponding to H_{1a2} and H_{1b2}. For H_{1a1}, the corresponding unconstrained model is the model with μ_{con}, $\mu_{amn} = \mu_{pat}$, and μ_{sim} as the three free mean parameters. Similarly, the corresponding unconstrained model for H_{1b1} is the model with μ_{con}, μ_{amn}, and $\mu_{pat} = \mu_{sim}$ as the three free mean parameters. For the model under the hypothesis H_{01}: $\mu_{con} = \mu_{amn} = \mu_{pat} = \mu_{sim}$, we may view the data from one group with $J = 1$. In this way, the formulas developed in Sections 8.3 and 8.4 can be directly applied to all six models under consideration. We then compare these six models using each of the four Bayesian criteria, namely L measure, DIC, LPML, and marginal likelihood. In all computations, a vague inverse gamma $\mathcal{IG}(0.0001, 0.0001)$ is specified for σ^2. In addition, several values of a_0, such as $a_0 = 0.01$, $a_0 = 0.0001$, and $a_0 = 0$, in the prior (8.6) for $\boldsymbol{\mu}$ are considered to investigate the robustness of these four Bayesian criteria to the specification of this key hyperparameter (a_0). Additional values of a_0 are also considered for the marginal likelihood, as this criterion is more sensitive to the specification of a_0. Recall that we assume $a_{01} = a_{02} = \cdots = a_{0J} = a_0$ and $\boldsymbol{y}_0 = \boldsymbol{0}$ in (8.6). Note that in Table 8.1 the models under H_{01} and H_2 are unconstrained and the models under H_{1a1}, H_{1b1}, H_{1a2}, and H_{1b2} are subject to inequality constraints. It can be shown that all of these four hypotheses are in accordance with the invariant assumption given by (8.10). Thus, the corresponding

Table 8.2. Detailed summaries of the L measures for the DID data ($a_0 = 0.0001$)

	Sum of $E\{\text{Var}(z_{ij}\|\boldsymbol{\theta})\|D\}$	Sum of $\text{Var}\{E(z_{ij}\|\boldsymbol{\theta})\|D\}$	Sum of $\text{Var}(z_{ij}\|D)$	Sum of Bias^2
H_{01}	2244.54	23.88	2268.42	2196.47
H_{1a1}	266.63	8.50	275.12	260.48
H_{1b1}	259.84	8.28	268.12	253.84
H_{1a2}	243.33	10.24	253.57	237.63
H_{1b2}	243.34	10.32	253.66	237.63
H_2	243.33	10.35	253.68	237.63

prior normalizing constant $C_0(\sigma^2, \boldsymbol{y}_0 = \boldsymbol{0}, \boldsymbol{a}_0)$ is free from σ^2 and a_0. In all computations, 100,000 Gibbs iterations with a burn-in of 2000 iterations were used to compute all the Bayesian criteria if the closed-form expressions are not available.

Table 8.2 shows the detailed breakdown summaries of the L measures in (8.15) under the six hypotheses based on the prior with $a_0 = 0.0001$. It is expected that the model under H_2 gives the best prediction and the model under H_{01} should lead to the worst prediction. This is exactly reflected in the L measures as the sums of Bias^2 $(= \sum_{j=1}^{J} \sum_{i=1}^{n_j} \{E(z_{ij}|D) - y_{ij}\}^2)$ are 237.63 under H_2 and 2196.47 under H_{01}, where 237.63 is the smallest value and 2196.47 is the largest value among the six models. When the inequality constraints are plausible, we expect that the sum of Bias^2 under the constrained model is comparable to that under the unconstrained model. From Table 8.2, we see that the sums of Bias^2 under the models corresponding to the hypotheses H_{1a2} and H_{1b2} are nearly identical to the one obtained under the hypothesis H_2. Moreover, if the inequality constraints are favorable, the sum of the variances of the conditional means, namely $\sum_{j=1}^{J} \sum_{i=1}^{n_j} \text{Var}\{E(z_{ij}|\boldsymbol{\theta})|D\}$, under the constrained model should be smaller than the one under the unconstrained model. From Table 8.2, we clearly observe this phenomenon by comparing these quantities under H_{1a2} and H_{1b2} to that under H_2. These results demonstrate that the L measure does indeed have a very nice statistical interpretation.

The L measures for the DID data based on various values of a_0 and ν are given in Table 8.3. From this table, we see that the values of L measures are sensitive to the choice of a_0. However, when a_0 is getting small, the L measure tends to be more robust. Interestingly, the order of the L measures under the six models does not change across all three values of a_0 as well as across three values of ν. Based on the L measure, the models under H_{1a2}, H_{1b2}, and H_2 fit the DID data equally well, with the model under H_{1a2} fitting the data slightly better. It is evident that these three models fit the DID data much better than the other three models. In particular, the model under H_{01} fits the data poorly. Although $\nu = \frac{1}{2}$ is a desirable choice in general, in comparing the six models given in Table 8.3, the L measures with $\nu = 0.1$ and $\nu = 0.9$

Table 8.3. The L measures for the DID data based on various priors

	$a_0 = 0.01$			$a_0 = 0.0001$			$a_0 = 0$		
	L(0.1)	L(0.5)	L(0.9)	L(0.1)	L(0.5)	L(0.9)	L(0.1)	L(0.5)	L(0.9)
H_{01}	2520.67	3399.38	4278.10	2488.07	3366.65	4245.24	2487.73	3366.32	4244.91
H_{1a1}	354.61	459.00	563.39	301.17	405.36	509.55	300.62	404.81	509.00
H_{1b1}	347.02	448.76	550.50	293.51	395.04	496.58	292.96	394.50	496.03
H_{1a2}	331.34	426.59	521.84	277.33	372.38	467.44	276.77	371.83	466.88
H_{1b2}	331.54	426.79	522.05	277.42	372.47	467.52	276.87	371.92	466.97
H_2	331.66	426.92	522.17	277.44	372.49	467.55	276.89	371.94	466.99

Table 8.4. The DIC values for the DID data

	$a_0 = 0.01$			$a_0 = 0.0001$			$a_0 = 0$		
	$d(\bar{\boldsymbol{\theta}})$	p_d	DIC	$d(\bar{\boldsymbol{\theta}})$	p_d	DIC	$d(\bar{\boldsymbol{\theta}})$	p_d	DIC
H_{01}	563.06	1.95	566.96	563.01	1.99	566.98	563.01	1.99	566.98
H_{1a1}	364.49	3.62	371.74	362.59	3.97	370.54	362.59	3.98	370.55
H_{1b1}	362.15	3.62	369.38	360.17	3.97	368.11	360.16	3.98	368.12
H_{1a2}	356.17	4.52	365.21	353.96	4.93	363.83	353.96	4.94	363.84
H_{1b2}	356.17	4.56	365.28	353.96	4.96	363.89	353.96	4.97	363.89
H_2	356.18	4.59	365.35	353.96	4.98	363.92	353.96	4.98	363.92

lead to the same conclusion as the L measure with $\nu = \frac{1}{2}$ for the DID data. In this sense, the L measure is quite robust in ν for the DID data.

Next, we computed the DIC values for all six models and the results are shown in Table 8.4. The DIC values are quite robust to the choice of a_0. Similar to the L measure, the top DIC models are those under H_{1a2}, H_{1b2}, and H_2, with the model under H_{1a2} achieving the smallest DIC value. Note that the DIC measure was computed based on the posterior mean of $\boldsymbol{\theta}$. To examine the robustness of the DIC measure in the choice of $\overline{\boldsymbol{\theta}}$ in (8.21), we computed the DIC values by taking $\overline{\boldsymbol{\theta}}$ to be the posterior mean, the posterior median, and the posterior mode of $\boldsymbol{\theta}$. As an illustration, the DIC values under these choices for $\overline{\boldsymbol{\theta}}$ are reported in Table 8.5 under the hypothesis H_{1a2} with $a_0 = 0.0001$. We see that those DIC values are very similar. Similar results were obtained but not reported here under other hypotheses with different values of a_0.

Table 8.6 shows the LPML values. Similar to the DIC measure, the LPML is extremely robust to the choice of a_0. Again, the top three LPML models are identical to those identified by the L measure and the DIC measure. The best LPML model is the one under H_{1a2}. The log marginal likelihoods were also computed and the results are presented in Table 8.7. When $a_0 = 0.01$, the top marginal likelihood model is H_{1a2}. When $a_0 = 0.0001$, the top marginal likelihood model becomes H_{1b1}. When a_0 is nearly 0, the worst model H_{01}

identified by the L measure, DIC, and LPML turns out to be the best based on the marginal likelihood criterion. This result is expected, as the model H_{01} has the smallest dimension. This phenomenon is known as Bartlett's or Lindley's paradox (see [1, 31]).

As the marginal likelihood-based criterion is not robust to the choice of a_0, a_0 needs to be more carefully elicited. When historical data or training data are available, a guide value of a_0 is discussed in [4, 23]. As for the DID data, there are no historical data available, we use an empirical Bayes method to choose a guide value of a_0. We observe that for the DID data, the log marginal likelihood is a concave function of a_0 under each of the six models in Table 8.1, which can be seen clearly from Figure 8.1, in which log marginal likelihoods as functions of a_0 are plotted under H_{1a1}, H_{1b1}, H_{1a2}, H_{1b2}, and H_2. Let $a_{0,H}$ denote the "best" value of a_0 that maximizes the log marginal likelihood under the hypothesis H. The log marginal likelihoods under the "best" values of a_0 are given in Table 8.8. According to the marginal likelihood-based criterion, H_{1b2} is the best model and H_{1a2} is the second best model for all six $a_{0,H}$ values. We notice that the difference in the log marginal likelihoods between H_{1b2} and H_{1a2} is quite small. In addition, under the "best" values of a_0, the phenomenon of the Bartlett's or Lindley's paradox disappears. In general, we recommend choosing a guide value of a_0 to be the smallest $a_{0,H}$ among all hypotheses under consideration. The prior under this guide value is least informative among the most "informative" priors under all the hypotheses being considered.

In view of the four Bayesian criteria considered in this example, the posterior predictive-based criteria, namely L measure, DIC, and LPML, consistently select H_{1a2} as the best model, whereas the log marginal likelihood-based criterion selects H_{1b2} as the best model. However, the difference between models H_{1a2} and H_{1b2} in all four criteria is nearly indistinguishable. Thus, these two models virtually fit the DID data equally well. Therefore, we recommend that both H_{1a2} and H_{1b2} should be considered for further investigation.

8.5.2 The Emotional Reactivity Data

We consider the data from a study that addressed the influence of depression severity on emotional reactivity after different types of peer evaluation feedback in preadolescent children (see [35]). As discussed in Chapter 2, 139 children, between the ages of 10 and 13 years, participated in this study. The

Table 8.5. The DIC values based on various estimates of $\boldsymbol{\theta}$ under H_{1a2} ($a_0 = 0.0001$)

Estimate	μ_{pat}	μ_{con}	μ_{sim}	μ_{amn}	σ^2	$d(\bar{\boldsymbol{\theta}})$	p_d	DIC
Mean	3.109	13.279	1.876	4.560	2.589	353.96	4.93	363.83
Median	3.106	13.279	1.877	4.560	2.551	353.94	4.96	363.85
Mode	3.189	13.289	1.776	4.589	2.434	354.18	4.72	363.62

Table 8.6. The LPML values for the DID data

	$a_0 = 0.01$	$a_0 = 0.0001$	$a_0 = 0$
H_{01}	-283.29	-283.29	-283.29
H_{1a1}	-185.75	-185.42	-185.42
H_{1b1}	-184.64	-184.31	-184.32
H_{1a2}	-182.53	-182.26	-182.26
H_{1b2}	-182.56	-182.28	-182.29
H_2	-182.60	-182.30	-182.31

response variable is the score on the positive affect scale of the Positive And Negative Affect Schedule (PANAS-P). The primary interest of this study was to examine the effect of peer evaluation in mood under three conditions and three levels of depression, as shown in Table 8.9. The hypotheses considered in this example are given in Table 8.10.

We use the same notion as discussed in Section 8.5.1 that each hypothesis defines a model. Thus, we compare the five models listed in Table 8.10 using each of the four Bayesian criteria, namely L measure, DIC, LPML, and marginal likelihood. Again, we assume $a_{01} = a_{02} = \cdots = a_{0J} = a_0$ and $\boldsymbol{y}_0 = \boldsymbol{0}$ in (8.6). Note that in Table 8.10, the models under H_0 and H_2 are unconstrained and the models under H_{1a}, H_{1b}, and H_{1c} are subject to in-

Table 8.7. The log marginal likelihoods for the DID data

	$a_0 = 0.01$	$a_0 = 0.001$	$a_0 = 0.0001$	$a_0 = 10^{-50}$
H_{01}	-294.70	-295.23	-296.32	-349.27
H_{1a1}	-205.07	-200.99	-203.61	-362.40
H_{1b1}	-204.07	-199.81	-202.40	-361.19
H_{1a2}	-203.16	-199.54	-203.23	-414.96
H_{1b2}	-203.15	-199.53	-203.23	-414.96
H_2	-205.63	-202.02	-205.71	-417.44

Table 8.8. The log marginal likelihoods under the "best" value of a_0 for the DID data

	"Best"	Hypothesis					
	Value of $a_{0,H}$	H_{01}	H_{1a1}	H_{1b1}	H_{1a2}	H_{1b2}	H_2
H_{01}	0.0074	-294.671	-203.452	-202.401	-201.513	-201.505	-203.988
H_{1a1}	0.0017	-295.015	-200.835	-199.664	-199.178	-199.174	-201.656
H_{1b1}	0.0016	-295.038	-200.835	-199.662	-199.200	-199.196	-201.678
H_{1a2}	0.0021	-294.945	-200.868	-199.706	-199.148	-199.144	-201.627
H_{1b2}	0.0020	-294.954	-200.861	-199.696	-199.149	-199.144	-201.627
H_2	0.0021	-294.945	-200.868	-199.706	-199.148	-199.144	-201.627

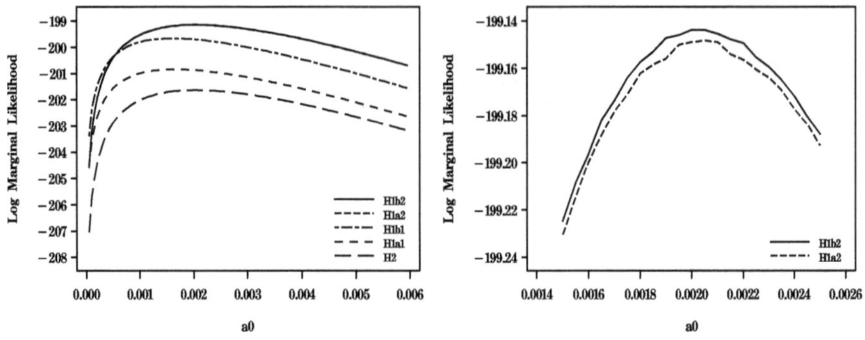

(a) (b)

Fig. 8.1. Plots of log marginal likelihoods versus a_0 under H_{1a1}, H_{1b1}, H_{1a2}, H_{1b2}, and H_2 (a) and under H_{1a2} and H_{1b2} (b), respectively, for the DID data

equality constraints. It can be shown that the constrained parameter space under each of three hypotheses H_{1a}, H_{1b}, and H_{1c} satisfies the invariant condition given by (8.10). Thus, the corresponding prior normalizing constant $C_0(\sigma^2, \boldsymbol{y}_0 = \mathbf{0}, \boldsymbol{a}_0)$ is free from σ^2 and a_0. Similar to Section 8.5.1, a vague inverse gamma $\mathcal{IG}(0.0001, 0.0001)$ is specified for σ^2 and several values of a_0, such as $a_0 = 0.01$, $a_0 = 0.0001$, and $a_0 = 0$, in the prior (8.6) for $\boldsymbol{\mu}$ are considered to investigate whether these four Bayesian criteria are robust to the specification of this key hyperparameter (a_0). Also, 100,000 Gibbs iterations with a burn-in of 2000 iterations were used to compute all the Bayesian criteria if the closed form expressions are not available.

Table 8.9. The design of experiment

	Condition		
Depress	Positive	Neutral	Negative
Low	μ_1	μ_2	μ_3
Moderate	μ_4	μ_5	μ_6
High	μ_7	μ_8	μ_9

Table 8.10. List of hypotheses

H_0 :	$\{\mu_7 - \mu_8\} = \{\mu_4 - \mu_5\} = \{\mu_1 - \mu_2\}$, $\{\mu_9 - \mu_8\} = \{\mu_6 - \mu_5\} = \{\mu_3 - \mu_2\}$
H_{1a}:	$\{\mu_7 - \mu_8\} < \{\mu_4 - \mu_5\} < \{\mu_1 - \mu_2\}$, $\{\mu_9 - \mu_8\} < \{\mu_6 - \mu_5\} < \{\mu_3 - \mu_2\}$
H_{1b}:	$\{\mu_7 - \mu_8\} < \{\mu_4 - \mu_5\} < \{\mu_1 - \mu_2\}$, $\{\mu_9 - \mu_8\} > \{\mu_6 - \mu_5\} > \{\mu_3 - \mu_2\}$
H_{1c}:	$\{\mu_7 - \mu_8\} > \{\mu_4 - \mu_5\} > \{\mu_1 - \mu_2\}$, $\{\mu_9 - \mu_8\} > \{\mu_6 - \mu_5\} > \{\mu_3 - \mu_2\}$
H_2 :	$\{\mu_7 - \mu_8\}, \{\mu_4 - \mu_5\}, \{\mu_1 - \mu_2\}$, $\{\mu_9 - \mu_8\}, \{\mu_6 - \mu_5\}, \{\mu_3 - \mu_2\}$

Table 8.11. Detailed summaries of the L measures for the emotional reactivity data ($a_0 = 0.0001$)

	Sum of $E\{\text{Var}(z_{ij}\|\boldsymbol{\theta})\|D\}$	Sum of $\text{Var}\{E(z_{ij}\|\boldsymbol{\theta})\|D\}$	Sum of $\text{Var}(z_{ij}\|D)$	Sum of Bias^2
H_0	4378.42	157.48	4535.90	4315.20
H_{1a}	4445.99	185.05	4631.04	4486.54
H_{1b}	4330.44	188.96	4519.40	4356.33
H_{1c}	4081.44	226.03	4307.47	4058.44
H_2	4103.38	265.66	4369.04	4044.09

Table 8.11 shows the detailed breakdown summaries of the L measures in (8.15) under the five hypotheses based on the prior with $a_0 = 0.0001$. Similar to Table 8.2, the unconstrained model under H_2 gives the best prediction and the most restrictive model under H_0 leads to the worst prediction, as the sums of Bias^2 $(= \sum_{j=1}^{J} \sum_{i=1}^{n_j} \{E(z_{ij}|D) - y_{ij}\}^2)$ are 4044.09 under H_2 and 4315.20 under H_0. Note that under H_2 with $a_0 = 0.0001$, the posterior means of $\mu_7 - \mu_8$, $\mu_4 - \mu_5$, and $\mu_1 - \mu_2$ are 6.87, 2.13, and -0.64, respectively, and the posterior means of $\mu_9 - \mu_8$, $\mu_6 - \mu_5$, and $\mu_3 - \mu_2$ are -2.32, -4.76 and -10.01, respectively. Thus, the inequality constraints under H_{1c} are most plausible for this dataset. As expected, the sum of Bias^2 under H_{1c} is most comparable to that under the unconstrained model under H_2. In addition, the sums of the variances of the conditional means, namely $\sum_{j=1}^{J} \sum_{i=1}^{n_j} \text{Var}\{E(z_{ij}|\boldsymbol{\theta})|D\}$, under all the constrained models are smaller than the one under the unconstrained model corresponding to the hypothesis H_2. These results are consistent with those obtained in Section 8.5.1.

The values of the L measures for the emotional reactivity data based on various values of a_0 ($a_0 = 0.01$, 0.0001, and 0) and ν ($\nu = 0.1$, 0.5, and 0.9) are given in Table 8.12. These L measures suggest that there is an overwhelming evidence in favor of the model with the favorable constraints under H_{1c}. In addition, for all values of a_0 and ν, the L measures consistently show that the

Table 8.12. The L measure for the emotional reactivity data based on various priors

	$a_0 = 0.01$			$a_0 = 0.0001$			$a_0 = 0$		
	L(0.1)	L(0.5)	L(0.9)	L(0.1)	L(0.5)	L(0.9)	L(0.1)	L(0.5)	L(0.9)
H_0	4988.82	6714.99	8441.15	4967.42	6693.50	8419.58	4967.20	6693.28	8419.36
H_{1a}	5101.41	6897.04	8692.67	5079.69	6874.31	8668.93	5080.28	6874.97	8669.65
H_{1b}	4969.64	6712.10	8454.55	4955.03	6697.57	8440.10	4955.03	6697.55	8440.08
H_{1c}	4735.73	6359.02	7982.31	4713.31	6336.69	7960.07	4713.20	6336.59	7959.97
H_2	4797.29	6415.02	8032.76	4773.45	6391.08	8008.72	4773.20	6390.84	8008.48

Table 8.13. The DIC values for the emotional reactivity data

	$a_0 = 0.01$			$a_0 = 0.0001$			$a_0 = 0$		
	$d(\bar{\boldsymbol{\theta}})$	p_d	DIC	$d(\bar{\boldsymbol{\theta}})$	p_d	DIC	$d(\bar{\boldsymbol{\theta}})$	p_d	DIC
H_0	872.02	5.94	883.90	872.00	6.00	884.00	872.00	6.00	884.00
H_{1a}	877.48	6.67	890.81	877.41	6.75	890.91	877.41	6.76	890.93
H_{1b}	873.30	6.96	887.23	873.31	7.06	887.42	873.31	7.06	887.43
H_{1c}	863.46	8.60	880.66	863.46	8.69	880.85	863.46	8.69	880.85
H_2	863.01	9.89	882.78	862.98	9.99	882.96	862.98	9.99	882.96

Table 8.14. The LPML values for the emotional reactivity data

	$a_0 = 0.01$	$a_0 = 0.0001$	$a_0 = 0$
H_0	-442.67	-442.74	-442.74
H_{1a}	-446.31	-446.37	-446.38
H_{1b}	-444.58	-444.71	-444.71
H_{1c}	-441.50	-441.63	-441.63
H_2	-442.65	-442.79	-442.79

model under H_{1c} fits the data best and the model under H_{1a} fits the data worst.

Table 8.13 shows the DIC values. We can see from this table that the DIC values are very robust to the choice of a_0. The best and worst DIC models are exactly the same as those based on the L measures. Among the models corresponding to hypotheses H_{1a}, H_{1b}, and H_{1c}, the model with least favorable constraints has the smallest dimensional penalty. The values of p_d also indicate that all constrained models (H_0, H_{1a}, H_{1b}, and H_{1c}) are less complex than the unconstrained model (H_2).

The LPML values for the emotional reactivity data were computed and the results are reported in Table 8.14. Similar to the L measure and DIC criteria, the best model under the LPML criterion is the one under H_{1c} and the model under H_{1a} is the worst. However, based on the LPML values, the models corresponding to H_0 and H_2 are nearly indistinguishable. In addition,

Table 8.15. The log marginal likelihoods for the emotional reactivity data

	$a_0 = 1$	$a_0 = 0.1$	$a_0 = 0.01$	$a_0 = 0.0001$	$a_0 = 10^{-50}$
H_0	-464.09	-455.59	-458.30	-469.44	-734.23
H_{1a}	-467.51	-463.15	-470.93	-491.48	-968.11
H_{1b}	-465.33	-460.01	-467.00	-487.29	-963.92
H_{1c}	-462.01	-454.41	-460.96	-481.21	-957.84
H_2	-463.72	-456.48	-463.09	-483.35	-959.98

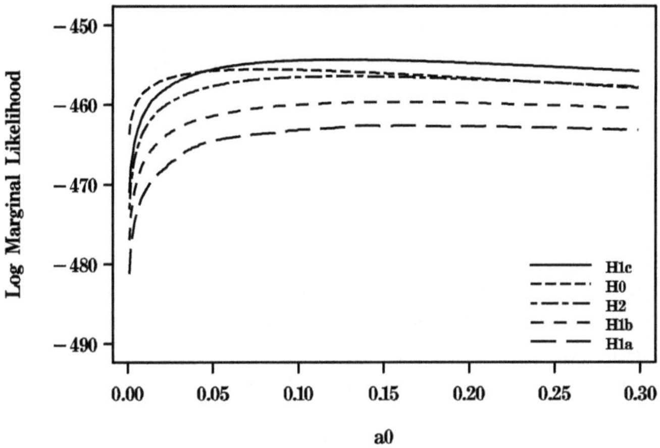

Fig. 8.2. Plots of log marginal likelihoods versus a_0 under H_0 , H_{1a}, H_{1b}, H_{1c}, and H_2 for the emotional reactivity data

we computed the log marginal likelihoods for all the five models, which are given in Table 8.15. When $a_0 = 1$, the top marginal likelihood model is H_{1c}. However, when $a_0 \leq 0.01$, the top marginal likelihood model becomes H_0. Thus, Bartlett's or Lindley's paradox kicks in much earlier for the emotional reactivity data than for the DID data, as shown in Table 8.7. Similar to Section 8.5.1, we also observe that the marginal likelihoods are concave functions of a_0 under all five hypotheses, as shown in Figure 8.2. We also compute the "best" values ($a_{0,H}$) of a_0 that maximize the log marginal likelihoods under H_0, H_{1a}, H_{1b}, H_{1c}, and H_2. The log marginal likelihoods under the "best" values of a_0 are given in Table 8.16. Under these five $a_{0,H}$ values, the marginal likelihood-based method consistently selects H_{1c} as the best model and the phenomenon of the Bartlett's or Lindley's paradox disappears. Similar to the

Table 8.16. The log marginal likelihoods under the "best" value of a_0 for the emotional reactivity data

	"Best"	Hypothesis				
	value of $a_{0,H}$	H_0	H_{1a}	H_{1b}	H_{1c}	H_2
H_0	0.078	-455.52	-463.68	-460.38	-454.66	-456.75
H_{1a}	0.165	-456.19	-462.61	-459.71	-454.48	-456.52
H_{1b}	0.148	-456.00	-462.67	-459.69	-454.39	-456.44
H_{1c}	0.124	-455.76	-462.85	-459.78	-454.33	-456.40
H_2	0.126	-455.78	-462.85	-459.76	-454.33	-456.40

L measure, DIC, and LPML, the worst model is H_{1a} for all values of $a_{0,H}$, as shown in Table 8.16. Based on the results given in Tables 8.12, 8.13, 8.14, and 8.16, all four criteria select the same best model, and, hence, the constrained model, $\{\mu_7-\mu_8\} > \{\mu_4-\mu_5\} > \{\mu_1-\mu_2\}$, $\{\mu_9-\mu_8\} > \{\mu_6-\mu_5\} > \{\mu_3-\mu_2\}$, is most suitable to the emotional reactivity data.

8.6 Discussion

In this chapter, we have discussed the four commonly used Bayesian selection criteria, including the L measure, DIC, and LPML and marginal likelihood, for the selection of constrained models. One noticeable difference between the marginal likelihood (or Bayes factor) and the Bayesian criteria based on the posterior predictive distribution is that the Bayes factor suffers the Bartlett's or Lindley's paradox and the other three criteria do not. This paradox makes the Bayes factor extremely sensitive to the specification of the prior distributions of model parameters. In Section 8.5.1, to select between the models corresponding to hypotheses H_{01} and H_2 for the DID data, we observe the paradox when the prior (8.6) for $\boldsymbol{\mu}$ with $a_0 = 0.0001$ is specified. For the emotional reactivity data discussed in Section 8.5.2, this paradox is observed even when a much less vague prior, namely the the prior (8.6) for $\boldsymbol{\mu}$ with $a_0 = 0.01$, is specified. In this sense, the other three criteria are more advantageous, as these criteria are quite robust to the specification of the prior distributions. However, when a_0 is carefully elicited, the Bartlett's or Lindley's paradox disappears and the marginal likelihood-based criterion is more in agreement with the other posterior predictive-based criteria, as illustrated in Sections 8.5.1 and 8.5.2.

Among the L measure, DIC, and LPML, the L measure has some nice statistical interpretations; that is, the L measure allows us to monitor how well a model fits the data in terms of the posterior predictive variance and bias. This property may aid us in determining whether certain inequality constraints are favorable for a given dataset. Specifically, a set of inequality constraints are more favorable than another set of constraints if the corresponding model produces a smaller sum of Bias^2 (viz. $\sum_{j=1}^{J}\sum_{i=1}^{n_j}\{E(z_{ij}|D) - y_{ij}\}^2$). The model with inequality constraints may be considered to be simpler than the unconstrained one. The dimensional penalty term, p_d, in DIC may be viewed as a measure of the model complexity. This is a more direct measure of model complexity than the one in the Bayes factor, which is the prior normalizing constant under the constrained model. However, it is not clear whether p_d can be analytically available under certain models with inequality constraints. It is of great theoretical interest to quantify the inequality constraints in terms of the reduction of the number of parameters.

From the computational point view, it is much easier to compute the L measure, DIC, and LPML than the Bayes factor. As long as we are able to sample from the posterior distribution under the constrained model, the

Monte Carlo estimates of these quantities discussed in Sections 8.4.1 to 8.4.3 are efficient and numerically stable. In Section 8.4.1, two Monte Carlo methods are discussed for computing the marginal likelihood. These methods work well for the models considered in Sections 8.5.1 and 8.5.2 for the DID and emotional reactivity data. However, when the prior probability that the parameters fall into the constrained parameter space under the unconstrained model is extremely small, the Monte Carlo estimator given in (8.35) may become very inefficient. In this case, the Gibbs stopper estimator given in (8.38) may be more efficient. However, due to the complexity induced by inequality constraints, the Gibbs stopper estimator may still not be satisfactory. Therefore, a much more sophisticated and yet efficient Monte Carlo estimator of the marginal likelihood or the Bayes factor needs to be developed. This is well deserved to be an important future research project.

As discussed in [3], the L measure is not well calibrated since this measure is not properly scaled. This is also true for the DIC and LPML measures. In Sections 8.5.1 and 8.5.2, we have seen that for certain constrained models, these measures are very similar numerically. Is a model with a smaller criterion value statistically better than another model? Or is the difference in the criterion values due to uncertainty in the random sample? One way to address this important issue is to calibrate these criteria under certain distributions so that they can be more easily interpreted. The idea of calibration is not totally new. In fact, a calibration is proposed in [3] for the L measure based on the prior predictive distribution. The extension of the calibration of [3] for the L measure to the DIC and LPML measures appears to be straightforward. However, the properties of the calibration for the models under inequality constraints need to be examined carefully.

When we compare two nested models with different dimensions based on the marginal likelihoods or the Bayes factor, we have observed the Bartlett's or Lindley's paradox for both the DID and emotional reactivity data. This paradox may partially be the direct consequence of the way to interpret the strength of evidence according to the value of Bayes factor. A popular rule for interpreting the strength of evidence is proposed in [27] and a slight modification of Jeffreys' proposal is given in [28]. Based on this rule, for example, if the Bayes factor for comparing model 1 to model 2 is greater than 150, there is a strong evidence in favor of model 1 over model 2. The decision rule of purely comparing the observed Bayes factor to a predetermined value, which is independent of the sampling distribution of the Bayes factor, called a "critical value", can be misleading, since the sampling distribution depends on the models being compared and the priors involved in deriving the posterior distributions. More importantly, as discussed in [20], the prior predictive distributions of the Bayes factor are asymmetric cross models. As a consequence, the decision rules for determining the strength of evidence based only on the observed value of the Bayes factor may be problematic. This problem of the Bayes factor was also discovered in [37]. To address this problem, a calibrating value of the Bayes factor based on the prior predictive distributions and the decision rule based on this calibrating value for selecting the model were proposed in [20]. Based on the calibrating value, the Bartlett's or Lindley's paradox can be alleviated although it may not disappear completely. In summary, similarly to the L measure, the Bayes factor should be properly calibrated as well.

References

[1] Bernardo, J.M., Smith, A.F.M.: Bayesian Theory. New York, Wiley (1994)

[2] Celeux, G., Forbes, F., Robert, C.P., Titterington, D.M.: Deviance information criteria for missing data models (with Discussion). Bayesian Analysis, **1**, 651–706 (2007)

[3] Chen, M.-H., Dey, D.K., Ibrahim, J.G.: Bayesian criterion based model assessment for categorical data. Biometrika, **91**, 45–63 (2004)

[4] Chen, M.-H., Ibrahim, J.G.: The relationship between the power prior and hierarchical models. Bayesian Analysis, **1**, 551–574 (2006)

[5] Chen, M.-H., Ibrahim, J.G.: Conjugate priors for generalized linear models. Statistica Sinica, **13**, 461–476 (2003)

[6] Chen, M.-H., Ibrahim, J.G., Yiannoutsos, C.: Prior elicitation, variable selection and and Bayesian computation for logistic regression models. Journal of the Royal Statistical Society, B, **61**, 223–242 (1999)

[7] Chen, M.-H., Shao, Q.-M.: Monte Carlo methods on Bayesian analysis of constrained parameter problems. Biometrika, **85**, 73–87 (1998)

[8] Chib, S.: Marginal likelihood from the Gibbs output. Journal of the American Statistical Association, **90**, 1313–1321 (1995)

[9] Clyde, M.A.: Bayesian model averaging and model search strategies. In: Bernardo, J.M., Berger, J.O., Dawid, A.P., Smith, A.F.M. (eds) Bayesian Statistics 6 (pp. 157–185). Oxford, Oxford University Press (1999)

[10] Clyde, M.A., George, I.E.: Model uncertainty. Statistical Science, **19**, 81–94 (2004)

[11] Gelfand, A.E., Dey, D.K.: Bayesian model choice: Asymptotics and exact calculations. Journal of the Royal Statistical Society, B, **56**, 501–514 (1994)

[12] Gelfand, A.E., Dey, D.K., Chang, H.: Model determinating using predictive distributions with implementation via sampling-based methods (with Discussion). In: Bernardo, J.M., Berger, J.O., Dawid, A.P., Smith, A.F.M. (eds) Bayesian Statistics 4 (pp. 147–167). Oxford, Oxford University Press (1992)

[13] Gelfand, A.E., Ghosh, S.K.: Model choice: A minimum posterior predictive loss approach. Biometrika, **85**, 1–13 (1998)

[14] Gelman, A. Meng, X.L., Stern, H.S.: Posterior predictive assessment of model fitness via realized discrepancies (with Discussion). Statistica Sinica, **6**, 733–807 (1996)

[15] Geisser, S.: Predictive Inference: An Introduction. London, Chapman & Hall (1993)

[16] George, E.I., McCulloch, R.E.: Variable selection via Gibbs sampling. Journal of the American Statistical Association, **88**, 881–889 (1993)

[17] George, E.I., McCulloch, R.E.: Approaches for Bayesian variable selection. Statistica Sinica, **7**, 339–374 (1997)
[18] George, E.I., McCulloch, R.E., Tsay, R.: Two approaches to Bayesian model selection with applications. In: Berry, D., K. Chaloner, K., Geweke, J. (eds) Bayesian Analysis in Statistics and Econometrics: Essays in Honor of Arnold Zellner (pp. 339–348). New York, Wiley (1996)
[19] Geweke, J.: Efficient simulation from the multivariate normal and Student-t distributions subject to linear constraints. In: Computing Science and Statistics: Proceedings of the Twenty-Third Symposium on the Interface (pp. 571–578). Fairfax Station, VA, Interface Foundation of North America Inc. (1991)
[20] Gonzalo García-Donato, G., Chen, M.-H.: Calibrating Bayes factor under prior predictive distributions. Statistica Sinica, **15**, 359–380 (2005)
[21] Hans, C., Dunson, D.B.: Bayesian inferences on umbrella orderings. Biometrics, **61**, 1018–1026 (2005)
[22] Huntjens, R.J.C., Peters, M.L., Woertman, L., Bovenschen, L.M., Martin, R.C., Postma, A.: Inter-identity amnesia in dissociative identity disorder: A simulated memory impairment? Psychological Medicine, **36**, 857–863 (2006)
[23] Ibrahim, J.G., Chen, M.-H and Sinha, D.: On optimality properties of the power prior. Journal of the American Statistical Association, **98**, 204–213 (2003).
[24] Ibrahim, J.G., Chen, M.-H and Sinha, D.: Criterion based methods for Bayesian model assessment. Statistica Sinica, **11**, 419–443 (2001).
[25] Ibrahim, J.G., Chen, M.-H., Sinha, D.: Bayesian Survival Analysis. New York, Springer-Verlag (2001)
[26] Ibrahim, J.G., Laud, P.W.: A predictive approach to the analysis of designed experiments. Journal of the American Statistical Association, **89**, 309–319 (1994).
[27] Jeffreys, H.: Theory of Probability (3rd ed.). Oxford, Clarendon Press (1961)
[28] Kass, R.E., Raftery, A.E.: Bayes factors. Journal of the American Statistical Association, **90**, 773–795 (1995)
[29] Klugkist, I., Kato, B., Hoijtink, H.: Bayesian model selection using encompassing priors. Statistica Neerlandica, **59**, 57–69 (2005)
[30] Laud, P.W., Ibrahim, J.G.: Predictive model selection. Journal of the Royal Statistical Society, B, **57**, 247–262 (1995)
[31] Lindley, D.V.: A statistical paradox. Biometrika, **44**, 187–192 (1957)
[32] Madigan, D., Raftery, A.E.: Model selection and accounting for model uncertainty in graphical models using Occam's window. Journal of the American Statistical Association, **89**, 1535–1546 (1994)
[33] Raftery, A.E., Madigan, D., Hoeting, J.A.: Bayesian model averaging for linear regression models. Journal of the American Statistical Association, **92**, 179–191 (1997)
[34] Raftery, A.E., Madigan, D., Volinsky, C.T.: Accounting for model uncertainty in survival analysis improves predictive performance. In: Bernardo, J.M., Berger, J.O., Dawid, A.P., Smith, A.F.M. (eds) Bayesian Statistics 5 (pp. 323–350). Oxford, Oxford University Press (1995)
[35] Reijntjes, A.: Emotion-Regulation and Depression in Pre-adolescent Children. Ph.D. thesis, Free University, Amsterdam (2004)

[36] Spiegelhalter, D.J., Best, N.G., Carlin, B.P., Linde, A. van der: Bayesian measures of model complexity and fit (with Discussion). Journal of the Royal Statistical Society, B, **64**, 583–639 (2002)
[37] Vlachos, P.K., Gelfand, A.E.: On the calibration of Bayesian model choice criteria. Journal of Statistical Planning and Inference, **111**, 223–234 (2003)
[38] Yu, J.Z., Tanner, M.A.: An analytical study of several Markov chain Monte Carlo estimators of the marginal likelihood. Journal of Computational and Graphical Statistics, **8**, 839–853 (1999)

9

Bayesian Versus Frequentist Inference

Eric-Jan Wagenmakers[1], Michael Lee[2], Tom Lodewyckx[3], and Geoffrey J. Iverson[2]

[1] Department of Psychology, University of Amsterdam, Roetersstraat 15, 1018 WB Amsterdam, the Netherlands ej.wagenmakers@gmail.com
[2] Department of Cognitive Sciences, University of California at Irvine, 3151 Social Science Plaza, Irvine CA 92697, USA mdlee@uci.edu and giverson@uci.edu
[3] Department of Quantitative and Personality Psychology, University of Leuven, Tiensestraat 102, 3000 Leuven, Belgium tom.lodewyckx@student.kuleuven.be

9.1 Goals and Outline

Throughout this book, the topic of order restricted inference is dealt with almost exclusively from a Bayesian perspective. Some readers may wonder why the other main school for statistical inference – *frequentist* inference – has received so little attention here. Isn't it true that in the field of psychology, almost all inference is frequentist inference?

The first goal of this chapter is to highlight why frequentist inference is a less-than-ideal method for statistical inference. The most fundamental limitation of standard frequentist inference is that it does not condition on the observed data. The resulting paradoxes have sparked a philosophical debate that statistical practitioners have conveniently ignored. What cannot be so easily ignored are the practical limitations of frequentist inference, such as its restriction to nested model comparisons.

The second goal of this chapter is to highlight the theoretical and practical advantages of a Bayesian analysis. From a theoretical perspective, Bayesian inference is principled and prescriptive and – in contrast to frequentist inference – a method that does condition on the observed data. From a practical perspective, Bayesian inference is becoming more and more attractive, mainly because of recent advances in computational methodology (e.g., Markov chain Monte Carlo and the WinBUGS program [95]). To illustrate, one of our frequentist colleagues had been working with the WinBUGS program and commented "I don't agree with the Bayesian philosophy, but the WinBUGS program does allow me to implement complicated models with surprisingly little effort." This response to Bayesian inference is diametrically opposed to the one that was in vogue until the 1980s, when statisticians often sympathized with the Bayesian philosophy but lacked the computational tools to implement models with a moderate degree of complexity.

The outline of this chapter is as follows: Section 9.2 introduces the Fisherian and the Neyman-Pearson flavors of frequentist inference and goes on to list a number of limitations associated with these procedures. Section 9.3 introduces Bayesian inference and goes on to lists a number of its advantages. Section 9.4 briefly presents our conclusions.

9.2 Frequentist Inference and Its Problems

Frequentist inference is based on the idea that probability is a limiting frequency. This means that a frequentist feels comfortable assigning probability to a repeatable event in which the uncertainty is due to randomness, such as getting a full house in poker (i.e., aleatory uncertainty [78]). When n hands are played and a full house is obtained in s cases, then, with n very large, the probability of a full house is just s/n. But a frequentist must refuse to assign probability to an event where uncertainty is also due to lack of knowledge, such as the event of Alexander Grischuk ever winning a major poker championship (i.e., epistemic uncertainty [34, 78]).

Because uncertainty about parameters is epistemic, frequentist inference does not allow probability statements about the parameters of a statistical process. For instance, the fact that a frequentist 95% confidence interval for the normal mean μ is $[-0.5, 1.0]$ does not mean that there is a 95% probability that μ is in $[-0.5, 1.0]$. Instead, what it means is that if the same procedure to construct confidence intervals was repeated very many times, for all kinds of different datasets, then in 95% of the cases would the true μ lie in the 95% confidence interval (cf. the example presented in Section 9.2.1).

Discussion of frequentist inference is complicated by the fact that current practice has become an unacknowledged amalgamation of the p-value approach advocated by Fisher [32] and the α-level approach advocated by Neyman and Pearson [76]. Hubbard and Bayarri [49, p. 176] summarized and contrasted the paradigms as follows:

> The level of significance shown by a p value in a Fisherian significance test refers to the probability of observing data this extreme (or more so) under a null hypothesis. This data-dependent p value plays an epistemic role by providing a measure of inductive evidence against H_0 in single experiments. This is very different from the significance level denoted by α in a Neyman-Pearson hypothesis test. With Neyman-Pearson, the focus is on minimizing type II, or β, errors (i.e., false acceptance of a null hypothesis) subject to a bound on type I, or α, errors (i.e., false rejections of a null hypothesis). Moreover, this error minimization applies only to long-run repeated sampling situations, not to individual experiments, and is a prescription for behaviors, not a means of collecting evidence.

Clearly then, Fisher's approach is very different from that of Neyman and Pearson. Yet, most researchers believe the paradigms have somehow merged and interpret the p-value both as a measure of evidence and as a repetitive error rate. It appears that the confusion between the two different procedures is now close to total, and it has been argued that this mass confusion "has rendered applications of classical statistical testing all but meaningless among applied researchers." [49, p. 171]. Additional references include [9, 18, 37, 38, 39, 40, 43, 91].

We now discuss several general problems of both the Fisherian and the Neyman-Pearson procedure (cf. [14, 26, 48, 53, 91, 97]). Although only the Neyman-Pearson procedure is truly frequentist (i.e., it requires knowledge about performance in long-run sampling situations), we will perpetuate the confusion and refer to both the Fisherian and the Neyman-Pearson procedure as "frequentist."

9.2.1 Frequentist Inference Generally Does Not Condition on the Observed Data

As argued by Berger and Wolpert [14], frequentist evidence is often pre-experimental or unconditional.[4] This means that "a particular procedure is decided upon for use, and the accuracy of the evidence from an experiment is identified with the long run behavior of the procedure, were the experiment repeatedly performed." [14, p. 5]. We illustrate the problem with this unconditional approach by an example that highlights the pathological properties of frequentist confidence intervals (cf. [15, p. 468]).

Consider a uniform distribution with mean μ and width 1. Draw two values randomly from this distribution, label the smallest one s and the largest one l, and check whether the mean μ lies in between s and l. If this procedure is repeated very many times, the mean μ will lie in between s and l in half of the cases. Thus, (s, l) gives a 50% frequentist confidence interval for μ. But suppose that for a particular draw, $s = 9.8$ and $l = 10.7$. The difference between these values is 0.9, and this covers 9/10th of the range of the distribution. Hence, for these particular values of s and l we can be 100% confident that $s < \mu < l$, even though the frequentist confidence interval would have you believe you should only be 50% confident.

This example shows why it is important to condition on the data that have actually been observed. The key problem is that frequentist methods do not do this, so that for data x, "(...) a procedure which looks great pre-experimentally could be terrible for particular x(...)" [14, p. 9]. Other examples of pathological behavior of frequentist confidence intervals can be found in [15, pp. 466–469], [14], and, in particular, [52].

[4] Frequentist' procedures sometimes do condition on important aspects of the data. Conditioning is always partial, however, and there exist situations in which it is unclear what aspects of the data on which to condition.

Table 9.1. Two different sampling distributions, $f(y)$ and $g(y)$ that lead to two different p-values for $y = 5$

	Data y					
Distribution	$y = 1$	$y = 2$	$y = 3$	$y = 4$	$y = 5$	$y = 6$
$f(y)\|H_0$	.04	.30	.31	.31	.03	.01
$g(y)\|H_0$	.04	.30	.30	.30	.03	.03

9.2.2 Frequentist Inference Depends on Data That Were Never Observed

The p-value is the probability under the null hypothesis of observing data *at least as extreme* as the data that were actually observed. This means that the p-value is partly determined by data that were never observed, as is illustrated in the following example (cf. [4, 14, 19, 97]).

Assume the data y can take on six integer values, $y \in \{1, 2, ..., 6\}$, according to one of the sampling distributions $f(y)$ or $g(y)$. Further assume that what is observed is $y = 5$. As can be seen from Table 9.1, the observed datum is equally likely under $f(y)$ and $g(y)$. Yet, a one-sided p-value is $.03 + .01 = .04$ under $f(y)$ and $.03 + .03 = .06$ under $g(y)$. This is solely due to the fact that the more extreme observation $y = 6$, which was never observed, is less likely under $f(y)$ than it is under $g(y)$. Jeffreys famously summarized the situation: "*What the use of P implies, therefore, is that a hypothesis that may be true may be rejected because it has not predicted observable results that have not occurred.* This seems a remarkable procedure" [55, p. 385, italics in original].

9.2.3 Frequentist Inference Depends on the Intention With Which the Data Were Collected

Because p-values are calculated over the sample space, changes in the sample space can greatly affect the p-value. For instance, assume that a participant answers a series of 17 test questions of equal difficulty; 13 answers are correct, 4 are incorrect, and the last question was answered incorrectly. Under the standard binomial sampling plan (i.e., "ask 17 questions"), the two-sided p-value is .049. The data are, however, also consistent with a negative binomial sampling plan (i.e., "keep on asking questions until the fourth error occurs"). Under this alternative sampling plan, the experiment could have been finished after four questions, or after a million. For this sampling plan, the p-value is .021.

What this simple example shows is that the intention of the researcher affects statistical inference – the data are consistent with both sampling plans, yet the p-value differs. Berger and Wolpert [14, pp. 30–33] discussed the resulting counterintuitive consequences through a story involving a naive scientist and a frequentist statistician.

In the story, a naive scientist has obtained 100 independent observations that are assumed to originate from a normal distribution with mean θ and standard deviation 1. In order to test the null hypothesis that $\theta = 0$, the scientist consults a frequentist statistician. The mean of the observations is 0.2, and hence the p-value is a little smaller than .05, which leads to a rejection of the null hypothesis. However, the statistician decides to probe deeper into the problem and asks the scientist what he would have done in the fictional case that the experiment had *not* yielded a significant result after 100 observations. The scientist replies that he would have collected another 100 observations. Thus, it may be hypothesized that the implicit sampling plan was not to collect 100 observation and stop; instead, the implicit sampling plan was to first take 100 observations and check whether $p < .05$. When the check is successful, the experiment stops, but when the check fails, another 100 observations are collected and added to the first 100, after which the experiment stops.

The statistician then succeeds in convincing the scientist that use of the implicit sampling plan requires a correction in order to keep the type I error rate at $\alpha = .05$ [81]. Unfortunately, this correction for planning multiple tests now leads to a p-value that is no longer significant. Therefore, the puzzled scientist is forced to continue the experiment and collect an additional 100 observations. Note that the interpretation of the data (i.e., significant or not significant) depends on what the scientist was planning to do in a situation that did not actually occur. If the very same data had been collected by a scientist who had answered the statistician's question by saying, whether truthfully or not, "I would not have collected any more observations," then the data would have been judged to be significant: Same data, different inference.

But the story becomes even more peculiar. Assume that the scientist collects the next 100 observations and sets up another meeting with the statistician. The data are now significant. The statistician, however, persists and asks what the scientist would have done in case the experiment had not yielded a significant result after 200 observations. Suppose that the scientist now answers "This would have depended on the status of my grant renewal. If my grant is renewed, I would have had enough funds to test another 100 observations. If my grant is not renewed, I would have had to stop the experiment. Not that this matters, of course, because the data were significant anyway."

The frequentist statistician then explains that the inference depends on the grant renewal; if the grant is not renewed, the sampling plan stands and no correction is necessary. But if the grant is renewed, the scientist could have collected more data, in the fictional case that the data would not have been significant after 200 observations. This calls for a correction for planning multiple tests, similar to the first one. Berger and Wolpert [14, p. 33] end their story: "The up-to-now honest scientist has had enough, and he sends in a request to have the grant renewal denied, vowing never again to tell the statistician what he could have done under alternative scenarios."

We believe that most researchers find it awkward that the conclusions from frequentist statistics depend critically on events that have yet to happen – events that, moreover, seem to be irrelevant with respect to the data that have actually been obtained.

9.2.4 Frequentist Inference Does Not Prescribe Which Estimator Is Best

Frequentist inference is not derived from a set of simple axioms that describe rational behavior. This means that any statistical problem potentially affords more than one frequentist solution, and it may be unclear which one is best. For instance, many different estimators may be proposed for a particular parameter θ. Which estimator for θ should we prefer? The common strategy is to narrow down the set of admissible estimators by considering only estimators that are *unbiased*. An estimator $t(\cdot)$ based on data y is unbiased when

$$\int_Y t(y)p(y|\theta)\ dy = \theta, \tag{9.1}$$

for all θ, where Y indicates the sample space (cf. [65]); that is, the only estimators taken into consideration are those that, averaged over the data that could arise, do not systematically overestimate or underestimate θ.

Although the criterion of unbiasedness has intuitive appeal, it is in fact highly contentious. First, the criterion is based on all possible datasets that could be observed (i.e., the sample space Y). This means that the intention of the researcher affects which estimators are unbiased and which are not. For instance, for the binomial sampling plan the unbiased estimator is s/n, where s is the number of correct responses out of a total of n questions, but for the negative binomial sampling plan the unbiased estimator is $(s-1)/(n-1)$. Second, an estimator that is unbiased for θ may well be biased for some nonlinear transformation of θ such as $\sqrt{\theta}$.

Finally, unbiased estimators may perform uniformly worse than biased estimators. Consider, for instance, the datum y distributed as $\mathcal{N}(\sqrt{\theta}, 1)$ with $\theta > 0$. The unbiased estimator for θ is $t(y) = y^2 - 1$. But when $|y| < 1$, $t(y)$ is negative, which conflicts with the knowledge that $\theta > 0$. A new estimator $t^{new}(y)$ may be proposed that is given by $t^{new}(y) = y^2 - 1$ when $|y| \geq 1$ and $t^{new}(y) = 0$ otherwise. The new estimator $t^{new}(y)$ is biased but does uniformly better than the unbiased estimator $t(y)$, this means that $t(y)$ is *inadmissible* (cf. [79]).

It should also be noted that in the above example, $t(y)$ is biased downward when $|y| < 1$, but biased upward when $|y| \geq 1$. Thus, an estimator may be unbiased for all possible datasets taken together, but it may – at the same time – be biased for every single dataset considered in isolation [79, p. 122].

Frequentist statisticians are aware of this problem, in the sense that they acknowledge that "(...) an overly rigid insistence upon unbiasedness may lead to difficulties" [96, p.432]. This statement highlights an important problem:

Frequentist inference does not specify a unique solution for every statistical problem. When unfortunate consequences of, say, "an overly rigid insistence upon unbiasedness" become apparent, adhoc estimators may be proposed that remedy the immediate problem – but this clearly is not a satisfactory state of affairs.

9.2.5 Frequentist Inference Does Not Quantify Statistical Evidence

According to the Fisherian tradition, p-values reflect the strength of evidence against the null hypothesis. General guidelines associate specific ranges of p-values with varying levels of evidence: A p-value greater than .1 yields "little or no real evidence against the null hypothesis," a p-value less than .1 but greater than .05 implies "suggestive evidence against the null hypothesis," a p-value less than .05 but greater than .01 yields "moderate evidence against the null hypothesis," and a p-value less than .01 constitutes "very strong evidence against the null hypothesis" [17, p. 9]; see also [101, p. 157].

If p-values truly reflect evidence, a minimum requirement is that equal p-values provide equal evidence against the null hypothesis (i.e., the p-postulate [97]). According to the p-postulate, $p = .05$ with 10 observations constitutes just as much evidence against the null hypothesis as does $p = .05$ after 50 observations.

It may not come as a surprise that Sir Ronald Fisher himself was of the opinion that the p-postulate is correct: "It is not true...that valid conclusions cannot be drawn from small samples; if accurate methods are used in calculating the probability [the p-value], we thereby make full allowance for the size of the sample, and should be influenced in our judgement only by the value of probability indicated" [31, p. 182], as cited in [91, p. 70].

Nevertheless, some researchers believe that the p-postulate is false and that $p = .05$ after 50 observations is more reliable than $p = .05$ after 10 observations. For instance, Rosenthal and Gaito [84] found that the confidence with which a group of psychologists were willing to reject the null hypothesis increased with sample size (cf. [75]). Consistent with the psychologists' intuition, an article co-authored by 10 reputable statisticians stated that "a given p-value in a large trial is usually stronger evidence that the treatments really differ than the same p-value in a small trial of the same treatments would be" [80, p. 593], as cited in [91, p. 71].

Finally, several researchers have argued that when the p-values are the same, studies with small sample size actually provide *more* evidence against the null hypothesis than studies with large sample size (e.g., [1, 3, 69, 75]). For a summary of the debate, see [90]. Abelson considered the very question of whether a researcher would be happier with a $p = .05$ after testing 10 cases per group or after testing 50 cases per group, and then firmly concluded "Undergraduates inevitably give the wrong reply: Fifty cases per group, because a bigger sample is more reliable." The appropriate answer is "ten cases per group, because if the p values are the same, the observed effect size has to be bigger with a smaller n" [1, p. 12].

In order to draw a firm conclusion about the veracity of the p-postulate, we first need to define what "evidence" is. The details of a rational (i.e., coherent or Bayesian) definition of evidence are presented in Section 9.3. Here it suffices to say that such an analysis must always reject the p-postulate (e.g., [26, 64, 98]): From a rational perspective, $p = .05$ after only 10 observations is more impressive than $p = .05$ after 1000 observations. In fact, it may happen that that for a large dataset, a frequentist analysis will suggest that the null hypothesis should be rejected, whereas a rational analysis will suggest that the null hypothesis is strongly supported.

9.2.6 Frequentist Inference Does Not Apply to Non-nested Models

Consider the study on Dissociative Identity Disorder (DID), introduced in Chapter 2 and discussed throughout this book. In this study, Huntjens et al. [50] set out to study memory processes in DID-patients. These patients often report *inter-identity amnesia* (i.e., impaired memory for events experienced by identities that are not currently present). For instance, the identity "lonely girl" may have limited or no knowledge of the events experienced by the identity "femme fatale." To test whether DID-patients were really affected by interidentity amnesia or whether they were simulating their amnesia, the authors assessed the performance of four groups of subjects on a multiple-choice recognition test. The dependent measure was the number of correct responses. The first group were the DID-patients, the second group were Controls, the third group were controls instructed to simulate interidentity amnesia (Simulators), and the fourth group were controls who had never seen the study list and were therefore True amnesiacs.

From the psychological theorizing that guided the design of the experiment, one can extract several hypotheses concerning the relative performance of the different groups. One hypothesis, H_{1a}, states that the mean recognition scores μ for DID-patients and True amnesiacs are the same and that their scores are higher than those of the Simulators: $\mu_{con} > \{\mu_{amn} = \mu_{pat}\} > \mu_{sim}$. Another hypothesis, H_{1b}, states that the mean recognition scores μ for DID-patients and Simulators are the same and that their scores are lower than those of the True amnesiacs: $\mu_{con} > \mu_{amn} > \{\mu_{pat} = \mu_{sim}\}$.

The hypotheses H_{1a} and H_{1b} are non-nested, and frequentist inference is not well suited for the comparison of such models (e.g., [60]). The main problem is that it is not clear whether H_{1a} or H_{1b} should be considered the null hypothesis. One might try both possibilities, but this runs the danger of simultaneously rejecting (or accepting) both H_{1a} and H_{1b}. Moreover, it is not clear how to interpret the hypothetical result of $p = .04$ when H_{1a} serves as the null hypothesis, and $p = .06$ when H_{1b} serves as the null hypothesis – even though H_{1a} is rejected and H_{1b} is not, this does not mean that H_{1b} is much better than H_{1a}.

9.2.7 Interim Conclusion

In the preceding analyses we have argued that frequentist procedures suffer from fundamental philosophical and practical problems. These problems are not some kind of well-kept secret, as statisticians have written about these frequentist flaws for many decades; the website `http://biology.uark.edu/coop/Courses/thompson5.html` documents some of their efforts by listing 402 articles and books that criticize the use of frequentist null hypothesis testing.

Indeed, the selection of problems mentioned in Sections 9.2.1 to 9.2.6 was certainly not exhaustive. Other problems include the fact that α-levels are arbitrary "surely, God loves the .06 nearly as much as the .05" [85, p. 1277], the fact that inference in sequential designs is overly complicated and conservative "Sequential analysis is a hoax" [2, p. 381], the fact that p-values are often misinterpreted, even by those teaching statistics [44], the fact that Fisherian frequentist inference does not allow one to obtain evidence in support of the null hypothesis, and the fact that frequentist inference is internally inconsistent or *incoherent*. The latter means that when statistical conclusions need to be backed up by betting on them, the frequentist will be a sure loser (for details, see Section 9.3.1).

All of this makes one may wonder why – despite the harsh criticism – the flogged horse of frequentist inference is still alive and well, at least in the field of psychology [1]. We believe the reason for this is most likely an unfortunate combination of several factors. Among the most important of these are ease of application, presumed lack of an appealing alternative, limited statistical knowledge among practitioners, faulty and one-sided teaching of statistics at universities, historical precedent, and – for a few special cases – exact numerical correspondence of frequentist "flogged horse" inference with rational inference to be discussed below.

This concludes our summary of frequentist inference and its problems. We now turn to a discussion of the other major statistical paradigms for statistical inference, which differs from frequentist inference in a few key assumptions. We will argue that, both philosophically and practically, this paradigm constitutes a superior alternative to frequentist inference.

9.3 Bayesian Inference and Its Advantages

In Bayesian inference, parameters are random variables. Uncertainty or degree of belief with respect to the parameters is quantified by probability distributions. For a given model, say H_1, the prior distribution $p(\theta|H_1)$ for a parameter θ is updated after encountering data y to yield a posterior distribution $p(\theta|y, H_1)$. The posterior information contains all of the relevant information about θ. Note that the posterior distribution is conditional on the data y that have been observed; data that could have been observed, but were not, do not affect Bayesian inference.

Specifically, Bayes' rule states that the posterior distribution $p(\theta|y, H_1)$ is proportional to the product of the prior $p(\theta|H_1)$ and the likelihood $f(y|\theta, H_1)$:

$$p(\theta|y, H_1) = p(\theta|H_1) f(y|\theta, H_1)/m(y|H_1). \tag{9.2}$$

In this equation, $m(y|H_1)$ is the marginal probability of the data; it is computed by integrating out the model parameters using the law of total probability:

$$m(y|H_1) = \int p(y, \theta|H_1)\ d\theta = \int p(\theta|H_1) f(y|\theta, H_1)\ d\theta. \tag{9.3}$$

This shows that $m(y|H_1)$ can also be interpreted as a *weighted average likelihood* where the weights are provided by the prior distribution $p(\theta|H_1)$. Because $m(y|H_1)$ is a number that does not depend on θ, $m(y|H_1)$ can be conveniently ignored when the goal is to estimate θ. However, when the goal is Bayesian hypothesis testing, $m(y|H_1)$ becomes critically important.

For concreteness, consider the choice between two possibly non-nested models, H_1 and H_2. The extension to more than two models is entirely possible and follows the same recipe. Bayes' rule dictates how the prior probability of H_1, $p(H_1)$, is updated through the data to give the posterior probability of H_1, $p(H_1|y)$:

$$p(H_1|y) = p(H_1)m(y|H_1)/\sum_t p(H_t)m(y|H_t). \tag{9.4}$$

In the same way, one can calculate the posterior probability of H_2, $p(H_2|y)$. The ratio of these posterior probabilities is given by

$$\frac{p(H_1|y)}{p(H_2|y)} = \frac{p(H_1)}{p(H_2)}\frac{m(y|H_1)}{m(y|H_2)}, \tag{9.5}$$

which shows that the posterior odds $p(H_1|y)/p(H_2|y)$ is equal to the product of the prior odds $p(H_1)/p(H_2)$ and the ratio of marginal probabilities $m(y|H_1)/m(y|H_2)$. Thus, the ratio of marginal probabilities – henceforth referred to as the Bayes factor [55] – quantifies the change from prior to posterior odds brought about by the data. The Bayes factor, or the log of the Bayes factor, is often interpreted as the *weight of evidence* coming from the data [42]. Thus, a Bayes factor hypothesis test prefers the model under which the observed data are most likely. For details see [13], [15, Chapter 6], [41, Chapter 7], [56], and [77]; for an introduction in Bayesian inference, see Chapters 3 and 4 of the present book.

Jeffreys [55] proposed labeling the evidence provided by the Bayes factor according to a classification scheme that was subsequently revised by Raftery [82, Table 6]. Table 9.2 shows the Raftery classification scheme. The first column shows the Bayes factor, and the second column shows the associated

Table 9.2. Interpretation of the Bayes factor in terms of evidence

Bayes factor BF_{12}	$p(H_1\|y)$	Evidence
1–3	.50–.75	Weak
3–20	.75–.95	Positive
20–150	.95–.99	Strong
> 150	$> .99$	Very strong

posterior probability when it is assumed that both H_1 and H_2 are a priori equally plausible. The third column shows the verbal labels for the evidence at hand, in the case of a comparison between two models. Note that these verbal labels are associated with the level of evidence that is provided by the Bayes factor (i.e., a comparison between two models). These verbal labels should not be associated with posterior model probabilities (PMPs) when the set of candidate models is larger than two, for example, consider the problem of finding the best set of predictors for a regression equation. By considering all possible combinations of predictors, the model space can easily comprise as many as 100,000 candidate models. When a single model out of such a large set has a posterior probability of, say, .50, this would constitute a dramatic increase over its prior probability of .000001, and hence the data provide "very strong" rather than "weak" evidence in its favor.

Bayesian procedures of parameter estimation and hypothesis testing have many advantages over their frequentist counterparts. Below is a selective list of 10 specific advantages that the Bayesian paradigm affords.

9.3.1 Coherence

Bayesian inference is *prescriptive*; given the specification of a model, there exists only one way to obtain the appropriate answer. Bayesian inference does not require adhoc solutions to remedy procedures that yield internally inconsistent results. Bayesian inference is immune from such inconsistencies because it is founded on a small set of axioms for rational decision making. Several axiom systems have been proposed, but they all lead to the same conclusion: Reasoning under uncertainty can only be coherent if it obeys the laws of probability theory (e.g., [15, 20, 22, 23, 30, 53, 66, 68, 83, 92]).

One of the famous methods to prove this far-reaching conclusion is due to Bruno de Finetti and involves a betting scenario [22]. Assume there exists a legally binding ticket that guarantees to pay 1 euro should a proposition turn out to be true. For instance, the proposition could be "In 2010, the Dutch national soccer team will win the world cup." Now you have to determine the price you are willing to pay for this ticket. This price is the "operational subjective probability" that you assign to the proposition.

The complication is that this scenario also features an opponent. The opponent can decide, based on the price that you determined, to either buy this ticket from you or to make you buy the ticket from him. This is similar to the "I cut, you choose" rule where the person who cuts a cake gets to choose the last piece; it is then in that person's own interest to make a fair judgment.

In the example of the ticket, it is obviously irrational to set the price higher than 1 euro, because the opponent will make you buy this ticket from him and he is guaranteed to make a profit. It is also irrational to set the price lower than 0 euro, because the opponent will "buy" the ticket from you at a negative price (i.e., gaining money) and is again guaranteed to make a profit.

Now suppose you have to determine the price of three individual tickets. Ticket A states "In 2010, the Dutch national soccer team will win the world cup"; ticket B states "In 2010, the French national soccer team will win the world cup"; and ticket C states "In 2010, either the Dutch or the French national soccer team will win the world cup." You can set the prices any way you want. In particular, there is nothing to keep you from setting the prices such that price(ticket A) + price(ticket B) $\neq$ price(ticket C). However, when you set the prices this way, you are guaranteed to lose money compared to your opponent; for instance, suppose you set price(ticket A) = 0.5 euro, price(ticket B) = 0.3 euro, and price(ticket C) = 0.6 euro. Then the opponent will buy ticket C from you, sell you tickets A and B, and he is guaranteed to come out ahead. A set of wagers that ensures that somebody will make a profit, regardless of what happens, is called a *Dutch book*.

Using betting scenarios such as the above, de Finetti showed that the only way to determine subjective values and avoid a certain loss is to make these values obey the rules of probability theory (i.e., the rule that probabilities lie between 0 and 1, the rule that mutually exclusive events are additive, and the rule of conditional probability); that is, the only way to avoid a Dutch book is to make your prices for the separate tickets *cohere* according to the laws of probability calculus.

The concept of coherence refers not just to the betting scenario, but more generally to the combination of information in a way that is internally consistent. For example, consider Bayesian inference in the case that the data arrive in two batches, y_1 and y_2 [79, pp. 64-65]. Following the adage "today's posterior is tomorrow's prior" [65, p. 2], we can update from the initial prior $p(\theta)$ to a posterior $p(\theta|y_1)$ and then update this posterior again, effectively treating $p(\theta|y_1)$ as a prior, to finally obtain $p(\theta|y_1, y_2)$. The crucial aspect is that when the data are conditionally independent, it does not matter whether we observe the dataset batch-by-batch, all at once, or in reverse order.

As an additional example of coherence, consider a set of three models: H_1 postulates that $\mu_a = \mu_b = \mu_c$, H_2 postulates that $\{\mu_a = \mu_b\} > \mu_c$, and H_3 postulates that $\mu_a > \{\mu_b = \mu_c\}$. Then, denoting the Bayes factor for model H_i over model H_j by BF_{ij}, we can deduce from the identity

$$\frac{m(y|H_1)}{m(y|H_2)} = \frac{m(y|H_1)}{m(y|H_3)} \frac{m(y|H_3)}{m(y|H_2)}, \tag{9.6}$$

that $BF_{12} = BF_{32} \times BF_{13}$. This means that if the data are twice as likely under H_1 than under H_3 and thrice as likely under H_3 than under H_2, we know that the data are six times more likely under H_1 than under H_2. If the Bayes factors would not commute like this, one could construct a situation in which one could hold intransitive beliefs – a situation that would violate the axioms of rational decision making upon which Bayesian inference rests.

9.3.2 Automatic Parsimony

In statistical hypothesis testing, the ideal model captures all of the replicable structure and ignores all of the idiosyncratic noise. Such an ideal model yields the best predictions for unseen data coming from the same source. When a model is too complex, it is said to *overfit* the data; the model mistakenly treats idiosyncratic noise as if it were replicable structure. When a model is too simple, it is said to *underfit* the data, which means that the model fails to capture all of the replicable structure in the data. Models that underfit or overfit the data provide suboptimal predictions and are said to generalize poorly (e.g., [73, 100]).

The main challenge of hypothesis testing or model selection is to identify the model with the best predictive performance. However, it is not immediately obvious how this should be done; complex models will generally provide a better fit to the observed data than simple models, and therefore one cannot simply prefer the model with the best "goodness-of-fit" – such a strategy would lead to massive overfitting. Intuition suggests that this tendency for overfitting should be counteracted by putting a premium on simplicity. This intuition is consistent with the law of parsimony or "Occam's razor" (cf. `http://en.wikipedia.org/wiki/Occam's_Razor`), which states that when everything else is equal, simple models are to be preferred over complex models [53, Chapter 20].

Formal model selection methods try to quantify the trade-off between goodness-of-fit and parsimony. Many of these methods measure a model's overall performance by the sum of two components: One that measures descriptive accuracy and one that places a premium on parsimony. The latter component is also known as the Occam factor [70, Chapter 28]. For many model selection methods, the crucial issue is how to determine the Occam factor. One of the attractive features of Bayesian hypothesis testing is that it automatically determines the model with the best predictive performance – Bayesian hypothesis testing therefore incorporates what is known as an automatic Occam's razor. In order to see why this is the case, we explore two lines of reasoning.

First, recall that Bayesian model selection is based on the marginal probability of the data given model t, $m(y|H_t)$. Now denote a sequence of n data points by $y^n = (y_1, ..., y_n)$; for example, y_{i-1} denotes the $(i-1)^{th}$ individual data point, whereas y^{i-1} denotes the entire sequence of observations ranging from y_1 up to and including y_{i-1}. Quantify predictive performance for a single data point by the logarithmic loss function $-\ln \widehat{p}_i(y_i)$: the larger the probability that $\widehat{p}_i$ (determined based on the previous observations y^{i-1}) assigns to the observed outcome y_i, the smaller the loss. From the definition of conditional probability (i.e., $p(y_i|y^{i-1}) = p(y^i)/p(y^{i-1})$), it then follows that the marginal probability of the data may be decomposed as a series of sequential, "one-step-ahead" probabilistic predictions (e.g., [21, 99]):

$$\begin{aligned} m(y^n|H_t) &= p(y_1, ..., y_n|H_t) \\ &= p(y_n|y^{n-1}, H_t)p(y_{n-1}|y^{n-2}, H_t)...p(y_2|y_1, H_t)p(y_1|H_t). \end{aligned} \tag{9.7}$$

Thus, (9.7) shows that the model with the highest marginal probability will also have the smallest sum of one-step-ahead prediction errors, as $-\ln m(y^n|H_t) = \sum_{i=1}^{n} - \ln p(y_i|y^{i-1}, H_t)$.

According to the second line of reasoning, every statistical model makes a priori predictions. Complex models have a relatively large parameter space and are therefore able to make many more predictions and cover many more eventualities than simple models. However, the drawback for complex models is that they need to spread out their prior probability across their entire parameter space. In the limit, a model that predicts almost everything has to spread out its prior probability so thinly that the occurrence of any particular event will not greatly add to that model's credibility. Formally, the marginal probability of the data is calculated by averaging the likelihood $f(y|\theta, H_t)$ over the prior $p(\theta|H_t)$. When the prior is very spread out, it will occupy a relatively large part of the parameter space in which the likelihood is almost zero, and this greatly decreases the average or marginal likelihood.

As a more concrete example, consider two people, Bart and Lisa, who each get 100 euros to bet on the winner of the 2010 world cup soccer. Bart decides to divide his money evenly over 10 candidate teams, including those from Brazil and Germany. Lisa divides her money over just two teams, betting 60 euros on the team from Brazil and 40 euros on the team from Germany. Now if either Brazil or Germany turn out to win the 2010 world cup, Lisa wins more money than Bart. By betting all her money on just two teams, Lisa was willing to take a risk, whereas Bart was just trying to keep his options open. For Bart, this means that even if his prediction of Brazil winning turns out to be correct, he will still lose the 90 euros he bet on the other countries to win. The point of the story is that, both at the betting office and in Bayesian inference, hedging your bets is not necessarily the best option, because this requires you to spread your resources – be it money or prior probability mass – thinly over the alternative options.

9.3.3 Extension to Non-nested Models

Bayesian hypothesis testing is based on the marginal probability of the data given model t, $m(y|H_t)$, and therefore it does not make a fundamental distinction between nested and non-nested models. This means that Bayesian hypothesis testing can be applied in many more situations than frequentist hypothesis testing. In cognitive psychology, for instance, important substantive questions concern the extent to which the law of practice follows a power function versus an exponential function or the extent to which category learning is best described by an exemplar model or a prototype model. For Bayesian inference, the substantive questions can be statistically tested in exactly the same way, whether the competing models are nested or not. For frequentist inference, however, the fact that the models are non-nested causes grave complications.

Another class of possibly non-nested models that are of great relevance for psychologists are those that incorporate order restrictions. For instance, consider again the case of the Huntjens et al. study on DID discussed in Section 9.2.6 and throughout this book. For the data from the study, hypothesis H_{1a} states that the mean recognition scores μ for DID-patients and True amnesiacs are the same and that their scores are higher than those of the Simulators: $\mu_{con} > \{\mu_{amn} = \mu_{pat}\} > \mu_{sim}$, whereas hypothesis H_{1b} states that the mean recognition scores μ for DID-patients and Simulators are the same and that their scores are lower than those of the True amnesiacs: $\mu_{con} > \mu_{amn} > \{\mu_{pat} = \mu_{sim}\}$. Within the frequentist paradigm, a comparison of these models is problematical. Within the Bayesian paradigm, however, the comparison is natural and elegant (e.g., [35, 47, 59, 60, 61, 62, 94]).

The general recipe, outlined in O'Hagan and Forster [79, pp. 70-71] is to carry out order restricted inference by first considering the posterior distribution of the unconstrained model and then restricting one's attention to the part of the posterior distribution that obeys the parameter constraints. In a Markov chain Monte Carlo (MCMC) simulation, for instance, this can be accomplished automatically by retaining only those samples that are in line with the constraints. The work reported in this book attests to the ability of Bayesian inference to address substantive psychological questions that involve order restrictions in a manner that is unattainable by frequentist means.

9.3.4 Flexibility

Bayesian inference allows for the flexible implementation of relatively complicated statistical techniques such as those that involve hierarchical nonlinear models (e.g., [71, 72, 74, 86, 87, 88, 89]). In hierarchical models, parameters for individual people are assumed to be drawn from a group-level distribution. Such multilevel structures naturally incorporate both the differences and the commonalities between people and therefore provide experimental psychology with the means to settle the age-old problem of how to deal with individual differences.

Historically, the field of experimental psychology has tried to ignore individual differences, pretending instead that each new participant is a replicate of the previous one [6]. As Bill Estes and others have shown, however, individual differences that are ignored can lead to averaging artifacts in which the data that are averaged over participants are no longer representative for any of the participants (e.g., [28, 29, 45]). One way to address this issue, popular in psychophysics, is to measure each individual participant extensively and deal with the data on a participant-by-participant basis.

In between the two extremes of assuming that participants are completely the same and that they are completely different lies the compromise of hierarchical modeling (cf. [63]). The theoretical advantages and practical relevance of a Bayesian hierarchical analysis for common experimental designs has been repeatedly demonstrated by Jeff Rouder and colleagues (e.g., [86, 88, 89]). Although hierarchical analyses can be carried out using orthodox methodology [46], there are strong philosophical and practical reasons to prefer the Bayesian methodology (e.g., [36, 68]).

9.3.5 Marginalization

Bayesian statistics makes it easy to focus on the relevant variables by integrating out so-called nuisance variables (e.g., [5, 12]). Consider, for instance, the case of the normal distribution, for which the likelihood function is given by

$$f(y|\mu, \sigma) = \frac{1}{\sigma\sqrt{2\pi}} \exp\left(-\frac{(y-\mu)^2}{2\sigma^2}\right). \tag{9.8}$$

For this example, we follow [70, Chapter 24] and propose *conjugate improper priors* for μ and σ. A prior is said to be *conjugate* when it is in the same distributional family as the posterior distribution. For instance, when the prior for μ is normal, the posterior for μ is also normal. Conjugate priors are often the only ones that allow analytical derivation of the posterior. A prior is said to be *improper* when it does not integrate to a finite number. For instance, when the prior for μ is a normal distribution with mean $\mu_0 = 0$ and standard deviation $\sigma_\mu \to \infty$, this yields a prior that is flat across the entire real line. For the present example, we use conjugate improper priors on μ and σ because they lead to elegant analytical results that correspond to results from frequentist inference.

In particular, we assume here that the prior on μ is normal with mean $\mu_0 = 0$ and standard deviation $\sigma_\mu \to \infty$. This flat prior simply states that all values of μ are equally likely a priori. Because σ is always greater than 0, but $\log \sigma$ covers the entire real line, a standard "uninformative" prior is flat on the log scale, which transforms to the prior $p(\sigma) = 1/\sigma$. Using these priors, one can analytically derive the joint posterior distribution of μ and σ given the data (i.e., $p(\mu, \sigma|y)$) (e.g., [70, Chapter 24]).

Now that we have defined the priors and know the joint posterior distribution of μ and σ, consider two scenarios in which one needs to eliminate a nuisance parameter. In the first scenario, we want to learn about the mean μ of a normal distribution with unknown standard deviation σ. Thus, μ is the parameter of interest, whereas σ is a parameter that one would like to ignore (i.e., a nuisance parameter).

Using the law of total probability, it is straightforward to marginalize over, or integrate out, σ, as $p(\mu|y) = \int p(\mu, \sigma|y)\, d\sigma$. The fact that this equation can be rewritten as $p(\mu|y) = \int p(\mu|\sigma, y)p(\sigma)\, d\sigma$ highlights the fact that the nuisance parameter σ can only be integrated out once it has been assigned a prior distribution. After integrating out σ, the resulting posterior marginal distribution for $p(\mu|y)$ turns out to be the Student-t distribution, the famous frequentist distribution for a test statistic that involves the mean of a normal distribution with unknown variance [54].

In the second situation, we want to learn about the standard deviation σ of a normal distribution with unknown mean μ. This means that σ is the parameter of interest, whereas μ is now the nuisance parameter. From the joint posterior distribution of μ and σ, we can again apply the law of total probability, this time to integrate out μ, as follows: $p(\sigma|y) = \int p(\sigma, \mu|y)\, d\mu = \int p(\sigma|\mu, y)p(\mu)\, d\mu$. As before, this equation shows that the nuisance parameter μ can only be integrated out when it has been assigned a prior distribution. After computing the marginal posterior distribution $p(\sigma|y)$, the Most Probable value for σ (given the data y) turns out to be $\sigma_{MP} = \sqrt{S^2/(n-1)}$, where n equals the number of observations and $S^2 = \sum_{i=1}^{n}(y_i - \bar{y})^2$. The factor $n-1$ (instead of n) also occurs in frequentist inference, where $S^2/(n-1)$ is the unbiased estimator for the variance of a normal distribution with unknown mean.

In sum, Bayesian inference allows the user to focus on parameters of interest by integrating out nuisance parameters according to the law of total probability. The resulting marginal posterior distributions may have matching frequentist counterparts, but this only holds in a few special cases.

9.3.6 Validity

Bayesian inference yields results that connect closely to what researchers want to know. To clarify this claim by analogy, Gerd Gigerenzer has suggested that for many researchers statistical inference involves an internal Freudian struggle amond the Superego, the Ego, and the Id (e.g., [37, 39]). In Gigerenzer's analogy, the Superego promotes Neyman-Pearson hypothesis testing, in which an α-level is determined in advance of the experiment. The Ego promotes Fisherian hypothesis testing, in which the precise value of p supposedly measures the strength evidence against the null hypothesis. Finally, the Id desires that the hypotheses under consideration are assigned probabilities, something that the Superego and Ego are unable and unwilling to do. As a result of this unconscious internal conflict, researchers often report results from frequentist procedures, but often believe – implicitly or even explicitly – that they have learned something about the probability of the hypotheses under consideration.

We agree with Gigerenzer that, deep down inside, what researchers really want is to draw Bayesian conclusions. Or, in the words of Dennis Lindley, "Inside every Non-Bayesian, there is a Bayesian struggling to get out," as cited in [53]. This assertion is supported by the fact that researchers often misinterpret frequentist concepts – and misinterpret them in a manner that is decidedly Bayesian (i.e., the interpretation would have been correct if the method of inference had been Bayesian) [44].

To illustrate the foregoing with a concrete example, consider a frequentist confidence interval for the normal mean μ: $\mu \in [-0.5, 1.0]$. As we have seen in Section 9.2.1, the correct but counterintuitive interpretation of this result is that when the frequentist procedure is applied very many times to all kinds of possible datasets, the different intervals cover the true value of μ in 95% of the cases. But why would this be relevant for the researcher who wants to learn about μ for his or her data? In contrast, consider the same $[-0.5, 1.0]$ interval for μ, but now assume it is a Bayesian 95% credible interval. Consistent with intuition and consistent with what researchers want to know, this Bayesian interval conveys that there is a .95 probability that μ lies in $[-0.5, 1.0]$. From the viewpoint of "operation subjective probability" discussed in Section 9.3.1, this confidence interval means that when a coherent researcher is asked to set a fair price for a ticket that promises to pay 1 euro should the assertion "μ is in $[-0.5, 1.0]$" turn out to be true, that researcher will set the price of the ticket at exactly 0.95 euro.

9.3.7 Subjectivity That Is Open to Inspection

A common objection to Bayesian inference is that it is subjective and therefore has no place in scientific communication. For instance, in an article entitled "Why Isn't Everyone a Bayesian?" Bradly Efron argued that "Strict objectivity is one of the crucial factors separating scientific thinking from wishful thinking" and concluded that "The high ground of scientific objectivity has been seized by the frequentists" [27, p. 4].

Efron's claims need to be amended for several reasons. First, from a subjective Bayesian perspective, there is no such thing as "strict objectivity," as reasoning under uncertainty is always relative to some sort of background knowledge. In this view, the search for "strict objectivity" is a quixotic ideal. Thus, subjective Bayesians might want to change Efron's claim to "The high ground of scientific objectivity is a concept that cannot be seized by anyone, because it does not exist."

Second, there exists a school of *objective* Bayesians, who specify priors according to certain predetermined rules [58]. Given a specific rule, the outcome of statistical inference is independent of the person who performs the analysis.

Examples of objective priors include the unit information priors, that is, priors that carry as much information as a single observation [57], priors that are invariant under transformations [55], and priors that maximize entropy [51]. Objective priors are generally vague or uninformative (i.e., thinly spread out over the range for which they are defined). Thus, objective Bayesians might want to change Efron's claim to "Although the high ground of scientific objectivity may *appear* to be seized by the frequentists, objective Bayesians have a legitimate claim to scientific objectivity also."

Third, frequentist inference is not as objective as one may (wishfully) think. As illustrated in Section 9.2.3, the intention with which an experiment is carried out can have a profound impact on frequentist inference. The undisclosed ideas and thoughts that guided experimentation are crucial for calculating frequentist measures of evidence. Berger and Berry concluded that the perceived objectivity of frequentist inference is largely illusionary [11]. Thus, critics of frequentist inference might want to change Efron's claim to "Although the high ground of scientific objectivity may *appear* to be seized by the frequentists, upon closer inspection this objectivity is only make-believe, as in reality frequentists have to rely on the honesty and introspective ability of the researchers who collected the data."

In contrast to frequentist inference, Bayesian inference generally does not depend on subjective intentions (cf. Section 9.2.3), or on data that were never observed (cf. Section 9.2.2) [67]. The posterior distribution of parameters θ is written $p(\theta|y)$, and the marginal probability of a model, say H_0, is given by $m(y|H_0)$; in both cases, y is the observed data, and it is irrelevant what other data could have been observed but were not.

In Bayesian inference, the subjectivity that Efron alluded to comes in through the specification of the prior distribution for the model parameters. Regardless of whether this specification occurs automatically, as in the case of objective priors, or whether it occurs through the incorporation of prior knowledge, as in the case of subjective priors, the crucial point is that the prior distribution is formally specified and available for all other researchers to inspect and criticize. This also means that Bayesian subjectivity can be analyzed by formal methods that quantify robustness to the prior (e.g., [8, 25]). Note how different the notion of subjectivity is for the two paradigms: Bayesian subjectivity is open to inspection, whereas frequentist subjectivity is hidden from view, carefully locked up in the minds of the researchers that collected the data. Therefore, a final adjustment of Efron's statement might read "Scientific objectivity is illusionary, and both Bayesian inference and frequentist inference have subjective elements; the difference is that Bayesian subjectivity is open to inspection, whereas frequentist subjectivity is not."

9.3.8 Possibility of Collecting Evidence in Favor of the Null Hypothesis

Bayesian hypothesis testing allows one to obtain evidence in favor of the null hypothesis. In the Fisherian paradigm, p-values can only be used to reject the null hypothesis. The APA task force on statistical inference stressed this point by issuing the warning "Never use the unfortunate expression *accept the null hypothesis*" [102, p. 599]. Of course, what is unfortunate here is not so much the expression, but rather the fact that Fisherian p-values are incapable of providing support for the null hypothesis. This limitation hinders scientific progress, because theories and models often predict the absence of a difference. In the field of visual word recognition, for instance, the entry-opening theory [33] predicts that masked priming is absent for items that do not have a lexical representation. Another example from that literature concerns the work by Bowers et al. [16], who have argued that priming effects are equally large for words that look the same in lowercase and uppercase (e.g., kiss/KISS) or that look different (e.g., edge/EDGE), a finding supportive of the hypothesis that priming depends on abstract letter identities.

A final example comes from the field of recognition memory, where Dennis and Humphreys' "bind cue decide model of episodic memory" (BCDMEM) predicts the absence of a list-length effect and the absence of a list-strength effect [24]. This radical prediction of a null effect allows researchers to distinguish between context-noise and item-noise theories of inference in memory. Within the Fisherian paradigm, support for such informative predictions can only be indirect.

In contrast to the Fisherian hypothesis test, the Bayesian hypothesis test quantifies evidence by comparing the marginal probability of the data given one hypothesis, say $m(y|H_A)$, to the the marginal probability of the data given another hypothesis, say $m(y|H_B)$. The null hypothesis has no special status in Bayesian inference, and evidence for it is quantified just as it is for any other hypothesis, in a way that automatically strikes a balance between goodness-of-fit and parsimony (cf. Section 9.3.2).

9.3.9 Opportunity to Monitor Evidence as It Accumulates

Bayesian hypothesis testing allows one to monitor the evidence as the data come in [10]. In contrast to frequentist inference, Bayesian inference does not require special corrections for "optional stopping" [97].

Consider, for instance, a hypothetical experiment on the neural substrate of dissociative identity disorder. In this experiment, the researcher Marge has decided in advance to use functional magnetic resonance imaging (fMRI) to test 30 patients and 90 normal controls in a total of 4 between-subjects conditions, using the same design as Huntjens et al. [50]. Marge inspects the data after 15 participants in each condition have been tested and finds that the results quite convincingly demonstrate the pattern she hoped to find. Unfortunately for Marge, she cannot stop the experiment and claim a significant result, as she would be changing the sampling plan halfway through and be guilty of "optional stopping." She has to continue the experiment, wasting not just her time and money, but also the time and efforts of the people who undergo needless testing.

Within the frequentist paradigm, it is possible to adopt special sampling plans that take into account the need or desire to monitor the data as they accumulate; however, these sampling plans yield conclusions that are much more conservative than the one that assumes a fixed sample size. Thus, the very same data may lead to a clearly significant result under a fixed sample size scheme, but to a clearly nonsignificant result under a variable sample size scheme; the difference is due to the fact that the variable sample size scheme incorporates a correction for the eventuality that the experiment could have ended at a different point in time than it actually did.

In contrast, for Bayesian hypothesis testing there is nothing wrong with gathering more data, examining these data, and then deciding whether or not to stop collecting new data – no special corrections are needed. As stated by Edwards et al., "(...) the rules governing when data collection stops are irrelevant to data interpretation. It is entirely appropriate to collect data until a point has been proven or disproven, or until the data collector runs out of time, money, or patience." [26, p. 193].

9.3.10 Possibility of Incorporating Prior Knowledge

Bayesian inference allows prior knowledge to influence conclusions [67]. Priors are not only tremendously useful for incorporating existing knowledge, they are also a prerequisite for rational inference: "If one fails to specify the prior information, a problem of inference is just as ill-posed as if one had failed to specify the data." [53, p. 373]. Another perspective on priors was put forward by Berger, who argued that "(...) when different reasonable priors yield substantially different answers, can it be right to state that there *is* a single answer? Would it not be better to admit that there is scientific uncertainty, with the conclusion depending on prior beliefs?" [7, p. 125]. Thus, rather than considering priors a nuisance, we believe they are useful [67], necessary [53], and informative with respect to the robustness of one's conclusions [7]. Priors are an integral part of rational inference; one can only enjoy the Bayesian omelet when one is prepared to break the Bayesian eggs [93, p. 578].

9.4 Concluding Comments

In experimental psychology, the dominance of frequentist inference is almost complete. The first goal of this chapter was to demonstrate that the frequentist framework, despite its popularity, has several serious deficiencies. The second goal of this chapter was to show how the Bayesian framework is both flexible and principled. Our conclusion is that the field of psychology can gain a lot by moving toward the Bayesian framework for statistical inference and by moving away from the frequentist framework.

Perhaps frequentist inference has survived for so long because researchers translate the frequentist statistical outcomes to informal Bayesian conclusions. For instance, most experienced experimental psychologists would take seriously a priming effect of 25 msec ($p = .03$, $N = 30$ subjects, $k = 20$ items per condition), whereas they would be skeptical of a priming effect of 4 msec ($p = .03$, $N = 257$ subjects, $k = 20$ items per condition). Such an informal Bayesian interpretation of frequentist results is another indication of the internal conflict between the frequentist Superego and Ego versus the Bayesian Id; see Section 9.3.6 and [37].

It is our hope that more and more psychologists will start to move away from frequentist inference and turn instead to formal Bayesian inference. It may take therapy, medication, or perhaps even surgery, but in the end, researchers will be happier people once they allow their inner Bayesian to come out.

References

[1] Abelson, R.P.: On the surprising longevity of flogged horses: Why there is a case for the significance test. Psychological Science, **8**, 12–15 (1997)

[2] Anscombe, F.J.: Sequential medical trials. Journal of the American Statistical Association, **58**, 365–383 (1963)

[3] Bakan, D.: The test of significance in psychological research. Psychological Bulletin, **66**, 423–437 (1966)

[4] Barnard, G.A.: The meaning of a significance level. Biometrika, **34**, 179–182 (1947)

[5] Basu, D.: On the elimination of nuisance parameters. Journal of the American Statistical Association, **72**, 355–366 (1977)

[6] Batchelder, W.H.: Cognitive psychometrics: Combining two psychological traditions. CSCA Lecture, Amsterdam, The Netherlands, October 2007.

[7] Berger, J.O.: Statistical Decision Theory and Bayesian Analysis (2nd ed.). New York, Springer (1985)

[8] Berger, J.O.: Robust Bayesian analysis: Sensitivity to the prior. Journal of Statistical Planning and Inference, **25**, 303–328 (1990)

[9] Berger, J.O.: Could Fisher, Jeffreys and Neyman have agreed on testing? Statistical Science, **18**, 1–32 (2003)

[10] Berger, J.O., Berry, D.A.: The relevance of stopping rules in statistical inference. In: Gupta, S.S., Berger, J.O. (eds) Statistical Decision Theory and Related Topics: Vol. 1. New York, Springer (1988)

[11] Berger, J.O., Berry, D.A.: Statistical analysis and the illusion of objectivity. American Scientist, **76**, 159–165 (1988)

[12] Berger, J.O., Liseo, B., Wolpert, R.L.: Integrated likelihood methods for eliminating nuisance parameters. Statistical Science, **14**, 1–28 (1999)

[13] Berger, J.O., Pericchi, L.R.: The intrinsic Bayes factor for model selection and prediction. Journal of the American Statistical Association, **91**, 109–122 (1996)

[14] Berger, J.O., Wolpert, R.L.: The Likelihood Principle. Institute of Mathematical Statistics (2nd ed.), Hayward, CA (1988)

[15] Bernardo, J.M., Smith, A.F.M.: Bayesian Theory. New York, Wiley (1994)

[16] Bowers, J.S., Vigliocco, G., Haan, R.: Orthographic, phonological, and articulatory contributions to masked letter and word priming. Journal of Experimental Psychology: Human Perception and Performance, **24**, 1705–1719 (1998)

[17] Burdette, W.J., Gehan, E.A.: Planning and Analysis of Clinical Studies. Charles C. Springfield, IL, Thomas (1970)

[18] Christensen, R.: Testing Fisher, Neyman, Pearson, and Bayes. The American Statistician, **59**, 121–126 (2005)

[19] Cox, D.R.: Some problems connected with statistical inference. The Annals of Mathematical Statistics, **29**, 357–372 (1958)

[20] Cox, R.T.: Probability, frequency and reasonable expectation. American Journal of Physics, **14**, 1–13 (1946)

[21] Dawid, A.P.: Statistical theory: The prequential approach. Journal of the Royal Statistical Society, Series A, **147**, 278–292 (1984)

[22] De Finetti, B.: Theory of Probability, Vols. 1 and 2. New York, Wiley (1974)

[23] DeGroot, M.-H.: Optimal Statistical Decisions. New York, McGraw-Hill (1970)

[24] Dennis, S., Humphreys, M.S.: A context noise model of episodic word recognition. Psychological Review, **108**, 452–477 (2001)

[25] Dickey, J.M.: Scientific reporting and personal probabilities: Student's hypothesis. Journal of the Royal Statistical Society, Series B, **35**, 285–305 (1973)

[26] Edwards, W., Lindman, H., Savage, L.J.: Bayesian statistical inference for psychological research. Psychological Review, **70**, 193–242 (1963)

[27] Efron, B.: Why isn't everyone a Bayesian? The American Statistician, **40**, 1–5 (1986)

[28] Estes, W.K.: The problem of inference from curves based on group data. Psychological Bulletin, **53**, 134–140 (1956)

[29] Estes, W.K.: Traps in the route to models of memory and decision. Psychonomic Bulletin & Review, **9**, 3–25 (2002)

[30] Fishburn, P.C.: The axioms of subjective probability. Statistical Science, **1**, 335–345 (1986)

[31] Fisher, R.A.: Statistical Methods for Research Workers (5th ed.). London, Oliver and Boyd (1934)

[32] Fisher, R.A.: Statistical Methods for Research Workers (13th ed.). New York, Hafner (1958)

[33] Forster, K.I., Mohan, K., Hector, J.: The mechanics of masked priming. In: Kinoshita, S., Lupker, S.J. (eds) Masked Priming: The State of the Art. New York, Psychology Press (2003)

[34] Galavotti, M.C.: A Philosophical Introduction to Probability. Stanford, CA, CSLI Publications (2005)

[35] Gelfand, A.E., Smith, A.F.M., Lee, T. M.: Bayesian analysis of constrained parameter and truncated data problems using Gibbs sampling. Journal of the American Statistical Association, **87**, 523–532 (1992)
[36] Gelman, A., Hill, J.: Data Analysis Using Regression and Multilevel/Hierarchical Models. Cambridge, Cambridge University Press (2007)
[37] Gigerenzer, G.: The superego, the ego, and the id in statistical reasoning. In: Keren, G., Lewis, C. (eds) A Handbook for Data Analysis in the Behavioral Sciences: Methodological Issues. Hillsdale, NJ, Erlbaum (1993)
[38] Gigerenzer, G.: We need statistical thinking, not statistical rituals. Behavioral and Brain Sciences, **21**, 199–200 (1998)
[39] Gigerenzer, G.: Mindless statistics. The Journal of Socio–Economics, **33**, 587–606 (2004)
[40] Gigerenzer, G., Krauss, S., Vitouch, O.: The null ritual: What you always wanted to know about significance testing but were afraid to ask. In: Kaplan, D. (ed) The Sage Handbook of Quantitative Methodology for the Social Sciences. Thousand Oaks, CA, Sage (2004)
[41] Gill, J.: Bayesian Methods: A Social and Behavioral Sciences Approach. Boca Raton, FL, CRC Press (2002).
[42] Good, I.J.: Weight of evidence: A brief survey. In: Bernardo, J.M., DeGroot, M.-H., Lindley, D.V., Smith, A.F.M. (eds) Bayesian Statistics 2. New York, Elsevier (1985)
[43] Goodman, S.N.: P values, hypothesis tests, and likelihood: Implications for epidemiology of a neglected historical debate. American Journal of Epidemiology, **137**, 485–496 (1993)
[44] Haller, H., Krauss, S.: Misinterpretations of significance: A problem students share with their teachers? Methods of Psychological Research, **7**, 1–20 (2002)
[45] Heathcote, A., Brown, S., Mewhort, D.J.K.: The power law repealed: The case for an exponential law of practice. Psychonomic Bulletin & Review, **7**, 185–207 (2000)
[46] Hoffman, L., Rovine, M.J.: Multilevel models for the experimental psychologist: Foundations and illustrative examples. Behavior Research Methods, **39**, 101–117 (2007)
[47] Hoijtink, H.: Confirmatory latent class analysis: Model selection using Bayes factors and (pseudo) likelihood ratio statistics. Multivariate Behavioral Research, **36**, 563–588 (2001)
[48] Howson, C., Urbach, P.: Scientific Reasoning: The Bayesian Approach (3rd ed.). Chicago, Open Court (2006)
[49] Hubbard, R., Bayarri, M.J.: Confusion over measures of evidence (p's) versus errors (α's) in classical statistical testing. The American Statistician, **57**, 171–182 (2003)
[50] Huntjens, R.J.C., Peters, M.L., Woertman, L., Bovenschen, L.M., Martin, R.C., Postma, A.: Inter-identity amnesia in dissociative identity disorder: A simulated memory impairment? Psychological Medicine, **36**, 857–863 (2006)
[51] Jaynes, E.T.: Prior probabilities. IEEE Transactions on Systems Science and Cybernetics, **4**, 227–241 (1968)
[52] Jaynes, E.T.: Confidence intervals vs Bayesian intervals. In: Harper, W.L., Hooker, C.A. (eds) Foundations of Probability Theory, Statistical Inference, and Statistical Theories of Science, Vol. 2. Dordrecht, Reidel (1976)

[53] Jaynes, E.T.: Probability Theory: The Logic of Science. Cambridge, Cambridge University Press (2003)

[54] Jeffreys, H.: On the relation between direct and inverse methods in statistics. Proceedings of the Royal Society of London. Series A: Mathematical and Physical Sciences, **160**, 325–348 (1937)

[55] Jeffreys, H.: Theory of Probability. Oxford, Oxford University Press (1961)

[56] Kass, R.E., Raftery, A.E.: Bayes factors. Journal of the American Statistical Association, **90**, 377–395 (1995)

[57] Kass, R.E., Wasserman, L.: A reference Bayesian test for nested hypotheses and its relationship to the Schwarz criterion. Journal of the American Statistical Association, **90**, 928–934 (1995)

[58] Kass, R.E., Wasserman, L.: The selection of prior distributions by formal rules. Journal of the American Statistical Association, **91**, 1343–1370 (1996)

[59] Klugkist, I., Kato, B., Hoijtink, H.: Bayesian model selection using encompassing priors. Statistica Neerlandica, **59**, 57–69 (2005)

[60] Klugkist, I., Laudy, O., Hoijtink, H.: Inequality constrained analysis of variance: A Bayesian approach. Psychological Methods, **10**, 477–493 (2005)

[61] Klugkist, I., Laudy, O., Hoijtink, H.: Bayesian eggs and Bayesian omelettes: Reply to Stern (2005). Psychological Methods, **10**, 500–503 (2005)

[62] Laudy, O., Zoccolillo, M., Baillargeon, R.H., Boom, J., Tremblay, R.E., Hoijtink, H.: Applications of confirmatory latent class analysis in developmental psychology. European Journal of Developmental Psychology, **2**, 1–15 (2005)

[63] Lee, M.D., Webb, M.R.: Modeling individual differences in cognition. Psychonomic Bulletin & Review, **12**, 605–621 (2005)

[64] Lindley, D. V.: A statistical paradox. Biometrika, **44**, 187–192 (1957)

[65] Lindley, D.V.: Bayesian Statistics, a Review. Philadelphia, PA, SIAM (1972)

[66] Lindley, D.V.: Scoring rules and the inevitability of probability. International Statistical Review, **50**, 1–26 (1982)

[67] Lindley, D.V.: The analysis of experimental data: The appreciation of tea and wine. Teaching Statistics, **15**, 22–25 (1993)

[68] Lindley, D.V.: The philosophy of statistics. The Statistician, **49**, 293–337 (2000)

[69] Lindley, D.V., Scott, W.F.: New Cambridge Elementary Statistical Tables. London, Cambridge University Press (1984)

[70] MacKay, D.J.C.: Information Theory, Inference, and Learning Algorithms. Cambridge, Cambridge University Press (2003)

[71] Morey, R.D., Pratte, M.S., Rouder, J.N.: Problematic effects of aggregation in zROC analysis and a hierarchical modeling solution. Journal of Mathematical Psychology (in press)

[72] Morey, R.D., Rouder, J.N., Speckman, P.L.: A statistical model for discriminating between subliminal and near-liminal performance. Journal of Mathematical Psychology, **52**, 21–36 (2008)

[73] Myung, I.J., Forster, M.R., Browne, M.W.: Model selection [Special issue]. Journal of Mathematical Psychology, **44**(1–2) (2000)

[74] Navarro, D.J., Griffiths, T.L., Steyvers, M., Lee, M.D.: Modeling individual differences using Dirichlet processes. Journal of Mathematical Psychology, **50**, 101–122 (2006)
[75] Nelson, N., Rosenthal, R., Rosnow, R.L.: Interpretation of significance levels and effect sizes by psychological researchers. American Psychologist, **41**, 1299–1301 (1986)
[76] Neyman, J., Pearson, E.S.: On the problem of the most efficient tests of statistical hypotheses. Philosophical Transactions of the Royal Society A, **231**, 289–337 (1933)
[77] O'Hagan, A.: Fractional Bayes factors for model comparison. Journal of the Royal Statistical Society, Series B, **57**, 99–138 (1997)
[78] O'Hagan, A.: Dicing with the unknown. Significance, **1**, 132–133 (2004)
[79] O'Hagan, A., Forster, J.: Kendall's Advanced Theory of Statistics Vol. 2B: Bayesian Inference (2nd ed.). London, Arnold (2004)
[80] Peto, R., Pike, M.C., Armitage, P., Breslow, N.E., Cox, D.R., Howard, S.V., Mantel, N., McPherson, K., Peto, J., Smith, P.G.: Design and analysis of randomized clinical trials requiring prolonged observation of each patient, I: Introduction and design. British Journal of Cancer, **34**, 585–612 (1976)
[81] Pocock, S.J.: Group sequential methods in the design and analysis of clinical trials. Biometrika, **64**, 191–199 (1977)
[82] Raftery, A.E.: Bayesian model selection in social research. In: Marsden, P.V. (ed) Sociological Methodology. Cambridge, Blackwells (1995)
[83] Ramsey, F.P.: Truth and probability. In: Braithwaite, R.B. (ed) The Foundations of Mathematics and Other Logical Essays. London, Kegan Paul (1926)
[84] Rosenthal, R., Gaito, J.: The interpretation of levels of significance by psychological researchers. The Journal of Psychology, **55**, 33–38 (1963)
[85] Rosnow, R.L., Rosenthal, R.: Statistical procedures and the justification of knowledge in psychological science. American Psychologist, **44**, 1276–1284 (1989)
[86] Rouder, J.N., Lu, J.: An introduction to Bayesian hierarchical models with an application in the theory of signal detection. Psychonomic Bulletin & Review, **12**, 573–604 (2005)
[87] Rouder, J.N., Lu, J., Morey, R.D., Sun, D., Speckman, P.L.: A hierarchical process dissociation model. Journal of Experimental Psychology: General (in press)
[88] Rouder, J.N., Lu, J., Speckman, P.L., Sun, D., Jiang, Y.: A hierarchical model for estimating response time distributions. Psychonomic Bulletin & Review, **12**, 195–223 (2005)
[89] Rouder, J.N., Lu, J., Sun, D., Speckman, P., Morey, R., Naveh-Benjamin, M.: Signal detection models with random participant and item effects. Psychometrika (in press)
[90] Royall, R.: The effect of sample size on the meaning of significance tests. The American Statistician, **40**, 313–315 (1986)
[91] Royall, R.M.: Statistical Evidence: A Likelihood Paradigm. London, Chapman & Hall (1997)
[92] Savage, L.J.: The Foundations of Statistics. New York, Wiley (1954)

[93] Savage, L.J.: The foundations of statistics reconsidered. In: Neyman, J. (ed) Proceedings of the Fourth Berkeley Symposium on Mathematical Statistics and Probability, Vol. 1. Berkely, CA, University of California Press (1961)
[94] Smith, A.F.M., Roberts, G.O.: Bayesian computation via the Gibbs sampler and related Markov chain Monte Carlo methods. Journal of the Royal Statistical Society, Series B, **55**, 3–23 (1993)
[95] Spiegelhalter, D.J., Thomas, A., Best, N., Lunn, D.: WinBUGS Version 1.4 User Manual. Medical Research Council Biostatistics Unit, Cambridge (2003)
[96] Stuart, A., Ord, J.K., Arnold, S.: Kendall's Advanced Theory of Statistics Vol. 2A: Classical Inference & the Linear Model (6th ed.). London, Arnold (1999)
[97] Wagenmakers, E.-J.: A practical solution to the pervasive problems of p values. Psychonomic Bulletin & Review, **14**, 779–804 (2007)
[98] Wagenmakers, E.-J., Grünwald, P.: A Bayesian perspective on hypothesis testing. Psychological Science, **17**, 641–642 (2006)
[99] Wagenmakers, E.-J., Grünwald, P., Steyvers, M.: Accumulative prediction error and the selection of time series models. Journal of Mathematical Psychology, **50**, 149–166 (2006)
[100] Wagenmakers, E.-J., Waldorp, L.: Model selection: Theoretical developments and applications [Special issue]. Journal of Mathematical Psychology, **50**, 99–214 (2006)
[101] Wasserman, L.: All of Statistics: A Concise Course in Statistical Inference. New York, Springer (2004)
[102] Wilkinson, L., the Task Force on Statistical Inference: Statistical methods in psychology journals: Guidelines and explanations. American Psychologist, **54**, 594–604 (1999)

Part III

Beyond Analysis of Variance

10

Inequality Constrained Analysis of Covariance

Irene Klugkist[1], Floryt van Wesel[1], Sonja van Well[2], and Annemarie Kolk[2]

[1] Department of Methodology and Statistics, University of Utrecht, P.O. Box 80140, 3508 TC Utrecht, the Netherlands i.klugkist@uu.nl and f.vanwesel@uu.nl

[2] Clinical Psychology, University of Amsterdam, Roeterstraat 15, 1018 WB Amsterdam, the Netherlands s.m.vanwell@uva.nl and a.m.m.kolk@uva.nl

10.1 Introduction

This chapter deals with analysis of covariance (ANCOVA) with inequality constrained adjusted means. Bayesian evaluation of inequality constrained ANCOVA models was previously discussed in [3].

In an ANCOVA, two or more groups are compared on one outcome variable after correcting for one or more covariates. The outcome as well as the covariates are continuous variables. For general introductions of the ANCOVA model, see, for instance, [9, 10]. As a small example, consider an experiment where respondents with an anxiety disorder are randomly assigned to Therapy 1 or 2. Anxiety is measured before and after therapy and a difference score is computed for each respondent. The outcome variable of interest is the decrease in anxiety. The expectation of the researcher is that Therapy 1 will have better effects than Therapy 2; that is, his hypothesis is $\mu_1 > \mu_2$, where μ_1 and μ_2 denote the average decrease in anxiety for Therapy 1 and 2, respectively.

Assume that before the experiment, motivation of respondents was measured. Since motivation differs among participants and it can be expected that motivation is related to the therapy effect, motivation is included in the analysis as a covariate. Denoting the outcome (decrease in anxiety) for the ith respondent ($i = 1, \ldots, N$) with y_i and the covariate (motivation) with x_i, the statistical model is

$$y_i = \mu_1^* d_{1i} + \mu_2^* d_{2i} + \beta x_i + \varepsilon_i, \tag{10.1}$$

where ε_i is assumed to be normally distributed with mean zero and variance σ^2. Note that the nonstandard dummy coding with one dummy for each subgroup (d_1 and d_2, respectively) is used (see also Chapter 3). As a consequence, the parameter μ_1^* represents the predicted decrease in anxiety for a respondent from Therapy 1 ($d_{1i} = 1$ and $d_{2i} = 0$) with a motivation score $x_i = 0$. Likewise, μ_2^* represents the predicted decrease in anxiety for a respondent in

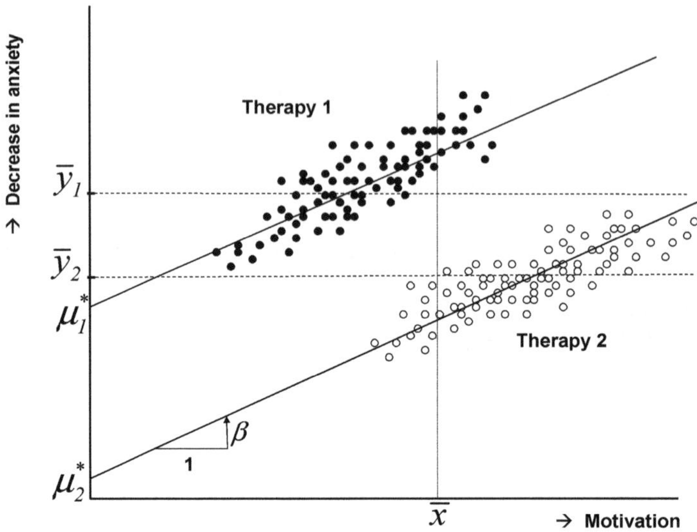

Fig. 10.1. Analysis of covariance

Therapy 2 ($d_{1i} = 0$ and $d_{2i} = 1$) with a motivation score 0. The relation between motivation and decrease in anxiety (in an ANCOVA model assumed to be equal for subgroups) is modeled by linear regression lines with slope β. For a graphical representation of this example, see Figure 10.1. It can be seen that, on average, the group receiving Therapy 1 has a larger decrease in anxiety than the group receiving Therapy 2 ($\bar{y}_1 > \bar{y}_2$). It can also be seen that the Therapy 1 group has, on average, a lower motivation than the Therapy 2 group. Furthermore, we see a positive relation between motivation on the x-axis and decrease in anxiety on the y-axis, as represented by the regression lines.

Compared to an analysis of variance (ANOVA), the incorporation of one or more covariates has two effects: (i) A correction is made for possible group differences on the covariate(s) and (ii) the residual variance is reduced. Let us start with the correction. By inclusion of the covariate, the average decrease in anxiety of the two groups is compared for therapy groups with equal average motivation. The relation between motivation (x) and decrease in anxiety (y) is used to compute the predicted average of y for a fixed value of x (for both groups the same). At the value $x = 0$ this provides μ_1^* and μ_2^* (i.e., the intercepts of the two regression lines). It is however more common to report so-called adjusted means. Adjusted means are the group mean outcomes (predictions on the regression lines) evaluated at the average score on the covariate(s). Using (10.1), the adjusted means can be obtained by filling in the average motivation score $\bar{x}$ (see also Figure 10.1). Another approach is centering the covariate(s) before it is entered in the model, in which case the μ^*'s in (10.1) are the adjusted means. In the remainder of this chapter as well as in the illustration, the covariate will always be centered before it is included in the analysis, simplifying the interpretation of the parameters.

The second effect of including covariates is that the residual variation will decrease. This can also be seen in Figure 10.1. In an ANOVA the residuals are the vertical distances between each observed score (the dots) and the corresponding subgroup's mean (the horizontal lines through $\bar{y}_1$ and $\bar{y}_2$). For many dots, these distances are relatively large. The residuals in the ANCOVA are the vertical distances between the dots and the regression lines. Many of these distances are much smaller. By decreasing the residuals and thus explaining a larger part of the variation, more powerful analyses are obtained.

So in the example, motivation is included in the model to correct for the difference in the average motivation of the two groups to obtain a fair comparison between the two therapies and to decrease the residual variance. Note, however, that the expectation of the researcher and the translation into a constrained hypothesis remains the same. After correcting for the effect of motivation, it is still expected that Therapy 1 will have a larger effect than Therapy 2 (i.e., $\mu_1^* > \mu_2^*$). The hypothesis now reflects an expected ordering in the effects of the two therapies *for subjects with the same motivation.*

In Section 10.2, a psychological experiment is introduced for which the researchers formulated several competing informative hypotheses. In Section 10.3, Bayesian evaluation for the general inequality constrained ANCOVA is presented. This section is relatively technical but readers can choose to skip it and continue with Section 10.4, where the results for the illustration presented in Section 10.2 are provided and discussed.

10.2 Illustration

The illustration is based on a psychological experiment conducted by Van Well, Kolk, and Klugkist [11] on the responsivity to stressors. Outcome variables contained several cardiovascular measures as well as a subjective stress response measure and were obtained at baseline and in three stress phases: anticipation, stressor, and recovery. Responsivity was investigated for two types of stressors, the Cold Pressor Test (CPT) and the N-Back task, but in this illustration only data from the CPT will be used. The CPT is a physiological stress task where participants put their right hand up to the wrist in a bucket of ice water for two minutes. Of interest were the effects of the gender relevance of the stressor, the respondent's sex, and gender role identification. Gender relevance was manipulated by varying the introduction to the stressor in masculine relevant, feminine relevant, or neutral terms. Gender role identification was measured with the GIAT [1, 12] and respondents were classified in the categories masculine or feminine. In total, 54 women and 40 men participated in the experiment.

Researchers have found different results concerning responses to gender relevant stressors. Lash, Eisler, and Southard [6] and Lash, Gillespie, Eisler, and Southard [7] found that men were more responsive than women when the stressor was presented as masculine relevant and that women were more responsive than men when the stressor was presented as feminine relevant. No sex differences were found when the stressor was presented as gender neutral. Other researchers investigated responses to gender role identification and gender relevance of the stressor. They also found stronger stress responses for participants with a masculine gender role identification in the masculine relevant condition, stronger responses for feminine role identification in the feminine relevant condition, and no difference between masculine and feminine role identification in the neutral condition [5, 8]. So both for sex and gender relevance of the manipulation and for gender role identification and gender relevance of the manipulation, there were indications for so-called match effects; that is, stronger responses for matching conditions (male-masculine, female-feminine). Furthermore, Kolk and van Well [4] tested sex and gender role identification match effects in one study and their results reveiled stronger gender match effects than sex match effects. Other researchers, however, also found opposite, so-called mismatch effects. In a study by Davis and Matthews [2] responses were stronger in situations that were relevant to the opposite sex or gender role identification. The results of another study suggested that easy tasks produce mismatch effects, whereas more difficult tasks produce match effects [13]. To further investigate the contradicting results, Van Well et al. [11] designed an experiment to examine the relationship among sex, gender role identification, and gender relevance of a stressor. They investigated sex (mis)match as well as gender role identification (mis)match effects.

In this chapter just a part of the data of Van Well et al. will be used. The illustration is limited to the outcome measures diastolic blood pressure (DBP) and systolic blood pressure (SBP) obtained during the stressor phase only. The two outcomes will be analyzed separately and in each analysis the baseline measurement of the corresponding outcome (DBP or SBP) is included as a covariate. This leads to an ANCOVA model with 12 subgroups (gender relevance by sex by gender role identification) and one covariate. Table 10.1 presents the group numbering that will be used in the formulation of the informative hypotheses.

Again using the dummy coding as explained before (one dummy variable for each group; see Chapters 3 and 4), the following model is obtained:

$$y_i = \sum_{j=1}^{12} \mu_j d_{ji} + \beta x_i + \varepsilon_i, \tag{10.2}$$

Table 10.1. Group labeling for the groups formed by sex (men, women), gender role identification (masculine, feminine), and gender relevance of the condition (masculine, feminine, neutral)

	Condition		
	Masculine	Feminine	Neutral
Men			
Masculine	1	2	3
Feminine	4	5	6
Women			
Masculine	7	8	9
Feminine	10	11	12

for $i = 1, \ldots, N$ respondents. Note that, to simplify notations, covariate adjusted means are denoted by μ_j (in Section 10.1 they were denoted by μ_j^* to emphasize the difference between a mean and an adjusted mean). Two separate analyses are performed. In the first analysis, y_i and x_i are the scores of the ith respondent on DBP in stressor phase and at baseline, respectively. The second analysis is an application of (10.2) with SBP measurements in stressor phase (y_i) and at baseline (x_i).

The sex (mis)match and gender role identification (mis)match theories are translated into statistical hypotheses that impose inequality constraints on the adjusted means. For instance, the sex match effect implies that men in the masculine relevant condition (i.e., groups 1 and 4) and women in the feminine relevant condition (i.e., groups 8 and 11) will have higher stress responses than respondents in the other groups. This leads to the first constrained hypothesis, H_{1a}, presented below. In a similar way, the sex mismatch (H_{1b}), the gender role identification match (H_{1c}), and the gender role identification mismatch (H_{1d}) hypotheses are formulated:

Sex effects

$$\begin{array}{ll} \text{Match:} & H_{1a} : \{\mu_1, \mu_4\} > \{\mu_2, \mu_3, \mu_5, \mu_6\}, \{\mu_8, \mu_{11}\} > \{\mu_7, \mu_9, \mu_{10}, \mu_{12}\}, \\ \text{Mismatch:} & H_{1b} : \{\mu_2, \mu_5\} > \{\mu_1, \mu_3, \mu_4, \mu_6\}, \{\mu_7, \mu_{10}\} > \{\mu_8, \mu_9, \mu_{11}, \mu_{12}\}. \end{array}$$

Gender role identification effects

$$\begin{array}{ll} \text{Match:} & H_{1c} : \{\mu_1, \mu_5\} > \{\mu_2, \mu_3, \mu_4, \mu_6\}, \{\mu_7, \mu_{11}\} > \{\mu_8, \mu_9, \mu_{10}, \mu_{12}\}, \\ \text{Mismatch:} & H_{1d} : \{\mu_2, \mu_4\} > \{\mu_1, \mu_3, \mu_5, \mu_6\}, \{\mu_8, \mu_{10}\} > \{\mu_7, \mu_9, \mu_{11}, \mu_{12}\}. \end{array}$$

Note that all hypotheses are nested in the unconstrained, encompassing hypothesis

$$H_2 : \mu_1, \mu_2, \mu_3, \mu_4, \mu_5, \mu_6, \mu_7, \mu_8, \mu_9, \mu_{10}, \mu_{11}, \mu_{12}.$$

The unconstrained hypothesis has no restrictions on the adjusted means and does not reflect a theory. It does, however, play a central role in the encompassing prior approach (see Chapter 4). For each constrained hypothesis H_t ($t \in \{1a, 1b, 1c, 1d\}$), the Bayes factor (BF) with the unconstrained hypothesis is computed and reflects to what extent the constraints are supported by the data.

Table 10.2 presents the means (M), standard deviations (SD), and sample sizes (N) for DBP and SBP at baseline and during the stressor phase for the 12 groups. Before moving to the results of the analyses, however, Bayesian model selection for inequality constrained ANCOVA models is introduced. In Section 10.3, the prior and posterior distributions are presented as well as the Gibbs sampler. This section is relatively technical and can be skipped by readers who are more interested in the practical example (the results of the illustration are discussed in Section 10.4). Note that, the Bayesian procedure that is applied in this chapter is very similar to the approach presented in Chapters 3 and 4 for the ANOVA model.

10.3 The Analysis of Covariance Model

10.3.1 The Model and Likelihood

The general ANCOVA model for $i = 1, \ldots, N$ respondents, outcome variable y_i, $k = 1, \ldots, K$ (centered) covariates x_{ki}, and $j = 1, \ldots, J$ groups with group membership denoted by d_{ji} ($d_{ji} = 1$ if the respondent is a member of group j, and zero otherwise) is given by

$$y_i = \sum_{j=1}^{J} \mu_j d_{ji} + \sum_{k=1}^{K} \beta_k x_{ki} + \varepsilon_i, \tag{10.3}$$

where μ_j is the covariate adjusted mean for group j, β_k is the slope parameter for the relation of the kth covariate with the outcome variable, and ε_i is assumed to be normally distributed with mean zero and variance σ^2. This leads to the likelihood function:

$$f(\boldsymbol{y}|\boldsymbol{\mu}, \boldsymbol{\beta}, \sigma^2, \boldsymbol{D}, \boldsymbol{X}) = \tag{10.4}$$

$$\prod_{i=1}^{N} \frac{1}{\sqrt{2\pi\sigma^2}} \exp \left\{ \frac{(y_i - \sum_{j=1}^{J} \mu_j d_{ji} - \sum_{k=1}^{K} \beta_k x_{ki})^2}{-2\sigma^2} \right\},$$

where $\boldsymbol{\mu}$ and $\boldsymbol{\beta}$ are vectors with adjusted means μ_j and slopes β_k, respectively, $\boldsymbol{D}$ is a $J \times N$ matrix with the (j, i)th element equal to d_{ji}, and $\boldsymbol{X}$ is a $K \times N$ matrix with elements x_{ki}.

Table 10.2. Sample means (M), standard deviations (SD), and sample sizes (N) for DBP and SBP at baseline and during the stressor phase

		Condition								
		Masculine			Feminine			Neutral		
DBP		M	SD	N	M	SD	N	M	SD	N
Men										
Masculine	baseline	79.25	22.06	5	73.64	9.13	9	66.93	6.57	6
	stress	99.04	26.93	5	90.53	9.13	9	81.01	9.53	6
Feminine	baseline	74.82	9.37	9	79.26	7.43	6	74.30	10.42	5
	stress	88.98	8.54	9	100.36	16.72	6	96.46	13.10	5
Women										
Masculine	baseline	75.33	13.61	9	75.15	7.57	8	78.85	9.71	10
	stress	92.69	11.89	9	95.13	14.08	8	91.83	15.14	10
Feminine	baseline	71.23	9.85	8	69.66	11.48	9	67.07	8.95	10
	stress	83.79	11.86	8	84.92	8.08	9	81.56	10.88	10
SBP		M	SD	N	M	SD	N	M	SD	N
Men										
Masculine	baseline	134.77	24.25	5	126.56	13.64	9	123.41	7.75	6
	stress	169.34	27.92	5	154.24	12.43	9	146.30	13.08	6
Feminine	baseline	127.27	12.04	9	134.57	17.71	6	130.04	21.08	5
	stress	152.04	15.04	9	165.93	24.66	6	163.36	19.78	5
Women										
Masculine	baseline	130.10	16.33	9	136.26	8.68	8	138.77	14.04	10
	stress	156.27	10.93	9	161.89	18.53	8	156.55	19.88	10
Feminine	baseline	128.78	11.52	8	131.44	14.28	9	127.30	16.66	10
	stress	145.06	11.88	8	152.01	12.93	9	146.13	17.87	10

10.3.2 Prior and Posterior

A prior distribution has to be specified for the model parameters. Let us assume that, other than the possible order constraints, there is no prior information. In that case, any noninformative prior can be used for estimation problems; for example a constant prior $p(\boldsymbol{\mu}, \boldsymbol{\beta}, \sigma^2) \propto 1$, a reference prior $p(\boldsymbol{\mu}, \boldsymbol{\beta}, \sigma^2) \propto \sigma^{-2}$, or a diffuse conjugate or semiconjugate prior. In terms of parameter estimation all noninformative priors will give (virtually) the same results, as was previously shown in Chapter 3. However, for model selection applications, the choice for a prior is more important. The encompassing prior approach introduced in Chapter 4 is also applied in this chapter.

The encompassing prior assumes prior independence of all parameters:

$$p(\boldsymbol{\mu}, \boldsymbol{\beta}, \sigma^2) = p(\sigma^2) \prod_{k=1}^{K} p(\beta_k) \prod_{j=1}^{J} p(\mu_j). \tag{10.5}$$

For each parameter a conjugate prior is specified; that is, a normal distribution with mean μ_0 and variance τ_0^2 for each μ_j, a normal distribution with mean β_{k0} and variance ω_{k0}^2 for each β_k, and a scaled inverse χ^2-distribution with degrees of freedom ν_0 and scale factor σ_0^2 for σ^2. Note that the same prior distribution is specified for all μ_j ($j = 1, \ldots, J$), since hypotheses with constraints on (some of) the μ_j are considered (for the specification rules used in the encompassing prior approach, see Chapter 4). This leads to the following specification of the general prior for a hypothesis H_t:

$$p(\boldsymbol{\mu}, \boldsymbol{\beta}, \sigma^2 | H_t) \propto$$

$$\text{Inv-}\chi^2(\sigma^2|\nu_0, \sigma_0^2) \prod_{k=1}^{K} \mathcal{N}(\beta_k|\beta_{k0}, \omega_{k0}^2) \prod_{j=1}^{J} \mathcal{N}(\mu_j|\mu_0, \tau_0^2) I_{\boldsymbol{\mu} \in H_t}, \tag{10.6}$$

where the indicator function $I_{\boldsymbol{\mu} \in H_t}$ equals 1 if $\boldsymbol{\mu}$ is in accordance with the constraints of H_t and 0 otherwise. Through this indicator function the prior information in the form of inequality constraints on (some of) the μ_j is incorporated: Areas that are not allowed according to the constraints receive zero prior density. The remaining area is proportional to the unconstrained prior.

If no constraints are imposed, the indicator function is not activated; that is, $I_{\boldsymbol{\mu} \in H_t}$ always has the value one. Stated differently, the encompassing prior is (10.6) without the indicator term. The specification of the parameters of the encompassing prior is data based and such that it is diffuse and thus low informative. As was previously explained in Chapter 4, a sample is drawn from the posterior distribution using a constant prior. The parameters of the scaled inverse χ^2-distribution are $\nu_0 = 1$ and σ_0^2 is equal to the posterior mean of the variance. The specification of μ_0 and τ_0^2 is done based on the broad interval that is a combination of the J 99.7% credibility intervals for each of the μ_j (see Sections 4.3.1 and 4.4.1 of Chapter 4). Finally, β_{k0} is equal to the posterior mean of β_k ($k = 1, \ldots, K$), and for the specification of ω_{k0}, we use three times the posterior standard deviation of β_k (so that, also for the β's, the prior is low informative compared to the information in the data).

The product of the joint prior distribution (10.6) and the density of the data (10.4) leads to the general posterior distribution:

$$p(\boldsymbol{\mu}, \boldsymbol{\beta}, \sigma^2 | \mathbf{y}, D, X, H_t) \propto f(\mathbf{y}|\boldsymbol{\mu}, \boldsymbol{\beta}, \sigma^2, D, X)$$

$$\text{Inv-}\chi^2(\sigma^2|\nu_0, \kappa_0^2) \prod_{k=1}^{K} \mathcal{N}(\beta_k|\eta_{k0}, \omega_{k0}^2) \prod_{j=1}^{J} \mathcal{N}(\mu_j|\alpha_0, \tau_0^2) I_{\boldsymbol{\mu} \in H_t}. \tag{10.7}$$

The encompassing posterior is (10.7) without the indicator function.

10.3.3 The Gibbs Sampler

The joint posterior (10.7) contains $J + K + 1$ parameters. To sample this multivariate posterior the Gibbs sampler is applied. In a Gibbs sampler, parameters are sampled iteratively from the univariate distribution of each parameter conditional upon the current values of the other parameters. This requires the specification of initial or starting values for all parameters and then repeatedly sampling each μ_j, β_k, and σ^2 from the corresponding conditional distributions. The sampling scheme for each iteration of the Gibbs sampler for the ANCOVA model therefore consists of the following steps:

1. For $j = 1, ..., J$, sample μ_j from $p(\mu_j|\boldsymbol{\mu}_{-j}, \boldsymbol{\beta}, \sigma^2, \boldsymbol{y}, X, D, H_t)$. The notation $\boldsymbol{\mu}_{-j}$ is used to denote the vector $\boldsymbol{\mu}$ with all except the jth element. The posterior for μ_j is equal to a truncated normal distribution with expectation
$$\frac{\frac{1}{\tau_0^2}\alpha_0 + \frac{1}{\sigma^2}(\sum_{i=1}^{N} d_{ji}y_i - \sum_{k=1}^{K}(\beta_k \sum_{i=1}^{N} d_{ji}x_{ki}))}{\frac{1}{\tau_0^2} + \frac{1}{\sigma^2}\sum_{i=1}^{N} d_{ji}} \tag{10.8}$$
and variance
$$\left(\frac{1}{\tau_0^2} + \frac{1}{\sigma^2}\sum_{i=1}^{N} d_{ji}\right)^{-1}. \tag{10.9}$$
The constraints of H_t combined with the current values of $\boldsymbol{\mu}_{-j}$ define possible lower and upper bounds for the values allowed for μ_j in the iteration at hand. The largest lower and smallest upper bound determine the truncation. Sampling from truncated normal distributions is straightforward via inverse probability sampling (see Chapter 3).
2. For $k = 1, ..., K$, sample β_k from $p(\beta_k|\boldsymbol{\mu}, \boldsymbol{\beta}_{-k}, \sigma^2, \boldsymbol{y}, \boldsymbol{X}, \boldsymbol{D}, H_t)$; that is, a normal distribution with expectation
$$\frac{\frac{1}{\omega_{k0}^2}\eta_{k0} + \frac{1}{\sigma^2}\left(\sum_{i=1}^{N} y_i x_{ki} - \sum_{i=1}^{N}\sum_{j=1}^{J} x_{ki}\mu_j d_{ji} - \sum_{i=1}^{N}\sum_{k'=1}^{K_{-k}} \beta_{k'}x_{ki}x_{k'i}\right)}{\frac{1}{\omega_{k0}^2} + \frac{1}{\sigma^2}\sum_{i=1}^{N} x_{ki}^2} \tag{10.10}$$
and variance
$$\left(\frac{1}{\omega_{k0}^2} + \frac{1}{\sigma^2}\sum_{i=1}^{N} x_{ki}^2\right)^{-1}. \tag{10.11}$$
The notation $\boldsymbol{\beta}_{-k}$ denotes all except the kth elements of $\boldsymbol{\beta}$, and K_{-k} denotes all except the kth element of $\{1, \ldots, \mathrm{K}\}$.
3. Sample σ^2 from $p(\sigma^2|\boldsymbol{\mu}, \boldsymbol{\beta}, \boldsymbol{y}, X, D, H_t)$; that is, a scaled inverse χ^2-distribution with degrees of freedom
$$N + \nu_0 \tag{10.12}$$
and scale parameter
$$\frac{\nu_0\kappa_0^2 + Ns^2}{\nu_0 + N}, \tag{10.13}$$
with $s^2 = \frac{1}{N}\sum_{i=1}^{N}(y_i - [\sum_{j=1}^{J}\mu_j d_{ji} + \sum_{k=1}^{K}\beta_k x_{ki}])^2$.

Note that the initial sampled values are affected by the arbitrary chosen starting values. Therefore, a first set of draws is discarded, the so-called burn-in period. The iterations after burn-in form the sample from the posterior distribution. Careful monitoring of the required size of the burn-in period and the total number of iterations from the posterior is important. In Chapter 3, graphical monitoring as well as a convergence diagnostic were discussed and illustrated and therefore this topic will not be further elaborated here.

10.3.4 Model Selection

Model selection using the encompassing prior approach, requires a sample from the unconstrained prior as well as a sample from the unconstrained posterior. With these two samples, any hypothesis H_t with inequality constraints imposed on (some of) the μ_j, can be evaluated using:

$$BF_{t2} = \frac{1/d_t}{1/c_t}, \tag{10.14}$$

where $1/c_t$ and $1/d_t$ are the proportions of the unconstrained prior and posterior in agreement with the constraints of H_t. To obtain a sample from the unconstrained posterior the Gibbs sampler as presented in the previous section can be used. Note that the indicator function is not active and therefore in Step 1 the conditional distributions are normal instead of truncated normal. To obtain an estimate for $1/c$, parameter values are sampled from the prior (10.6) again with the indicator function not activated (i.e., for the unconstrained prior it always has value 1). Because of independence of all parameters in the unconstrained prior, sampling from (10.6) is straightforward.

With the samples from unconstrained prior and posterior, BF_{t2} for each constrained hypothesis H_t with the unconstrained hypothesis H_2 can be estimated. The Bayes factor for the comparison of two constrained hypotheses H_t and $H_{t'}$ can be computed using

$$BF_{tt'} = \frac{BF_{t2}}{BF_{t'2}}. \tag{10.15}$$

Finally, assuming equal prior probabilities for all hypotheses under consideration, posterior model probabilities (PMP) for a finite set of T hypotheses can be computed using

$$\text{PMP}(H_t) = \frac{BF_{t2}}{\sum_{t' \in T} BF_{t'2}}. \tag{10.16}$$

A PMP provides the relative support for each hypothesis. For a more elaborate presentation and discussion of PMP values and their interpretation see Chapters 4 and 15.

10.4 Results

In this section we will discuss the results of the experiment described in Section 10.2 of this chapter. This psychological experiment consisted of 12 groups formed by sex, gender role identification, and gender relevance of the stressor. The dependent variables of interest were diastolic blood pressure (DBP) and systolic blood pressure (SBP) measured in the stressor phase. To correct for baseline DBP and SBP measurement, each of these variables was used as covariate. The researchers were interested in hypotheses involving sex and gender role identification match and mismatch effects.

In order to perform a Bayesian analysis, prior distributions need to be specified for all model parameters $\mu_1, ..., \mu_{12}, \beta$, and σ^2. Using the encompassing prior approach introduced in Chapters 3 and 4, only a prior distribution for the unconstrained model H_2 needs to be specified. Priors for constrained hypotheses are derived from the encompassing prior by truncation of the parameter space. In the encompassing prior approach, the unconstrained prior is specified to be relatively vague (i.e., low informative). Since the data will dominate the prior, objective estimates (not or hardly influenced by the prior) are obtained for the model parameters. Also the model selection (i.e., the comparison of the different hypotheses) is (virtually) objective for the inequality constrained hypotheses considered in this illustration, as was shown in Chapter 4.

The procedure to construct priors as described in Section 10.3.2 leads to the following encompassing prior for DBP:

$$p(\boldsymbol{\mu}, \boldsymbol{\beta}, \sigma^2|H_2) \propto \text{Inv-}\chi^2(\sigma^2|1, 68.2)\, \mathcal{N}(\beta|0.94, 0.07) \prod_{j=1}^{J} \mathcal{N}(\mu_j|122.85, 222.61)$$

and for SBP:

$$p(\boldsymbol{\mu}, \boldsymbol{\beta}, \sigma^2|H_2) \propto \text{Inv-}\chi^2(\sigma^2|1, 136.1)\, \mathcal{N}(\beta|0.85, 0.07) \prod_{j=1}^{J} \mathcal{N}(\mu_j|157.60, 844.48).$$

For each outcome measure, the prior is updated with the empirical data leading to the unconstrained posterior. The posterior represents the knowledge with respect to the model parameters after seeing the data.

10.4.1 Posterior Estimates

Posterior estimates for all model parameters $\mu_1, ..., \mu_{12}, \beta$, and σ^2 are obtained by drawing a sample from the unconstrained posterior distribution, using a Gibbs sampler (see Chapter 3 and Section 3 of this chapter). The number of iterations used was 20,000 after a burn-in of 1000. This sample provides a

Table 10.3. Posterior means (PM), posterior standard deviations (PSD), and lower (LB) and upper bound (UB) of the 95% CCIs for all parameters under the unconstrained model H_2 for both DBP and SBP

	DBP				SBP			
			95% CCI				95% CCI	
	PM	PSD	LB	UB	PM	PSD	LB	UB
μ_1	93.56	3.56	86.63	100.53	165.54	5.04	155.69	175.58
μ_2	90.54	2.70	85.27	95.84	157.82	3.86	150.30	165.37
μ_3	87.54	3.35	80.96	94.16	152.73	4.73	143.44	162.22
μ_4	87.95	2.78	82.73	93.20	155.04	3.85	147.45	162.67
μ_5	94.86	3.31	88.36	101.32	162.52	4.64	153.41	171.62
μ_6	95.52	3.57	88.43	102.45	163.73	5.08	153.82	173.74
μ_7	91.05	2.70	85.76	96.35	156.83	3.84	149.30	164.44
μ_8	93.56	2.88	87.91	99.20	157.30	4.08	149.24	165.38
μ_9	86.98	2.59	81.92	92.07	149.96	3.69	142.68	157.23
μ_{10}	86.23	2.86	80.69	91.89	147.06	4.03	139.28	155.04
μ_{11}	88.70	2.68	83.47	93.93	151.59	3.83	144.09	159.22
μ_{12}	87.79	2.60	82.69	92.94	149.25	3.65	142.17	156.36
β	0.94	0.08	0.78	1.10	0.85	0.08	0.69	1.01
σ^2	67.39	10.74	49.61	91.58	134.76	21.66	99.26	183.44

discrete representation of the posterior distribution and can be summarized using, for instance, the posterior mean, posterior standard deviation, and 95% Central Credibility Interval (CCI) for all model parameters of interest. The results are presented in Table 10.3. For example, the parameter μ_1 for DBP has a posterior mean of 93.56 with a posterior standard deviation of 3.56. The 95% CCI for μ_1 (DBP) has lower and upper bounds of 86.63 and 100.53, respectively. Furthermore, it can be seen that the 95% CCIs for the β parameters for both DBP and SBP (i.e., the regression coefficients representing the relation between covariate and outcome), do not include the value zero. This shows that the baseline measurements have a positive relationship with the measurements in the stressor phase and that it was a good idea to take them into account as covariates in the statistical model.

Note, again, that the estimates presented in Table 10.3 are computed using the unconstrained prior (i.e., without taking any inequality constraints into account). The question of interest is, however, which of the informative hypotheses presented in Section 10.2 is mostly supported by the data. The results are presented in the next section.

10.4.2 Evaluation of the Informative Hypotheses

The informative hypotheses formulated by the researchers represent a sex match effect (H_{1a}), a sex mismatch effect (H_{1b}), a gender role identification match affect (H_{1c}), and, finally, a gender role identification mismatch effect

Table 10.4. Bayes factors (BF_{t2}) and posterior model probabilities (PMP) for both DBP and SBP

	DBP		SBP	
	BF_{t2}	PMP	BF_{t2}	PMP
H_{1a}	0.13	.04	0.32	.04
H_{1b}	0.09	.03	0.06	.01
H_{1c}	1.97	.62	5.85	.81
H_{1d}	0.01	.00	0.00	.00
H_2	1.00	.31	1.00	.14

(H_{1d}). To evaluate these hypotheses the Bayes factor will be used as the selection criterion.

Model selection based on the encompassing prior approach requires samples from both unconstrained prior and posterior (see Chapter 4 and Section 10.3.4 in this chapter). The amount of iterations used was 500,000 after a burn-in of 1000. The BF for each informative hypothesis (H_t) against the unconstrained hypothesis (H_2) is calculated and denoted by BF_{t2}. It provides a measure of support for the constrained hypothesis (e.g., the value $BF_{1c2} = 2$ implies that H_{1c} is 2 times better than H_2 according to the data from this experiment). From the BFs, posterior model probabilities (PMPs) can be derived. A PMP represents the relative support for a hypothesis within a set of hypotheses (see Chapter 4). In this illustration, again equal prior model probabilities are assumed.

The resulting BFs and PMPs for the competing hypotheses regarding DBP and SBP are presented in Table 10.4. What conclusions can be drawn from these results? The BFs for H_{1a}, H_{1b}, and H_{1d} are all smaller than 1 for both DBP and SBP. This means that the incorporation of the constraints is not supported by the data; that is, the unconstrained hypothesis is a better hypothesis than each of these three informative hypothesis. The BF for H_{1c} for DBP is 1.97, which means that the support for this hypothesis is almost 2 times stronger than the support for the unconstrained hypothesis. The BF for H_{1c} for SBP is 5.85, concluding that for SBP, H_{1c} is considered to be almost 6 times better than the unconstrained hypothesis. This is also reflected in the PMP values. For both DBP and SBP, H_{1c} received the highest PMP (.62 and .81, respectively); that is, within this set of hypotheses the gender role identification match hypothesis receives most support from the data.

The conclusion for the data described in this illustration is that participants are most responsive to the CPT when their gender role identification matches the gender relevance of the stressor. Note, once more, that the research on which this illustration is based involves much more (i.e., multiple outcome measures, two different stressor tasks, and additional measurement moments (anticipation and recovery)). The interested reader is referred to Van Well et al. [11].

10.5 Conclusion

Comparison of the match and mismatch hypotheses H_{1a}, H_{1b}, H_{1c}, and H_{1d} using a standard frequentist approach would have been quite complicated. First of all, the point of departure in classical methods is almost invariably the formulation of one or more null hypotheses. In this example, for instance, the hypothesis stating that all 12 means are equal, or null hypotheses about certain subsets of marginal means (i.e., stating no main effect of sex) would have been formulated. The usual approach would be to test against uninformative alternative hypotheses (basically stating "not H_0"), and in the case of a significant result (rejection of the null) subsequent testing of several pairwise comparisons combined with informal (subjective) evaluation of the estimated means. Several disadvantages of this approach to the evaluation of order constrained informative hypotheses have been discussed in previous chapters (cf. Chapters 2 and 5). Also a second disadvantage has been mentioned before: Even if one succeeds in evaluating each informative hypothesis against a null hypothesis, no direct confrontation of the informative hypotheses with each other is made. Possible outcomes are that each, none, or several of the order constrained hypotheses are preferred over the null. It would be very hard, if not impossible, to draw conclusions about the match and mismatch theories in such a case. The Bayesian model selection approach, however, directly evaluates each of the informative hypotheses and provides a measure for the relative support for each of them. The resulting posterior model probabilities have an easy and straightforward interpretation. The interested reader is referred to `http://www.fss.uu.nl/ms/informativehypotheses` for software for inequality constrained analysis of covariance.

References

[1] Aidman, E.V., Carroll, S.M.: Implicit individual differences: Relationships between implicit self-esteem, gender identity, and gender attitudes. European Journal of Personality, **17**, 19–36 (2003)

[2] Davis, M.C., Matthews, K.A.: Do gender-relevant characteristics determine cardiovascular reactivity? Match versus mismatch of traits and situation. Journal of Personality and Social Psychology, **71**, 527–535 (1996)

[3] Klugkist, I., Laudy, O., Hoijtink, H.: Inequality constrained analysis of variance: A Bayesian approach. Psychological Methods, **10**, 477–493 (2005)

[4] Kolk, A.M.M., Well, S. van: Cardiovascular responses across stressor phases: The match of gender and gender-role identification with the gender relevance of the stressor. Journal of Psychosomatic Research, **62**, 197–205 (2007)

[5] Lash, S.J., Eisler, R.M., Schulman, R.S.: Cardiovascular reactivity to stress in men: Effects of masculine gender role stress appraisal and masculine performance challenge. Behavior Modification, **14**, 3–20 (1990)

[6] Lash, S.J., Eisler, R.M., Southard, D.R.: Sex differences in cardiovascular reactivity as a function of the appraised gender relevance of the stressor. Behavioral Medicine, **21**, 86–94 (1995)

[7] Lash, S.J., Gillespie, B.L., Eisler, R.M., Southard, D.R.: Sex differences in cardiovascular reactivity: Effects of the gender relevance of the stressor. Health Psychology, **10**, 392–398 (1991)

[8] Martz, D.M., Handley, K.B., Eisler, R.M.: The relationship between feminine gender role stress, body image, and eating disorders. Psychology of Women Quarterly, **19**, 493–508 (1995)

[9] Rutherford, A.: Introducing ANOVA and ANCOVA: A GLM Approach. London, Sage Publications (2001)

[10] Stevens, J.: Applied Multivariate Statistics for the Social Sciences (3rd ed.). Mahwah, NJ, Lawrence Erlbaum (1996)

[11] Well, S. van, Kolk, A.M., Klugkist, I.: The relationship between sex, gender role identification, and gender relevance of a stressor on stress response: Sex and gender (mis)match effects. Behavior Modification, **32**, 427–449 (2008)

[12] Well, S. van, Kolk, A.M., Oei, N.Y.L.: Direct and indirect assessment of gender role identification. Sex Roles, **56**, 617–628, (2007)

[13] Wright, R.A., Murray, J.B., Storey, P.L., Williams, B.J.: Ability analysis of gender relevance and sex differences in cardiovascular response to behavioral challenge. Journal of Personality and Social Psychology, **73**, 405–417 (1997)

11

Inequality Constrained Latent Class Models

Herbert Hoijtink[1] and Jan Boom[2]

[1] Department of Methodology and Statistics, Utrecht University, P.O. Box 80140, 3508 TC Utrecht, the Netherlands h.hoijtink@uu.nl
[2] Department of Developmental Psychology, Utrecht University, P.O. Box 80140, 3508 TC Utrecht, the Netherlands j.boom@uu.nl

11.1 Latent Class Analysis

11.1.1 Introduction

This chapter deals with inequality constrained latent class analysis. As will be exemplified, researchers often have competing theories that can be translated into inequality constrained latent class models. After this translation it is rather straightforward to evaluate these theories.

In this section latent class models will be introduced. Introductions to latent class analysis are given in [17, 18, 27]. The specification using inequality constraints will be elaborated in the next section and is discussed further in [8, 9, 11, 13, 14].

The data that will be analyzed in this chapter are the responses $x_{ij} \in \{0,1\}$ of $i = 1, ..., N$ persons to $j = 1, ..., J$ items. The item responses are dichotomous and can represent responses like {no, yes}, {incorrect, correct}, {disagree, agree}, or {does not apply, does apply}. The main goal of latent class analysis is to determine groups of persons with similar item responses. Since it is unknown who belongs to which group, these groups are called latent classes. Two subgoals can be distinguished: determination of the number of latent classes and characterization of each latent class.

A simple example of latent class analysis is presented in Figure 11.1, where the responses of $N = 20$ persons to $J = 8$ items are presented. The meaning of the items is displayed at the top of the data matrix. As can be seen, there are two types of items: arithmetic exercises that increase in difficulty and questions with respect to the meaning of words that are also increasing in difficulty. If this data matrix is subjected to a latent class analysis, the persons are subsequently grouped into two, three, etc. classes such that persons within groups are rather homogeneous, that is having rather similar item responses, and that persons between groups are rather heterogeneous. There are several methods that can be used to decide which number of classes is optimal. Bayesian model selection is discussed in Section 11.2.2. Bayesian

12x22	32.5x34	80/(12x7)	3.7/(17x6.3)	giggle	gently	giddy	gibberish	
0	1	0	0	0	0	0	1	$\omega_1 = .10$, $\pi = \{...\}$
0	0	1	0	1	0	0	1	
1	1	1	0	0	0	0	1	$\omega_2 = .25$, $\pi = \{.8,.8,.8,.6,.4,.0,.2,.4\}$
1	0	1	1	0	0	1	0	
1	1	0	1	1	0	0	0	
1	1	1	0	1	0	0	0	
0	1	1	1	0	0	0	1	
0	0	0	0	1	1	1	0	$\omega_3 = .50$, $\pi = \{.3,.3,.1,.1,.9,.8,.7,.6\}$
1	0	0	0	1	1	1	1	
0	1	0	0	1	0	1	0	
1	1	0	0	1	1	1	1	
0	0	0	0	0	1	1	1	
1	0	0	0	1	1	0	1	
0	0	1	0	1	1	0	1	
0	0	0	0	1	0	1	1	
0	1	0	0	1	1	1	0	
1	0	0	1	1	1	0	0	
0	1	1	1	1	1	1	1	$\omega_4 = .15$, $\pi = \{...\}$
1	0	1	1	0	1	1	1	
1	1	1	1	1	1	1	0	

Fig. 11.1. A simple example of latent class analysis

goodness of fit testing is discussed in [9], and classical model selection and classical goodness of fit testing are discussed in [6] and [16], respectively.

For the simple example $Q = 4$ classes are optimal. Each of these $q = 1, \ldots, Q$ classes can be described using a class weight ω_q and class-specific probabilities π_{qj}. The class weight is an estimate of the proportion of persons belonging to this class in the population of interest. As can be seen in Figure 11.1, the first class is rather small (a weight of .10) and the third class is rather large (a weight of .50). The class-specific probability π_{qj} denotes the proportion of persons in class q responding 1 to item j. To give an example, in the third class the probabilities of responding correctly to the arithmetic exercises is rather small (smaller than .3) and the probabilities of correctly explaining the words are rather high (larger than .6). This class appears to contain the persons that have problems with arithmetic, but find it relatively easy to explain the meaning of words. An interpretation of the other classes can be found in the same way. For example, the fourth class contains persons that have both arithmetic and word explanation abilities.

The latent class analysis described in the previous paragraphs is exploratory. This means that before the analysis, both the number of classes and the meaning of the classes (in terms of the class-specific probabilities) is unknown. Although exploratory latent class analysis is a valuable technique, there are a few unresolved issues:

1. The number of classes can only be determined approximately. The interested reader is referred to studies in [6, 11, 16]. The implication is that a theoretically important latent class may not be found or that unimportant classes are included.
2. The meaning of the classes is determined after execution of the analysis. Since the human mind is flexible, even classes that may not reappear in the analysis of new data will receive an interpretation. This will lead to an incorrect description of the population from which the data were sampled.
3. Often researchers have (competing) theories with respect to the number and meaning of the latent classes. However, the results of an exploratory latent class analysis may not correspond with any of these theories. This makes it complicated to decide which of the theories is the best.

As will be elaborated in the next section, these issues can be avoided through the use of confirmatory latent class analysis, that is, latent class models enhanced via the addition of inequality constraints among the class-specific probabilities. Of course this gives rise to new issues. These will be discussed both in the next section and in Section 11.4.

11.1.2 Informative Hypotheses Specified Using Inequality Constraints

Inequality constraints can be added to a latent class model in order to specify informative hypotheses. These models were first discussed in [5] with a sequel in [26], in which inequality constraints to specify non-parametric item response models were used. Both authors use maximum likelihood to obtain estimates of the restricted parameters of the latent class model and likelihood ratio tests with bootstrapped p-values to test the fit of each model. In [8, 9, 11] Bayesian statistics is used for estimation, goodness of fit testing, and model selection of inequality constrained latent class models. The range of models discussed by these authors is more encompassing than nonparametric item response models (cf. [13, 14]). The technical details of this approach will be presented in the next section.

The inequality constraints are usually of the form

$$\pi_{qj} > \pi_{q'j'}, \text{ or, } \pi_{qj} < \pi_{q'j'}, \text{ for } j \neq j' \text{ and/or } q \neq q', \qquad (11.1)$$

although more elaborate constraints are possible [9]. Using these constraints, four competing models can be specified for the data displayed in Figure 11.1.

The first model is the latent class counterpart of a one-dimensional nonparametric item response model, also known as the model of monotone homogeneity [19, Chapters 3, 4, and 5]. This model assumes that each person can be characterized by one ability and that for each of the items the probability of a correct response increases with ability (monotone homogeneity). This model can be specified using the following constraints:

$$\pi_{1j} < \pi_{2j} < \pi_{3j} < \pi_{4j}, \text{ for } j = 1, \ldots, J; \tag{11.2}$$

stated in words, the probability of responding correctly increases with class number for each of the items. The two-dimensional counterpart (see [8]) of the monotone homogeneity model can also be formulated. Applied to the simple example, this model assumes that persons are characterized by two abilities: arithmetic ability and the ability to explain words. This model can be specified using

$$\{\pi_{1j}, \pi_{2j}\} < \{\pi_{3j}, \pi_{4j}\}, \text{ for } j = 1, \ldots, 4; \tag{11.3}$$

that is, classes 1 and 2 contain persons with a smaller arithmetic ability than classes 3 and 4. Similarly,

$$\{\pi_{1j}, \pi_{3j}\} < \{\pi_{2j}, \pi_{4j}\}, \text{ for } j = 5, \ldots, 8; \tag{11.4}$$

that is, classes 1 and 3 contain persons with a smaller ability to explain words than classes 2 and 4.

A closer look at the items reveals a possibility to elaborate the models specified in the previous paragraph. Since the items appear to differ in difficulty, the following constraints could be added to both models (see also the discussion with respect to double monotony in Chapter 6 of [19]):

$$\pi_{q1} > \pi_{q2} > \pi_{q3} > \pi_{q4}, \text{ for } q = 1, \ldots, 4 \tag{11.5}$$

and

$$\pi_{q5} > \pi_{q6} > \pi_{q7} > \pi_{q8}, \text{ for } q = 1, \ldots, 4; \tag{11.6}$$

that is, in each latent class item 1 is easier than item 2, etc.

This section provided a first illustration of the translation of competing theories into inequality constrained latent class models. The resulting models represent hypotheses with respect to whether persons use one or two abilities to answer the items and with respect to the difficulty of the items. Bayesian computational statistics can be used to estimate the parameters of each constrained model and to select the best of the models specified. The next section contains a statistical elaboration of Bayesian estimation and model selection. Readers who are less interested in the technical details can skip the next section and continue in Section 11.3.

11.2 Parameter Estimation and Model Selection

11.2.1 Likelihood, Prior, and Posterior Distribution

The first step in Bayesian estimation and model selection is the specification of the likelihood function of the statistical model at hand and the specification of the prior distribution of the parameters of this model. The multiplication of both renders the posterior distribution, which is the point of departure for both parameter estimation and model selection.

The likelihood function of the latent class model introduced before is

$$f(\boldsymbol{x}_1, \ldots, \boldsymbol{x}_N \mid \boldsymbol{\omega}, \boldsymbol{\pi}_1, \ldots, \boldsymbol{\pi}_Q) = \prod_{i=1}^{N} \sum_{q=1}^{Q} P(\boldsymbol{x}_i \mid \theta_i = q)\omega_q, \tag{11.7}$$

where θ_i denotes the class membership of person i and

$$P(\boldsymbol{x}_i \mid \theta_i = q) = \prod_{j=1}^{J} \pi_{qj}^{x_{ij}} (1 - \pi_{qj})^{1-x_{ij}}. \tag{11.8}$$

The prior distribution of the latent class model is

$$p(\boldsymbol{\omega}, \boldsymbol{\pi}_1, \ldots, \boldsymbol{\pi}_Q \mid H_t) \propto I_{\boldsymbol{\omega}, \boldsymbol{\pi}_1, \ldots, \boldsymbol{\pi}_Q \in H_t}, \tag{11.9}$$

where H_t denotes the set of inequality constraints used to specify the t-th model. The indicator function $I_{\boldsymbol{\omega}, \boldsymbol{\pi}_1, \ldots, \boldsymbol{\pi}_Q \in H_t} = 1$ if the values of the class weights and class-specific probabilities are in accordance with the restrictions in model t, and 0 otherwise. This prior is uninformative because a priori all parameter vectors in agreement with the constraints of model t are equally likely. This prior is chosen for convenience and less well founded than the prior distributions discussed in Chapters 4, 6, 7, and 8. Those priors were chosen after careful thinking (encompassing priors), using a part of the data (intrinsic priors), or teststatistics are used to compute the Bayes factor, thus avoiding the need for prior distributions. Although [9] and [11] show that inferences based on the prior in (11.9) have good properties, more research into the choice of priors for inequality constrained latent class models is needed.

The posterior distribution of the latent class model is proportional to the product of likelihood and prior: $p(\boldsymbol{\omega}, \boldsymbol{\pi}_1, \ldots, \boldsymbol{\pi}_Q \mid \boldsymbol{x}_1, \ldots, \boldsymbol{x}_N) \propto f(\cdot)p(\cdot)$. A data-augmented Gibbs sampler (cf. [28]) can be used to obtain a sample from the posterior distribution of the latent class model. Using this sample, parameter estimates, posterior standard deviations, and central credibility intervals can easily be computed (cf. [10]). The Gibbs sampler is an iterative sequence across three steps preceded by an initialization:

- Initialization: Assign initial values to the class weights and class-specific probabilities. Any set of values that is in agreement with the constraints imposed by the model at hand is allowed.

- Step 1: Data augmentation. For $i = 1, ..., N$, sample class membership θ_i from its posterior distribution conditional upon the current values of $\boldsymbol{\omega}, \boldsymbol{\pi}_1, \ldots, \boldsymbol{\pi}_Q$:

$$p(\theta_i = q \mid \boldsymbol{\omega}, \boldsymbol{\pi}_1, \ldots, \boldsymbol{\pi}_Q, \boldsymbol{x}_1, \ldots, \boldsymbol{x}_N) = \frac{P(\boldsymbol{x}_i \mid \theta_i = q)\omega_q}{\sum_{q=1}^{Q} P(\boldsymbol{x}_i \mid \theta_i = q)\omega_q}, \quad (11.10)$$

  that is, a multinomial distribution with probabilities $p(\theta_i = q \mid \cdot)$ for $q = 1, \ldots, Q$.
- Step 2: Sample π_{qj} for $j = 1, \ldots, J$ and $q = 1, \ldots, Q$ from its posterior distribution conditional on the current values of θ_i for $i = 1, \ldots, N$, and the constraints imposed by the model at hand:

$$p(\pi_{qj} \mid \boldsymbol{\theta}, x_{1j}, \ldots, x_{Nj}, L, U) \propto Beta(s_{qj} + 1, N_q - s_{qj} + 1, L, U), \quad (11.11)$$

  where s_{qj} denotes the number of persons currently allocated to class q that responded 1 to item j, N_q denotes the number of persons currently allocated to class q, and L and U denote the lower and upper bound, respectively, on π_{qj} resulting from the constraints to which π_{qj} is subjected in the model at hand. Note that this conditional distribution is independent of $\boldsymbol{\omega}$ because of the conditioning on $\boldsymbol{\theta}$. Using inverse probability sampling, it is easy to sample a deviate from this truncated beta distribution: (a) sample a random number ν from a uniform distribution on the interval [0,1]; (b) compute the proportions α and β that are not admissible due to L and U:

$$\alpha = \int_0^L Beta(\pi_{qj} + 1, N_q - S_{qj} + 1) d\pi_{qj} \quad (11.12)$$

  and,

$$\beta = \int_U^1 Beta(\pi_{qj} + 1, N_q - S_{qj} + 1) d\pi_{qj}; \quad (11.13)$$

  (c) compute π_{qj} such that it is the deviate associated with the ν-th percentile of the admissible part of the posterior of π_{qj}:

$$\alpha + \nu(1 - \alpha - \beta) = \int_0^{\pi_{qj}} Beta(\pi_{qj} + 1, N_q - S_{qj} + 1) d\pi_{qj}. \quad (11.14)$$

- Step 3: Sample $\boldsymbol{\omega}$ from its distribution conditional on the current value of $\boldsymbol{\theta}$:

$$p(\boldsymbol{\omega} \mid \boldsymbol{\theta}) \propto Dirichlet(N_1 + 1, \ldots, N_Q + 1). \quad (11.15)$$

  Note that this conditional distribution is independent of $\boldsymbol{\pi}_1, \ldots, \boldsymbol{\pi}_Q$ because of the conditioning on $\boldsymbol{\theta}$. Using algorithm DIR-2 from [20], it is relatively easy to sample from a Dirichlet distribution subject to the constraint $\sum_{q=1}^{Q} \omega_q = 1$: (a) for $q = 1, \ldots, Q$, sample a random variable z_q from a gamma distribution with parameters $N_q + 1$ and 1; (b) compute $\omega_q = z_q / \sum_{q=1}^{Q} z_q$ for $q = 1, \ldots, Q$.

As explained in Chapter 3, after convergence the Gibbs sampler renders a sample from the posterior distribution of interest. This implies that after the deletion of a burn-in period, the sample can be used to: estimate the parameters of the latent class model (for each parameter the average of the values sampled); compute central credibility intervals (e.g., for each parameter the 5-th and 95-th percentile of values sampled); and, as will be elaborated in the next section, compute the marginal likelihood for each model under investigation. For all examples in this chapter the Gibbs sampler is run for 110,000 iterations. The first 10,000 iterations serve as a burn-in period and are discarded; subsequently the parameters sampled in every 100-th iteration are saved and used for parameter estimation and computation of the marginal likelihood.

11.2.2 Marginal Likelihood and Posterior Probabilities

Like in most of the other chapters, model selection will be based on the marginal likelihood and posterior probabilities. The method used to compute the marginal likelihood (and subsequently the Bayes factor and posterior probabilities) is unlike the methods presented in Chapters 4, 6, 7, and 8. Let $\boldsymbol{\xi}_t = (\boldsymbol{\omega}, \boldsymbol{\pi}_1, \ldots, \boldsymbol{\pi}_Q)$; then the marginal likelihood of a (constrained) latent class model is given by

$$m(\boldsymbol{x}_1, \ldots, \boldsymbol{x}_N \mid H_t) = \int_{\xi_t} f(\boldsymbol{x}_1, \ldots, \boldsymbol{x}_N \mid \boldsymbol{\xi}_t) Pr(\boldsymbol{\xi}_t \mid H_t) d\boldsymbol{\xi}_t. \tag{11.16}$$

The computation of the marginal likelihood is based on an idea that can be found in [12] and [21]. They suggested using importance sampling to approximate the integral in (11.16) using a large sample (e.g. 99%) of parameter vectors from the posterior distribution and to imagine that a small sample (e.g., 1%) comes from the prior distribution each with a density equal to the marginal likelihood. An estimate $\log \widehat{m}$ of $\log m(\boldsymbol{x}_1, \ldots, \boldsymbol{x}_N \mid H_t)$ can then be obtained via a simple iterative procedure (see [11] for a fast algorithm with known precision) based on the following implicit equation:

$$\log \widehat{m} = \log \frac{.01B\widehat{m} + \sum_b^{.99B} \frac{f(\boldsymbol{x}_1,\ldots,\boldsymbol{x}_N \mid \boldsymbol{\xi}_t)}{.01+.99f(\boldsymbol{x}_1,\ldots,\boldsymbol{x}_N \mid \boldsymbol{\xi}_t)/\widehat{m}}}{.01B + \sum_b^{.99B} \frac{1}{.01+.99f(\boldsymbol{x}_1,\ldots,\boldsymbol{x}_N \mid \boldsymbol{\xi}_t)/\widehat{m}}}, \tag{11.17}$$

where $b = 1, \ldots, B$ denotes the number of iterations of the Gibbs sampler after burn-in.

If each of the models under investigation, $t = 1, \ldots, T$, has an equal prior probability of $1/T$, then posterior probabilities for each of the models are easily computed using

$$P(H_t \mid \boldsymbol{x}_1, \ldots, \boldsymbol{x}_N) = \frac{BF_{t1}}{BF_{11} + \ldots BF_{T1}}, \text{ for } t = 1, \ldots, T, \tag{11.18}$$

where the Bayes factor BF_{t1} is defined by

$$BF_{t1} = \frac{m(\cdot \mid H_t)}{m(\cdot \mid H_1)}, \text{ if } t \neq 1, \tag{11.19}$$

and $BF_{11} = 1$.

11.2.3 Performance of the Estimate of the Marginal Likelihood

In this section a small simulation study will be used to obtain an indication of the performance of the marginal likelihood. A more elaborate study can be found in [11]. There exist numerous examples of competing models where the marginal likelihood will almost always select the correct model. This happens if theories are not nested; that is, this happens if it is not so that one theory equals the other theory with extra restrictions added. In the next section, two examples will be given. In these cases it is very easy to determine which theory is correct, since usually one of the sets of constraints is clearly wrong.

However, consider the following three nested models:

1. An unconstrained latent class model with three classes (subsequently denoted by U3).
2. A monotone homogeneous three class model like (11.2), that is, $\pi_{1j} < \pi_{2j} < \pi_{3j}$ for $j = 1, \ldots, 10$ (subsequently denoted by M3).
3. A double monotonous three class model like the combination of (11.2) and (11.5), that is, $\pi_{1j} < \pi_{2j} < \pi_{3j}$ for $j = 1, \ldots, 10$, and $\pi_{q1} > \ldots > \pi_{q10}$ for $q = 1, \ldots, 3$ (subsequently denoted by D3).

If a dataset is used to determine the empirical support for each of these models, it is rather difficult to determine which is the best model. If the third model is correct, the second and first models are also correct (but less parsimonious, less informative, and thus less desirable). If the second model is correct, so is the first model. Stated otherwise, finding the best of a number of nested models is a real challenge for the estimate of the marginal likelihood (11.17). If the marginal likelihood performs satisfactorily in this context, it is likely that it will also perform satisfactorily in the context of non-nested or partly nested models.

In Table 11.1 the population used to generate 10 data matrices containing the responses of 500 persons to 10 items is described. As can be seen, the population is in agreement with hypothesis M3. Each of the 10 data matrices was analyzed using U3, M3, and D3. The results are displayed in Table 11.2.

The goal of this simulation was to determine if M3 is preferred over U3 and whether D3 is disqualified. It is clear that according to $\log \widehat{m}$, D3 has to be disqualified: It has clearly smaller values for $\log \widehat{m}$ than the other two models. For 7 data matrices, $\log \widehat{m}$ was between 1 and 2.5 larger for M3 than for U3 (indicating preference for M3); for the other 3 data matrices it was between 1 and 1.5 smaller.

Table 11.1. Description of the population used to simulate 10 data matrices

	π_{i1}	π_{i2}	π_{i3}
1	.20	.55	.75
2	.25	.50	.80
3	.20	.55	.75
4	.25	.50	.80
5	.20	.55	.75
6	.25	.50	.80
7	.20	.55	.75
8	.25	.50	.80
9	.20	.55	.75
10	.25	.50	.80
ω	.20	.30	.50

From this simulation study the following conclusions can be obtained (see [11] for a further elaboration):

1. If a constrained model is not in accordance with the population from which the data were generated, it will have a smaller $\log \widehat{m}$ than the corresponding unconstrained model (compare the results for U3 and D3 in Table 11.2).
2. If a constrained model is in accordance with the population from which the data were generated, $\log \widehat{m}$ is often larger for the constrained model (compare the results for U3 and M3 in Table 11.2).
3. If one of two nested constrained models is correct, it will have a larger $\log \widehat{m}$ than the other model (compare the results for M3 and D3 in Table 11.2).
4. Based on experience with applications so far (see also the two examples presented in the next section), $\log \widehat{m}$ will clearly indicate which of two non-nested constrained models is the best.

Table 11.2. Log $\widehat{m}$ for analyses of the 10 simulated data matrices

Matrix	U3	M3	D3
1	−3189.31	−3188.80	−3203.70
2	−3177.96	−3179.07	−3189.74
3	−3122.71	−3121.62	−3145.39
4	−3204.65	−3202.39	−3216.77
5	−3165.45	−3164.31	−3178.37
6	−3163.79	−3164.55	−3176.30
7	−3188.22	−3184.88	−3203.91
8	−3114.56	−3113.69	−3139.60
9	−3144.34	−3141.95	−3151.72
10	−3141.23	−3142.69	−3152.11

As can be seen, $\log \widehat{m}$ performs rather good as a model selection criterion. Only the comparison of two nested models that are both correct is less than perfect. This is caused by the manner in which $\log \widehat{m}$ is computed: If, for example, an unconstrained and a constrained model are both correct (like U3 and M3 in the simulation), the constrained model ought to be preferred since it is more parsimonious. However, looking at (11.17), it can be seen that $\log \widehat{m}$ is computed using a sample from the posterior distribution of the model at hand. If the constrained model is correct, the sample from the posterior distribution of the *unconstrained model* will not often contain parameter values that are not in agreement with the *constrained model* (these regions of the parameter space will have a rather small posterior density if the constrained model is correct). Consequently, the samples obtained for both models will be rather similar and thus the values for $\log \widehat{m}$ will be rather similar. Stated otherwise, in this situation $\log \widehat{m}$ is biased against the constrained model.

Consequently, if $\log \widehat{m}$ is used to compare two nested models, the following decision rule can be used:

1. If $\log \widehat{m}$ is larger (even if only slightly) for the most constrained model, it should be preferred to the less constrained model.
2. If $\log \widehat{m}$ is only slightly smaller for the most constrained model, it should still be preferred to the less constrained model. In the simulation study "slightly" should be about 2.

Currently, research is in progress that will render an estimator of the marginal likelihood that is not biased in the situation discussed above. However, if the above is carefully considered, researchers should be able to select the best of a number of (un)constrained latent class models using (11.17).

11.3 Examples

11.3.1 Introduction

In the previous section the technical details of Bayesian estimation and selection of inequality constrained latent class models were elaborated. In this introduction to the example section, a nontechnical elaboration of Bayesian model selection will be given such that readers who skipped the previous section will understand the meaning of two quantities that will be used in the sequel: the marginal likelihood and posterior probabilities. Both will be introduced in the context of a simple example with two probabilities. The interested reader is referred to [11] for a simple example with only one probability.

Table 11.3 contains the responses of 10 hypothetical persons to two hypothetical items x_1 and x_2. The probabilities of a positive response will be denoted by π_1 and π_2, respectively. The hypotheses under investigation are $H_1 : \pi_1, \pi_2$ (i.e., there is no theory about the relative sizes of both probabilities) and $H_2 : \pi_1 > \pi_2$.

Table 11.3. Responses of 10 hypothetical persons to two hypothetical items

Items	Responses									
x_1	1	1	0	1	1	0	1	1	0	0
x_2	0	0	1	1	0	0	0	1	0	1

The marginal likelihood is a measure of the degree of support for a hypothesis provided by the data. In order to be able to compute a marginal likelihood, two ingredients are needed: the prior distribution and the likelihood of both probabilities. The square in Figure 11.2 represents the prior distribution of H_1. It is a uniform distribution; that is, a priori (before observing the data) each combination of values of π_1 and π_2 is equally likely. The lower right-hand triangle denotes the prior distribution for H_2; that is, a priori each combination of values of π_1 and π_2 is equally likely, as long as $\pi_1 > \pi_2$.

The ellipses in Figure 11.2 are so-called isodensity contours of the likelihood of π_1 and π_2. If $\pi_1 = .6$ and $\pi_2 = .4$ (i.e.,if both probabilities are equal to the probabilities of a positive response observed in the data) the maximum of the likelihood is obtained (the center of the smallest ellipse). The further the values of π_1 and π_2 are located from the maximum, the smaller the likelihood (the larger an ellipse the smaller the likelihood), that is, the smaller the support in the data for the values of π_1 and π_2 at hand.

Stated in words, the marginal likelihood (see also (11.16)) is the likelihood integrated with respect to the prior distribution of the hypothesis at hand.

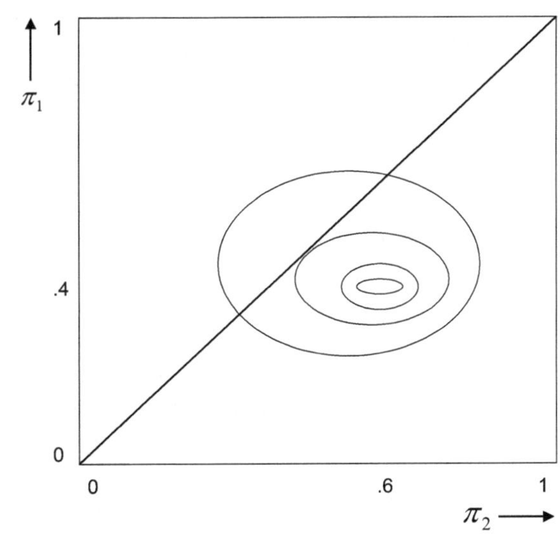

Fig. 11.2. A visual elaboration of the marginal likelihood

If uniform prior distributions are used (as is the case for H_1 and H_2), this statement can be simplified to the following: The marginal likelihood is the average height of the likelihood with respect to the prior of the hypothesis at hand. As can be seen in Figure 11.2, for H_1 the marginal likelihood is an average over the whole square. This implies an average based on many relatively large values (most of the likelihood is located in the lower right-hand triangle), but also many relatively small values (only a small part of the likelihood is located in the upper left-hand triangle). For H_2 the marginal likelihood is based on an average over the likelihood values in the lower right-hand triangle only; that is, the marginal likelihood $m(\boldsymbol{x} \mid H_2)$ of H_2 is larger than the marginal likelihood $m(\boldsymbol{x} \mid H_1)$ of H_1. Consequently, H_2 is a better model than H_1 because it is better supported by the data.

The marginal likelihood is an automatic Occam's razor (see also [12]). As can be seen in Figure 11.2, neither hypothesis is in contradiction with the data; that is, the fit of both hypotheses is similar. However, H_2 is more parsimonious than H_1 because due to the inequality constraint in H_2 a part of the parameter space is excluded. To determine the support of the data for a set of hypotheses, the marginal likelihood considers both fit and parsimoniousness of the hypotheses. Since both hypotheses in this simple example have a similar fit, H_2 is better than H_1 because it is more informative; that is, because it is more parsimonious.

The values of two (or more) marginal likelihoods can only be interpreted in relation to each other. With respect to the example at hand, $m(\boldsymbol{x} \mid H_2) = .060$ and $m(\boldsymbol{x} \mid H_1) = .035$; that is, H_2 is a better model than H_1. Assuming that a priori both hypotheses are considered to be equally likely, the Bayes factor (11.19) of both models is $BF_{21} = .060/.035 = 1.71$; that is, a posteriori (after observing the data) H_2 has become 1.71 times as likely as H_1. The easiest way to interpret marginal likelihoods is to translate them to posterior probabilities; that is, the probability that the hypothesis at hand is the best of the set of hypotheses under consideration after observing the data. For the hypotheses at hand the posterior probabilities (11.18) are $P(H_2 \mid \boldsymbol{x}) = 1.71/(1+1.71) = .63$ and $P(H_1 \mid \boldsymbol{x}) = 1.71/(1+1.71) = .37$. In the sequel both the marginal likelihood and posterior probabilities will be used to quantify the support in the data for the hypotheses under investigation.

11.3.2 Masculinity and Femininity

A contribution to an ongoing discussion with respect to the sex role stereotypes "masculinity" and "femininity" was given by [23]. Traditionally, masculinity and femininity are considered to be different extremes of the same (bipolar) dimension [15, 25]. However, this was criticized in [4] because there were no relevant variables that were related to either this bipolar dimension, masculinity or femininity. The author suggested that masculinity and femininity should be measured in a different way.

Table 11.4. Two theories translated in inequality constrained latent class models

Bipolar					Two-Dim.			
1	2	3	4	Item	N	M	F	A
++	+	−	−−	Vulnerable (F)	−	−	+	+
++	+	−	−−	Changeable (F)	−	−	+	+
++	+	−	−−	Intuitive (F)	−	−	+	+
++	+	−	−−	Emotional (F)	−	−	+	+
−−	−	+	++	Selfassured (M)	−	+	−	+
−−	−	+	++	Exact (M)	−	+	−	+
−−	−	+	++	Individualistic (M)	−	+	−	+
−−	−	+	++	Rational (M)	−	+	−	+

Ten stereotypical masculine and 10 feminine characteristics from the Groninger Androgyny Scale [7] were analyzed in [23]. This scale was constructed such that masculinity and femininity could be measured as separate dimensions (implying that persons could be both, none, or one of both), and not as the endpoints of a bipolar dimension. Eight of these characteristics (the top four labeled (F) are feminine, the bottom four labeled (M) are masculine) can be found in Table 11.4. As can be seen, the Groninger Androgyny Scale contains many masculine characteristics that are usually positively evaluated and feminine characteristics that are usually evaluated less positive or even negative. This is in accordance with the findings in [3] with respect to the feminine and masculine stereotype.

The items in Table 11.4 were scored by 166 students from the University of Groningen for a benefit of 10 guilders. The students had an average age of 22.4 years, 88 of the students were female, and 78 were male. The responses were originally scored on a 4-point scale. For the purpose of this chapter the responses were recoded to 0: the characteristic does not apply to me, and 1: the characteristic does apply to me.

In Table 11.4 the theories that "femininity and masculinity are the end points of a bipolar dimension" and "feminity and masculinity are two separate dimensions" are represented by an inequality constrained latent class model. Class-specific probabilities are represented by $--$, $-$, $+$, and $++$ such that $-- < - < + < ++$.

In the panel labeled "Bipolar" the restrictions are only within items (that is, within a row of the table) and not between items (that is, not between rows of the table). The numbers 1 through 4 refer to four latent classes that are ordered along a bipolar dimension ranging from "feminine" to "masculine". The class-specific probabilities are restricted such that the probability of choosing feminine characteristics is decreasing from class 1 to 4 and the probability of choosing masculine characteristics is increasing from class 1 to 4. This implies that feminine persons will be allocated to class 1 and have relatively high probabilities of choosing the feminine items and relatively small probabilities

Table 11.5. Log marginal likelihood and posterior model probabilities (PMPs)

Theory	$\log \widehat{m}$	PMP
Bipolar	−783.11	.013
Two-dimensional	−778.74	.987

of choosing the masculine items and that masculine persons will be allocated to class 4 with relatively high probabilities of choosing masculine items and relatively small probabilities of choosing feminine items.

In the panel labeled "Two-Dim." the first class contains the persons that are neither feminine nor masculine. The second class contains the masculine persons (the probabilities for the masculine characteristics are larger than the probabilities for the feminine characteristics in the same class and larger than all the probabilities in the first class). The third class contains the feminine person, and the fourth class the androgynous persons.

As can be seen in Table 11.5 the posterior probabilities show clearly that the support in the data is larger for the two-dimensional model than for the bipolar model. This is in accordance with the results obtained (using different methods) in [23]. In Table 11.6 the estimates of the parameters of the two-dimensional model are displayed. As can be seen, the estimates are in accordance with the constraints: In the masculine class the masculine items have high class-specific probabilities, in the feminine class the feminine items have high class-specific probabilities, and in the androgenous class all items have high class-specific probabilities. The number of persons in the androgenous class is rather high (55%). Since the sample of persons consisted of students who (at least in 1990) have less inhibition to show both feminine and masculine characteristics than the average person, this is not surprising.

Table 11.6. Estimates for the two-dimensional theory

Item	None	Masc.	Fem.	Andro.
Vulnerable (F)	.23	.30	.84	.84
Changeable (F)	.36	.27	.64	.65
Intuitive (F)	.41	.30	.82	.71
Emotional (F)	.24	.31	.91	.85
Selfassured (M)	.50	.78	.50	.76
Exact (M)	.21	.82	.26	.69
Individualistic (M)	.32	.83	.54	.86
Rational (M)	.25	.89	.35	.87
ω	.04	.17	.24	.55

Table 11.7. The eight balance items

Left Side				Item	Right Side			
Dist.	Weight	Add	Torque		Dist.	Weight	Add	Torque
2	3	5	6	1	4	1	5	4
2	4	6	8	2	4	1	5	4
3	1	4	3	3	1	4	5	4
3	2	5	6	4	4	1	5	4
1	3	4	3	5	3	1	4	3
1	4	5	4	6	2	2	4	4
3	2	5	6	7	2	3	5	6
4	1	5	4	8	2	2	4	4

11.3.3 The Balance Scale Task of Piaget

A well-known experiment from the psychological literature is the balance scale task of Piaget (cf. [1]). A picture of a simplified balance scale is shown to children. While the beam is fixed, a number of identical weights are placed on each side at certain distances from the fulcrum. For each of a number of balances (the items) the children have to predict which side will tip, if any. The weights on the balance differ with respect to their number and distance to the center. The formal (torque) rule to obtain the correct answer is that the balance is in equilibrium when the product of the number of weights and the distance from the center is equal at both sides of the balance. The interested reader can surf to `http://zap.psy.utwente.nl/english` and execute this task himself.

Here 8 items from the 19 balance scale items previously analyzed in [14] are used to compare 2 theories with respect to the strategies that children use to come up with a solution for each of the balance items: Siegler's theory [24] and an adjustment by Normandeau et al. [22]. The items are described in Table 11.7. Two types of items can be found in this table:

- Items 1 through 4 are "conflict weight" items. These are items for which both sides differ in distance and weight, and the correct answer is that the side with the most weight goes down.
- Item 5 through 8 are "conflict balance" items. For these items both sides also differ with respect to distance and weight. However, overall, the balance is in equilibrium.

The items were responded to by 887 Dutch children ranging in age from 4 until 16 years. The data were collected by students from the Department of Developmental Psychology of Utrecht University using a strict protocol of how to administer the balance scale task to the children.

According to Siegler's theory [24], children use one of three strategies to respond to these items:

Table 11.8. Two theories translated in inequality constrained latent class models

Siegler				Normandeau et al.			
Rule 1	Rule 3	Torque	Item	Rule 1	Add	Qual.	Torque
+	±	+	1,4	+	−	−	+
+	±	+	2,3	+	+	−	+
−	±	+	6,8	−	−	+	+
−	±	+	5,7	−	+	+	+

- Siegler's "Rule 1" states that children will look at the number of weights on each arm and will predict that the arm with the largest number of weights will go down. This will lead to correct answers to items 1 through 4 and to incorrect answers to items 5 through 8.
- Siegler's "Rule 3" states that children look at both distance and weight. However, if both vary, they will give a random prediction. This will lead to random answers to each of the eight items presented in Table 11.7.
- Children can also use the "Torque Rule"; in that case they will provide the correct answer to each of the items.

Note that Siegler's Rule 2 does not apply to the item set displayed in Table 11.7. To distinguish Rule 2 from Rule 1, items are needed that have an equal weight at each side of the balance. Consequently, Rule 2 will not be considered in the sequel.

According to Normandeau et al. [22] and Boom and ter Laak [2], Siegler's Rule 3 should be replaced by two other rules:

- The "Addition Rule" states that children will add the distance and weight for each side of the balance and predict that the side with the largest sum will go down. As can be seen in Table 11.7 this will render the correct response for items 2, 3, 5 and 7.
- The "Qualitative Proportion Rule" states that children understand that a heavy weight at a small distance from the fulcrum compensates for a small weight at a large distance from the fulcrum. The children will predict "balance" for all conflict problems and, consequently, give the correct answer to items 5 through 8.

As displayed in Table 11.8 the theories of Siegler and Normandeau et al. can be translated into inequality constrained latent class models. The class-

Table 11.9. Log marginal likelihood and posterior model probabilities (PMPs)

Theory	$\log \widehat{m}$	PMP
Siegler	−2834.61	.001
Normandeau et al.	−2793.78	.999

Table 11.10. Estimates for the theory of Normandeau et al.

Item	Rule 1	Add	Qual.	Torque
1	.96	.11	.17	.84
2	.99	.93	.17	.96
3	.97	.85	.13	.82
4	.94	.06	.17	.46
5	.00	.38	.27	.23
6	.03	.19	.34	.34
7	.05	.60	.28	.64
8	.07	.12	.26	.61
ω	.57	.11	.05	.27

specific probabilities for combinations of items and classes are represented by $-, \pm$, and $+$ with the restriction that $- < \pm < +$. To give an example, the children in class Rule 1 have a higher probability to respond correctly to the conflict weight items than to the conflict distance items. Table 11.9 presents the marginal likelihood and posterior probabilities for both models. As can be seen the support in the data is clearly in favor of the model proposed by Normandeau et al. [22]. Stated otherwise, according to data the Addition and Qualitative Proportion rules are better representations of the strategies that children use to answer to the items of the balance scale task than Rule 3. Estimates of the class-specific probabilities according to the restriction listed in the right-hand panel of Table 11.8 can be found in Table 11.10. It would have been nice if some of the large probabilities would have been larger (e.g., the probabilities for items 5, 6, 7, and 8 in the Qualitative Proportion class, and the probabilities of items 5 and 6 in the Torque class). Apparently four classes/response strategies are not yet enough to render a clear Torque class (high probabilities for all the items) and a somewhat more pronounced Qualitative Proportion class. The interested reader is referred to [14], in which more pronounced classes were obtained via the addition of two unrestricted classes (that clarify the structure in the data by taking out some of the noise) to the model of Normandeau et al.

11.4 Discussion: Exploratory and Informative Latent Class Analysis

In the first section of this chapter three unresolved issues related to exploratory latent class analysis were listed: The number of classes can only be determined approximately; the meaning of the classes has to be determined after execution of the analysis; and the results of an exploratory analysis may not correspond with any of the theories a researcher has in mind.

These issues are to a large extend solved using informative latent class analysis. Here a researcher will translate each of a set of competing theories into an inequality constrained latent class model. This avoids the third issue: A researcher does not have to figure out which theory has the closest correspondence with the results of an exploratory analysis; he straightforwardly gets the support of the data for each theory expressed as a posterior probability. It also avoids the first and the second issue because the researcher determines both the number of classes needed to adequately represent a theory and, using inequality constraints among the class-specific probabilities, he characterizes each class.

Of course this does not mean that there are no unresolved issues if a researcher decides to use informative latent class analysis. Here the focus will be on two methodological issues. The first issue can be phrased in questions like "What if my set of models does not contain the best model?", or "Can I evaluate as many models as I like?". The fear expressed in these questions is that an informative analysis with a limited number of competing models may not render the model that gives the best description of *the data at hand.* The fear is justified; an informative analysis will not render the best model for the data at hand. However, the questions should not address *the data at hand*, but *the population from which the data were sampled.* The answer to the question whether an informative analysis will render the best model for the population from which the data were sampled is clear: no. The best model is a model with the correct number of classes, and the class weights and the class-specific probabilities fixed at their population value. However, informative analysis will render a good model for the population from which the data were sampled; that is, if one believes that the following assumption is true: researchers know their research domain and are able to come up with a number of competing theories of which at least one is a more or less adequate description of the truth.

The second issue is the degree of precision with which a theory has to be specified in an inequality constrained latent class model. To give an example, the latent class containing the children applying the Torque rule in the balance scale task could be specified more precisely if the restriction $\pi_{4j} > .90$ for $j = 1, \ldots, J$ is added; that is, in the Torque class, the children have probabilities larger than .90 to correctly respond to each of the items. The question then becomes which specification better represents the theory of [22]. The answer to the question depends on the researcher executing an informative analysis (and of course on what his peers think). First of all, a theory should be clearly recognizable from the constraints imposed on a latent class model. Second, researchers should translate their theories into constrained latent class models such (that is, make them precise enough) that the data can be used to select the best of the models constructed. This makes translation of theories into constrained latent class models a rather subjective activity. Stated otherwise, the prior distributions formulated for each model are subjective and this makes selection of the best of a set of informative hypotheses a truly Bayesian way of making inferences.

Consequently, the choice between exploratory and informative latent class analysis is at the heart a choice between the classical and Bayesian way of making inferences. It is up to each researcher to decide whether he finds it easier to deal with the issues associated with exploratory or informative latent class analysis. Software for inequality constrained latent class analysis can be found at `http://www.fss.uu.nl/ms/informativehypotheses`.

References

[1] Boom, J., Hoijtink, H., Kunnen, S.: Rules in the balance. Classes, strategies or rules for the balance scale task. Cognitive Development, **16**, 717–735 (2001)

[2] Boom, J., Laak, J.F. ter: Classes in the balance. Latent class analysis and the balance scale task. Developmental Review, **27**, 127–149 (2007)

[3] Broverman, I.K., Vogel, S.R., Broverman, D.M., Clarkson, F.E., Rosenkrantz P.S.: Sex role stereotypes: A current appraisal. Journal of Social Issues, **28**, 59–78 (1972)

[4] Constantinople, A.: Masculinity-Femininity: An exception to a famous dictum? Psychological Bulletin, **80**, 389–407 (1973)

[5] Croon, M.A.: Latent class analysis with ordered latent classes. British Journal of Mathematical and Statistical Psychology, **43**, 171–192 (1990)

[6] Everitt, B.S.: A Monte Carlo investigation of the likelihood ratio test for number of classes in latent class analysis. Multivariate Behavioral Research, **23**, 531–538 (1988)

[7] Graaf, A. de: Konstructie van een verbeterde versie van de Groninger Androgynie Schaal (GRAS) [Construction of an improved version of the Groninger Androgyny Scale]. Heijmans Bulletin, HB-74-710-EX, University of Groningen, Faculty of Social Sciences (1984)

[8] Hoijtink, H., Molenaar I.W.: A multidimensional item response model: Constrained latent class analysis using the Gibbs sampler and posterior predictive checks. Psychometrika, **62**, 171–190 (1997)

[9] Hoijtink, H.: Constrained latent class analysis using the Gibbs sampler and posterior predictive p-values: Applications to educational testing. Statistica Sinica, **8**, 691–712 (1998)

[10] Hoijtink, H.: Posterior inference in the random intercept model based on samples obtained with Markov chain Monte Carlo methods. Computational Statistics, **3**, 315–336 (2000)

[11] Hoijtink, H.: Confirmatory latent class analysis: model selection using Bayes factors and (pseudo) likelihood ratio statistics. Multivariate Behavioral Research, **36**, 563–588 (2001)

[12] Kass, R.E., Raftery, A.E.: Bayes factors. Journal of the American Statistical Association, **90**, 773–795 (1995)

[13] Laudy, O., Zoccolillo, M., Baillargeon, R., Boom, J., Tremblay, R., Hoijtink, H.: Applications of confirmatory latent class analysis in developmental psychology. European Journal of Developmental Psychology, **2**, 1–15 (2005)

[14] Laudy, O., Boom, J., Hoijtink, H.: Bayesian computational methods for inequality constrained latent class analysis. In: Ark, A. van der, Croon, M., Sijtsma, K. (eds) New Developments in Categorical Data Analysis for the Social and Behavioral Sciences. Mahwah, NJ, Erlbaum (2004)
[15] Lewin, M.: In the Shadow of the Past: Psychology Portrays the Sexes. New York, Columbia University Press (1984)
[16] Lin, T.H., Dayton, C.M.: Model selection information criteria for non-nested latent class models. Journal of Educational and Behavioral Statistics, **22**, 249–268 (1997)
[17] Magidson, J., Vermunt, J.K.: Latent class models. In: Kaplan D, (ed) The Sage Handbook of Quantitative Methodology for the Social Sciences. London, Sage (2004)
[18] McCutcheon, A.L.: Latent Class Analysis. London, Sage (1987)
[19] Molenaar, I.W., Sijtsma, K.: Introduction to Nonparametric Item Response Theory. London, Sage (2002)
[20] Narayanan, A.: Computer generation of Dirichlet random vectors. Journal of Statistical Computation and Simulation, **36**, 19–30 (1990)
[21] Newton, M.A., Raftery, A.E.: Approximate Bayesian inference with the weighted likelihood bootstrap. Journal of the Royal Statistical Society, B, **56**, 3–48 (1994)
[22] Normandeau, S., Larivee, S., Roulin, J., Longeot, F.: The balance scale dilemma: Either the subject or the experimenter muddles through. Journal of Genetic Psychology, **150**, 237–250 (1989)
[23] Sanders, K., Hoijtink, H.: Androgynie bestaat [Androgyny exists]. Nederlands Tijdschrift voor de Psychologie, **47**, 123–133 (1992)
[24] Siegler, R.S.: Developmental sequences within and between concepts. Monographs of the Society for Research in Child Development, **46**, 2 (Serial No. 189) (1981)
[25] Terman, L., Miles, C.C.: Sex and Personality. Studies in Masculinity and Femininity. New York, McGraw-Hill (1936)
[26] Vermunt, J.K.: The use of restricted latent class models for defining and testing nonparametric and parametric IRT models. Applied Psychological Measurement, **25**, 283–294 (2001)
[27] Vermunt, J.K., Magidson, J.: Latent class analysis. In: Lewis-Beck, M., Bryman, A., Liao, T.F. (eds) The Sage Encyclopedia of Social Science Research Methods. London, Sage (2004)
[28] Zeger, S.L., Karim, M.R.: Generalized linear models with random effects: A Gibbs sampling approach. Journal of the American Statistical Association, **86**, 79–86 (1991)

12

Inequality Constrained Contingency Table Analysis

Olav Laudy

Department of Methodology and Statistics, Utrecht University, P.O. Box 80140, 3508 TC Utrecht, the Netherlands o.laudy@uu.nl

12.1 Introduction

In recent years, there has been growing interest in statistical models incorporating inequality constraints on model parameters. This is because the omnibus hypotheses can be replaced by more specific inequality constrained hypotheses [2]. In the extensive review in [3], literature is discussed on order restricted statistical models for contingency tables. What becomes clear in this review is that many order restricted models can be estimated and tested – however, not without thorough technical knowledge of the matter. Even if the software is provided, still the (applied) researcher is required to know a great deal about parameterizations of log-linear models. In the approach discussed in this chapter, models for contingency tables are presented in terms of cell probabilities or odds ratios rather than log-linear parameters. This allows researchers to test inequality constrained hypotheses in a format that is directly related to the data and in a way that researchers are used to think of hypotheses.

Being able to provide parameter estimates for various inequality constrained hypotheses gives rise to two questions: (1) Which of the constrained hypotheses fits the data? and (2) Which of the constrained hypotheses fits best? The former question requires to compare the inequality constrained hypothesis with the unconstrained alternative. The latter shows which constrained hypothesis out of a set of hypotheses is mostly supported by the data. In this chapter both questions are answered using Bayes factors and posterior model probabilities.

Throughout this chapter, one dataset is used to illustrate the approach. The research questions are first answered using log-linear models, a standard noninequality constrained approach that is widely available in statistical software packages. It is then shown how the inequality constrained approach is able to provide more specific answers to the research questions than the standard available methods. Note that it may seem "unfair" to compare a long existing method like log-linear analysis to a state-of-the-art Bayesian approach,

Table 12.1. Frequencies of service level, customer satisfaction (satis), and attitude toward next purchase

		Attitude toward next purchase	
Service level	Satis	unlikely	likely
No account manager	dissatisfied	23,174	23,140
	satisfied	67,895	81,356
Telephonic account manager	dissatisfied	2,104	2,223
	satisfied	4,585	6,781
Account manager	dissatisfied	331	257
	satisfied	599	977
Account team	dissatisfied	34	25
	satisfied	60	80

especially if one considers the many extensions that have been developed for log-linear models to handle a great variety of hypotheses, including inequality constrained hypotheses. However, as mentioned before, many of those specialized models require a great deal of technical knowledge, apart from the fact that they are not generally available in software. Thus, the comparison shows the differences between what can be done using the "standard", readily available log-linear models and a conceptually simple to understand Bayesian method. The example concerns the relation between customer satisfaction and attitude toward the next purchase. Data are taken from a business market customer satisfaction questionnaire conducted by a large telecom operator in the Netherlands. The data are presented in Table 12.1. Customer satisfaction was measured on a five-point Likert scale and then condensed to a dichotomous item by removing the neutral (3) category and by summing the category "very dissatisfied" and "dissatisfied" into a "dissatisfied" category, and by summing the category "very satisfied" and "satisfied" into a "satisfied" category. The question regarding attitude toward the next purchase is treated similarly and results in a category "unlikely" and a category "likely". The variable service level indicates how the customer is served. "No account manager" customers are served by general mailings, advertisements, and call centers. In general, these customers are small companies and home offices. The next category of customers are mostly medium enterprises and are served by telephonic account managers; each telephonic account manager serves about 300 customers. The major companies are served by an account manager who visits the company. Each account manager serves about 15 customers. The large enterprises are served by an account team; three to eight account managers serve exactly one customer. Note that although the size of the company is highly correlated with service level; the primary distinction between the categories is the way the customers are served. Depending on the complexity of the portfolio, yearly recurring revenues, and revenue potential, customers are appointed to a service level. In the sequel the variable "customer satisfaction" is abbreviated

Table 12.2. Examples of inequality constrained hypotheses

H_0	$\theta_1, \theta_2, \theta_3, \theta_4$
H_1	$\theta_1 > 1, \theta_2 > 1, \theta_3 > 1, \theta_4 > 1$
H_2	$\theta_1 < \theta_2 < \theta_3 < \theta_4$

as "satisfaction" or "satis"; the variable "attitude toward next purchase" is abbreviated as "attitude" or "att." The variable "service level" is abbreviated as "service" or "serv." Moreover, service level can be perceived as an ordinal variable; by increasing one service level is meant an increase toward a more dedicated service (i.e., toward the category "account team").

There are two research questions regarding these data. The first question concerns customer satisfaction: How satisfied are customers, and how does this relate to service level? The second question is: How is satisfaction related to attitude toward next purchase, and how is this association related to service level? These are fairly general questions and are specified in more detail in Section 12.4. However, as the following sections deal with computational issues regarding inequality constrained hypotheses, some examples are elaborated below. The odds ratio is used as a measure of association between customer satisfaction and attitude toward next purchase, and in each category of service level defined as

$$\theta_k = \frac{\pi_{11k}\pi_{22k}}{\pi_{12k}\pi_{21k}},$$

where π_{ijk} denote the cell probabilities in the contingency table, $i = 1, ..., I$ denote the categories of satisfaction, $j = 1, ..., J$ denote the categories of attitude and $k = 1, ..., K$ denote the categories of service level. In this example $I = J = 2$ and $K = 4$. Thus θ_k denotes the association between satisfaction and attitude for the k-th service level. An odds ratio of one indicates independence between satisfaction and attitude. As will be shown in the sequel, the odds ratio θ_1 equals 1.2, which indicates that for customers without an account manager, the odds of answering "likely" on the attitude question is 1.2 times larger for satisfied customers than for dissatisfied customers. For a detailed review of the odds ratio, see [1]. In Table 12.2, three hypotheses are displayed. Hypothesis H_0 does not impose any structure on the odds ratios. This hypothesis always fits the data perfectly, but it is not informative, nor parsimonious. In search for interpretable structure in the data, inequality constraints are specified: Hypothesis H_1 specifies a positive association between satisfaction and attitude for each service level. Hypothesis H_2 specifies that with increasing service level, the association between satisfaction and attitude also increases. It is of interest to investigate which of the hypotheses fits the data and which hypothesis fits the data best.

The chapter is built up as follows. First, in Section 12.2, the interpretation and parametrization of classical log-linear models is discussed, as this serves as comparison to the method proposed in this chapter. In Section 12.3, a

Bayesian approach to the evaluation of inequality constrained hypotheses for contingency tables is presented. It is shown how to obtain an estimate of Bayes factors and how to calculate the posterior model probability using Bayes factors. In Section 12.4, the various research questions are evaluated, both using the classical log-linear approach and the Bayesian approach proposed in this chapter.

12.2 Log-linear Models

A log-linear model decomposes a contingency table in an ANOVA-like manner into main and interaction effects [1]. We use standard notation for contingency tables with three variables A, B, and C with respective indexes $i = 1, ..., I$, $j = 1, ..., J$, and $k = 1, ..., K$. Suppose there is a multinomial sample of size N over the $I \times J \times K$ cells of the contingency table. Let f_{ijk} denote the observed frequency for cell (i, j, k). The cell probabilities π_{ijk} for the multinomial distribution form the joint distribution of the three categorical responses. Those responses are independent when $\pi_{ijk} = \pi_{i++}\pi_{+j+}\pi_{++k}$, for $i = 1, ..., I$, $j = 1, ..., J$, and $k = 1, ..., K$, where $\pi_{i++} = \sum_{j,k} \pi_{ijk}$. The expression for the independence on the scale of the frequencies is $f_{ijk} = N\pi_{i++}\pi_{+j+}\pi_{++k}$. On a logarithmic scale, independence has the additive form

$$\log(f_{ijk}) = \log(N) + \log(\pi_{i++}) + \log(\pi_{+j+}) + \log(\pi_{++k}).$$

The log expected frequencies for cell (i, j, k) is an additive effect of the i-th effect of variable A, the j-th effect of variable B and the k-th effect of variable C. Identifiability constraints have to be included and the usual notation for a log-linear independence model is

$$\log(f_{ijk}) = \mu + \lambda_i^A + \lambda_j^B + \lambda_k^C. \tag{12.1}$$

A common choice to identify the model is by setting the λ parameter of the last category of each variable to 0, thereby implying that the effect of the other parameters of that variable have to be interpreted as differences with respect to the last category. The parameters are not as easy and straightforward to interpret as one might like (e.g., in comparison to an ANOVA). The μ parameter can be viewed as the categorical counterpart of the grand mean in an ANOVA. The λ parameters are interpreted as additions to the log cell frequencies, or multipliers of the frequencies of an effect, however, it is more common to look at the signs and transform the parameters back to simple probabilities. Examples will be given in Section 12.4.4

The saturated model, the model that always perfectly fits the data, has the form

$$\log(f_{ijk}) = \mu + \lambda_i^A + \lambda_j^B + \lambda_k^C + \lambda_{ij}^{AB} + \lambda_{ik}^{AC} + \lambda_{jk}^{BC} + \lambda_{ijk}^{ABC}. \tag{12.2}$$

A two-way interaction effect (e.g., λ_{ij}^{AB}) captures the association between two variables. The three-way interaction effect(λ_{ijk}^{ABC}) shows how the association between the two variables varies over a third variable. With respect to the interpretation of a two-way effect, in the above explained parametrization, in a two-way table with $I = J = 2$, the exponent of the parameter λ_{ij}^{AB} equals the odds ratio as defined in Section 12.1. In tables where $\max(I, J) > 2$, the two-way effects are multiplication factors of the underlying one-way effects; however, it is more common to look at the sign and the significance of each two-way association term and use these to refer to the estimated or observed odds ratios. This is illustrated in the example in Section 12.4. Also, a three-way effect can be translated into a multiplication factor that shows how the two-way odds ratio is modified by each category of the third variable; however, it is more common to interpret the sign and refer to the estimated or observed odds ratios. As an example, suppose that the three-way interaction odds ratio equals 2. This indicates that for any two out of three variables, the odds ratio increases by a factor of 2 when moving one category up over the third variable.

As becomes clear, the interpretation of the parameters of a log-linear model is not straightforward. There is, however, another good use of log-linear models, and that is locating the source of dependencies. First, a shorthand notation for log-linear models is introduced. The independence model (12.1) is given by (A, B, C) and the saturated model (12.2) by (ABC), thus referring to the highest (interaction) effect in the model. Since the saturated model (ABC) allows for all effects, it always describes the data perfectly. Consider a saturated three-way model with exactly one three-way interaction parameter (this is the case when $I = J = K = 2$) that appears not to be significant. Then there is reason to estimate the model again, but without the three-way interaction term. The now highest interaction terms in the model are $\lambda_{ij}^{AB}, \lambda_{ik}^{AC}, and \lambda_{jk}^{BC}$, leading to the short notation (AB, AC, BC). This model allows for association between each pair of variables, but requires the association to be the same for each level of the third variable. The model (A, BC) shows an association between variables B and C, while this association is equal on each level of A. An alternative description is that variable A is jointly independent of B and C. There is a set of similarly structured models (e.g., (AB, C)) that have similar interpretations as (A, BC). The model (AB, BC) requires variables A and C to be independent for each level of B. This is called conditional independence. Note that conditional independence does not imply marginal independence; that is, A and C are independent for each level of B, but not when summed over B.

A likelihood ratio test can be used to compare models with different effects (i.e., different sets of λ parameters). The likelihood ratio test takes twice the difference of the log-likelihood values of two models and evaluates this in a chi-square distribution, with the degrees of freedom given by the difference in the number of parameters in the two models. The null hypothesis states that the smaller model does not fit worse than the larger model, and a significant effect indicates that the smaller model does indeed fit worse. Note that models do have to be nested; that is, the larger model contains all effects from the smaller model. The notation $LR(10) = 675, p = .000$ shows that a likelihood ratio test yields a value of 675 with 10 parameters difference between the two models, resulting in a significant effect. It is concluded that the smaller model fits worse.

Summarizing, the main objective for log-linear analysis is to find the smallest (i.e., fewest parameters) model that still fits the data. The various models are compared using likelihood ratio tests. After the smallest model is found that still fits the data, the observed odds ratios can be interpreted as being significant or not significant and references can be made to positive and negative associations.

One extension to log-linear models that can be performed using the standard statistical software is discussed. By crafting a special vector in the data matrix, and including this as a covariate, linear effects can be included in the log-linear model. This linear effect captures any linear trend that can be found over the consecutive categories of a variable, in one parameter. For example, suppose that in the saturated model (ABC), the two-way effect parameters λ_{ij}^{AB} take the values (2, 4, 6) over the third variable for $k = 1, 2, 3$; then a linear term $\beta^{AB} = 2$ would perfectly capture the trend. The values λ_{ij}^{AB} for each category of k can be calculated by multiplying the β^{AB} by the category of k; that is, $k \cdot \beta^{AB}$ for category $k = 3$ equals $3 \cdot 2 = 6$. When the data show a perfect linear trend, the linear effect model results in the same fit as the corresponding log-linear model, but it is more parsimonious (i.e., less parameters). Formally the example above is denoted as

$$\log(f_{ijk}) = \mu + \lambda_i^A + \lambda_j^B + \lambda_k^C + \beta^{AB} u_i v_j. \tag{12.3}$$

The terms u_i and v_j are the respective category numbers of variables A and B. To include such an effect using standard log-linear models, an extra column in the data has to be added, containing the product of the category numbers of variables A and B. For example, the value for $u_1 = 1$ and $v_2 = 2$ results in the value 2 in the linear effect column. See [1] for a thorough review of log-linear models.

12.3 Bayesian Analysis of Contingency Tables

In this section it is shown how a Bayesian analysis on contingency tables is performed. First, it will be presented how to obtain parameter estimates, subsequently estimation of Bayes factors and how they can be used to evaluate competing hypotheses as shown in Table 12.2 will be discussed. With respect to the former, Bayesian estimates are often obtained by a sampling procedure, whereas the classical approach mainly uses optimization. The optimization is performed with respect to the likelihood of the model. In Bayesian statistics, one element is added to the likelihood, namely the prior distribution. It is convenient to assume that before observing the data, the parameters in the model do not have one fixed value in the population, but rather have a variety of possible values, some values a priori more plausible than others, which leads to a prior distribution of parameters, or short, the prior. This prior is multiplied with the likelihood and results in what is called the posterior distribution: the distribution of the parameters after observing the data. As the posterior distribution can be quite difficult to optimize (i.e., find the maximum), a common solution is to take a sample (of parameters) and compute estimates using this sample. Note that the remainder of Section 12.3 is fairly technical and can be skipped if one is only interested in the application. A good introduction to Bayesian statistics can be found in [8].

12.3.1 Posterior Distribution

First, to enhance readability, vector notation is introduced. Let $\boldsymbol{f} = \{f_{ijk}|i = 1,...,I, j = 1,...,J, k = 1,...,K\}$ and $\boldsymbol{\pi} = \{\pi_{ijk}|i = 1,...,I, j = 1,...,J, k = 1,...,K\}$. Similar to the log-linear section, it is assumed that $\boldsymbol{f}$ follows a multinomial distribution, $\boldsymbol{f}|\boldsymbol{\pi} \sim M(\boldsymbol{\pi}, N)$. Let $\boldsymbol{\alpha}$ denote the parameters of the prior distribution, $p(\boldsymbol{\pi}|\boldsymbol{\alpha}) \sim Dirichlet(\boldsymbol{\alpha}) \propto \prod_{i=1}^{I}\prod_{j=1}^{J}\prod_{k=1}^{K}\pi_{ijk}^{\alpha_{ijk}-1}$. Before the analysis, a choice has to be made with respect to $\boldsymbol{\alpha}$. A common choice is a constant for each element of the vector $\boldsymbol{\alpha}$. In the examples, $\boldsymbol{\alpha} = \mathbf{1}$ is used, which leads to the estimate $\pi_{ijk} = f_{ijk}/N$ for the posterior mode.

Denote inequality constraint z as $r_z(\boldsymbol{\pi})$ for $z = 1,...,Z$. Examples of inequality constrained hypotheses are given in Table 12.2. The joint constraints are $R(\boldsymbol{\pi}) = (r_1(\boldsymbol{\pi}),...,r_Z(\boldsymbol{\pi}))$. The inequality constraints are accounted for in the prior distribution as follows:

$$p(\boldsymbol{\pi}|R(\boldsymbol{\pi}),\boldsymbol{\alpha}) = \frac{p(\boldsymbol{\pi}|\boldsymbol{\alpha})I_{\boldsymbol{\pi}\epsilon R(\boldsymbol{\pi})}}{\int p(\boldsymbol{\pi}|\boldsymbol{\alpha})I_{\boldsymbol{\pi}\epsilon R(\boldsymbol{\pi})}\,d\boldsymbol{\pi}},$$

where $I_{\boldsymbol{\pi}\in R(\boldsymbol{\pi})}$ is an indicator function that has the value one if $\boldsymbol{\pi}$ is in accordance with $R(\boldsymbol{\pi})$, and zero otherwise. The likelihood is given by $L(\boldsymbol{f}|\boldsymbol{\pi})$. It follows that the posterior distribution $p(\boldsymbol{\pi}|\boldsymbol{f}) \propto L(\boldsymbol{f}|\boldsymbol{\pi})p(\boldsymbol{\pi}|R(\boldsymbol{\pi}),\boldsymbol{\alpha})$ is

$$p(\boldsymbol{\pi}|\boldsymbol{f}) \propto \prod_{i=1}^{I}\prod_{j=1}^{J}\prod_{k=1}^{K}\pi_{ijk}^{f_{ijk}+\alpha_{ijk}-1}I_{\boldsymbol{\pi}\in R(\boldsymbol{\pi})},$$

restricted such that $\sum_{ijk}\pi_{ijk} = 1$.

Since the cell probabilities are restricted to sum to one, the cell probabilities cannot be sampled successively. Narayanan [13] showed that a Dirichlet distribution can be parameterized into a gamma distribution in such a way that the sampling procedure is simplified. Let $\gamma_{ijk} \sim Gamma(f_{ijk} + \alpha_{ijk}, 1)$; then the vector $(\pi_{111},...,\pi_{IJK})$ where $\pi_{ijk} = \gamma_{ijk}/\gamma_{+++}$ is distributed as

$Dirichlet(f_{111}+\alpha_{111},...,f_{ijk}+\alpha_{ijk},...,f_{IJK}+\alpha_{IJK})$. Under this parameterization the posterior becomes

$$P(\boldsymbol{\pi}|\boldsymbol{f}) \propto \prod_{i=1}^{I}\prod_{j=1}^{J}\prod_{k=1}^{K}\left(\frac{\gamma_{ijk}}{\gamma_{+++}}\right)^{f_{ijk}+\alpha_{ijk}-1} I_{\boldsymbol{\gamma}\in R(\boldsymbol{\gamma})}, \tag{12.4}$$

where $\gamma_{+++} = \sum_{ijk}\gamma_{ijk}$. Note that the inequality constraints on $\boldsymbol{\pi}$ are now also parameterized into inequality constraints on $\boldsymbol{\gamma}$, since $R(\boldsymbol{\pi}) = R(\frac{\boldsymbol{\gamma}}{\gamma_{+++}}) = R(\boldsymbol{\gamma})$. For example, suppose we have the following ordering: $\pi_{111} > \pi_{112} > \pi_{113}$ which equals $\frac{\gamma_{111}}{\gamma_{+++}} > \frac{\gamma_{112}}{\gamma_{+++}} > \frac{\gamma_{113}}{\gamma_{+++}}$. This reduces to $\gamma_{111} > \gamma_{112} > \gamma_{113}$.

It is now explained how the model parameters are sampled from the constrained posterior distribution. A sample is taken from $\gamma_{ijk} \sim Gamma(f_{ijk} + \alpha_{ijk}, 1)$ for $i = 1,...,I, j = 1,...,J, k = 1,...,K$. Without constraints, the parameters are independent, and an i.i.d. sample from the successive gamma distributions provides a draw from the posterior distribution. If inequality constraints are incorporated in the model, the successive draws of the gamma distribution are dependent, and we resort to the Gibbs sampler [7, 8].

The Gibbs sampler is an iterative procedure. Suppose we want to draw samples from the joint posterior distribution of γ under an inequality constraint $R(\gamma)$. Taking the full conditionals required for the Gibbs sampler reduces the multivariate constraints to univariate constraints. In the Gibbs sampler, the gamma distribution of a parameter thus has a lower and an upper bound, conditional upon the constraints and current values of all the other parameters. These values are denoted by $bounds(\gamma_{ijk}^{(s)}) = (l, u)$, where l denotes the maximum lower bound and u denotes the minimum upper bound over all Z constraints in iteration s. To keep the notation simple, we omit indexes for l and u.

In iteration $s = 0$, initial values have to be provided for γ_{ijk}. Any set of values that is in agreement with the constraints imposed upon the parameters can be used. Each iteration $s = 1, \ldots, S$ consists of the following steps:

(1) Cycle step 1 $\forall\ i, j, k$

  (1_a) Calculate $bounds(\gamma_{ijk}^{(s)}) = (l, u)$ given the current values of the parameters and $R(\gamma)$.

  (1_b) Sample $\gamma_{ijk}^{(s+1)} \sim Gamma(f_{ijk} + \alpha_{ijk}, 1|l, u)$.

(2) Compute $\pi_{ijk} = \gamma_{ijk}/\gamma_{+++}\ \forall\ i, j, k$ and deliver $(\pi_{111}, ..., \pi_{ijk}, ..., \pi_{IJK})$ as a draw of the correct truncated posterior.

Gelfand et al. [7] showed that in iteration s as $s \to \infty$ under mild conditions, the Gibbs sampler provides parameters that come from the correct constrained joint posterior distribution.

The naive way to sample from a truncated gamma distribution is to sample from the nontruncated gamma distribution until a deviate is sampled that satisfies the constraints. However, this is quite inefficient if only a small range of the distribution is admissible. Inverse probability sampling solves this problem. Details can be found in Chapter 3.

With regard to the sampling procedure, in our experience 9000 iterations and a burn-in of 1000 iterations generally lead to stable estimates. The required burn-in period can be longer, depending on how flexibly the sampled parameters can move through the parameter space, due to the constraints. The mixing of the Gibbs sampler is visually inspected by plotting $\boldsymbol{\pi}^{(s)}$ against s, for $s = 1, \ldots, S$ iterations [5].

For all the restrictions used in this chapter, the estimation procedure is basically the same. The author developed a software package that allows the users to specify a model and input the inequality constraints as text. To evaluate the inequality constraints in each sample of the posterior, a function parser (`http://www.its.uni-karlsruhe.de/~schmehl/functionparser.html`) is used. This parser reads the inequality constraints as text elements and translates it into equations. Afterward, the equations are solved numerically, using a simple root finder to obtain upper and lower bounds. Once the bounds are obtained, the estimation procedure is as explained above. More details can be found in [12].

12.3.2 Parameter Estimates, Posterior Standard Deviations, and Credibility Intervals

After removing the burn-in, the sample of S iterations from the constrained posterior distribution can be summarized to obtain parameter estimates, posterior standard deviations, and credibility intervals for each model parameter. By taking the averages over the S values, the Expected A Posteriori estimates (EAP) are obtained [8]. The posterior standard deviations are obtained by taking the standard deviations over the S values. The 90% central credibility intervals can be calculated by taking the 5th and 95th percentiles of the posterior sample. The posterior distribution may be skewed, in which case the credibility interval then correctly provides an asymmetric interval.

Furthermore, the summary measures can also be calculated for functions of the parameters. Let $g(\boldsymbol{\pi})$ describe a function of interest; then the S iterations of the posterior of the cell probabilities can be transformed according to $g(\boldsymbol{\pi})$. Summary measures can be calculated for this newly created vector. For example, suppose a contingency table with $I = J = K = 2$ is estimated using the constraint that the odds ratio of the collapsed table $(\pi_{11+}\pi_{22+})/(\pi_{21+}\pi_{12+}) > 1$. Let $g(\boldsymbol{\pi}) = (\pi_{11+}\pi_{22+})/(\pi_{21+}\pi_{12+})$. The vector $g^{(s)}(\boldsymbol{\pi})$ for $s = 1, ..., S$ represents the posterior distribution of this odds ratio. The EAP, the posterior standard deviation, and a central credibility interval can subsequently be calculated.

12.3.3 Bayes Factors and Posterior Model Probabilities

In the previous section, the sampling procedure has been explained. For each of the inequality constrained hypotheses, such a sampling procedure results in a set of (constrained) estimates. Subsequently, the question arises: Is the hypothesis compatible with the data and which of the hypotheses is best? Both questions can be evaluated using posterior model probabilities. Three ingredients are required to compute posterior model probabilities: a finite set of models or hypotheses, a prior model probability (not to be confused with the prior distribution for the parameters) for each hypothesis in the set, and marginal likelihoods or Bayes factors. The first ingredient requires careful considerations of the researcher: Which theories or expectations should be included? Should the unconstrained model be part of the set of hypotheses? With respect to the second ingredient, throughout this chapter (the conventional) equal prior model probabilities are used. So, for a set of T models the prior probability for model t ($t = 1, \ldots, T$) equals $1/T$. The last ingredient, the marginal likelihood, can be interpreted as the likelihood that the data are observed given that the hypothesis at hand is true. The fit of two hypotheses or models can be compared by examining the ratio of the marginal likelihoods: the Bayes factor [9, 10]. A great deal of literature shows that the computation of marginal likelihoods can be burdensome [4, 6, 14]. However, the hypotheses considered in this chapter are all constrained versions of the unconstrained model. Stated otherwise, the (in)equality constrained hypotheses are nested in the unconstrained model. Klugkist et al. [11] and Laudy and Hoijtink [12] showed that the calculation of a Bayes factor for nested hypotheses (that is, the Bayes factor for any constrained model with respect to the unconstrained model) is greatly simplified and does not require the computation of marginal likelihoods (see also Chapter 4).

Denote the unconstrained hypothesis as $H_0 : \boldsymbol{\pi}$ and the t-th constrained hypothesis as $H_t : R_t(\boldsymbol{\pi})$, where R_t is a function that imposes restrictions on $\boldsymbol{\pi}$, as introduced in Section 12.3.1. Note that in the sequel, the unconstrained hypothesis is always denoted by H_0. The Bayes factor BF_{t0} can be written as

$$BF_{t0} = c_t/d_t, \tag{12.5}$$

where $1/c_t$ denotes the proportion of the unconstrained prior that is in accordance with the constraints R_t, and $1/d_t$ denotes the proportion of the unconstrained posterior that is in accordance with the constraints R_t. An estimate of $1/c_t$ is obtained by sampling from the unconstrained prior distribution and calculating the proportion of parameter vectors that is in agreement with hypothesis H_t. An estimate of $1/d_t$ is obtained by sampling from the unconstrained posterior distribution and calculating the proportion of samples that is in agreement with the hypothesis H_t. The proportion of samples in agreement with the constraints after observing the data ($1/d_t$) is compared to (or stated differently, penalized by) the proportion of samples in agreement with the constraints a priori ($1/c_t$), rendering an estimate for the Bayes factor in (12.5). With respect to the interpretation, suppose a Bayes factor of a constrained hypothesis versus the unconstrained hypothesis takes the value 2;

Table 12.3. Approximating hypotheses for the estimation of BF_{10}

H_0:	$\pi_{11}, \pi_{12}, \pi_{13}$ (unconstrained)
$H_{1,1}$:	$\pi_{13} > .5$
$H_{1,2}$:	$\pi_{12} > \pi_{13} > .5$
H_1:	$\pi_{11} > \pi_{12} > \pi_{13} > .5$

then given the data at hand, the constrained hypothesis is twice as likely as the unconstrained hypothesis.

This approach works for any set of inequality constrained hypotheses and does not need a sampling procedure that can handle inequality constraints. However, when restrictions only allow a small part of the parameter space, the procedure is not very efficient. The proposed method can be extended such that also in complex restrictions, the procedure efficiently yields a Bayes factor. The extension makes use of the inequality constrained sampling procedure, as explained in Section 12.3.1.

To illustrate the computation of the Bayes factor for a "complex" constrained hypothesis, $H_1 : \pi_{11} > \pi_{12} > \pi_{13} > .5$, and the unconstrained $H_0 : \pi_{11}, \pi_{12}, \pi_{13}$ are considered. Note that straightforward application of (12.5) is possible; however, the following procedure is more efficient. The Bayes factor (BF_{10}) can be approximated by a stepwise procedure, by breaking up the restriction in parts, such that in each subsequent step a more restricted parameter space is taken into account. In several subsequent steps, the parameter space is more restricted (see Table 12.3). Note that the set of new hypotheses is constructed such that $H_1 \subset H_{1,2} \subset H_{1,1} \subset H_0$.

The Bayes factor for hypotheses $H_{1,1}$ and H_0 (denoted by $BF_{(1,1)0}$) is given by (12.5); that is, samples from both the unconstrained prior and posterior provide the proportions of these samples that are in agreement with $H_{1,1}$. The number of samples from both prior and posterior in agreement with $H_{1,1}$ will not be zero and the procedure is rather efficient.

In the next step, the Bayes factor for hypotheses $H_{1,2}$ and $H_{1,1}$ (denoted by $BF_{(1,2)(1,1)}$) is obtained by sampling from both the constrained prior and posterior of hypothesis $H_{1,1}$ and calculating the proportions of samples that are in agreement with hypothesis $H_{1,2}$. This procedure is repeated also for the Bayes factor for hypotheses H_1 and $H_{1,2}$. The Bayes factor of interest, BF_{10} is computed applying the product rule:

$$BF_{10} = BF_{1(1,2)} \times BF_{(1,2)(1,1)} \times BF_{(1,1)0}.$$

A Bayes factor provides the posterior odds of two hypotheses. For a finite set of hypotheses, representing a set of competing theories or expectations, posterior model probabilities for all models in the set can be computed from the Bayes factors.

Consider a set of just the unconstrained hypothesis (H_0) and one alternative hypothesis (H_1). The question is whether hypothesis H_1 is supported by the data. The posterior model probability $P_{H_1|H_0,H_1}$ – that is, the probability that hypothesis H_1 is true given the set of hypotheses H_0, H_1 – is given by

$$P_{H_1|H_0,H_1} = \frac{BF_{10}}{1 + BF_{10}}.$$

The probability of H_0 is $1 - P_{H_1|H_0,H_1}$. Note that a priori both hypotheses are equally likely. When $P_{H_1|H_0,H_1} = .9$, there is much evidence that the hypothesis is true. However, when $P_{H_1|H_0,H_1} = .6$, the evidence for hypothesis H_1 is much weaker.

If the set of hypotheses of interest consists of the unconstrained hypothesis H_0 and several alternative hypotheses H_t $(t = 1, \ldots, T)$, the question is which of the hypotheses supports the data best. The posterior model probabilities (PMPs) are computed using all Bayes factors of the constrained models with the unconstrained model (BF_{t0}), applying

$$\text{PMP}_{H_t|H_0,\ldots,H_T} = \frac{BF_{t0}}{1 + BF_{10} + \cdots + BF_{T0}}.$$

To obtain the PMP of the unconstrained model ($P_{H_0|H_0,\ldots,H_T}$) the numerator in the previous equation is replaced by the value 1.

Note that it is not always interesting to incorporate the unconstrained hypothesis into the set of hypotheses. Incorporating the unconstrained hypothesis H_0 shows whether the constrained hypotheses in the set provide a good description of the data: If the posterior probability of H_0 is large, none of the constrained hypotheses is supported by the data. However, it may be of interest to choose between the best of two restricted hypotheses. In that case, the unrestricted model (H_0) is not included in the set. Note that it is still part of the analyses because the unconstrained model is used to compute all the Bayes factors BF_{t0}. The PMPs of models H_t $(t = 1, \ldots, T)$, exclusive of the unconstrained model, are computed using

$$\text{PMP}_{H_t|H_1,\ldots,H_T} = \frac{BF_{t0}}{BF_{10} + \cdots + BF_{T0}}.$$

12.4 Example

In this section, both the classical log-linear models and the Bayesian inequality constrained procedure are illustrated using the data concerning the relation among service level, customer satisfaction, and attitude toward the next purchase. For the log-linear models, the model parameters are interpreted, and using a likelihood ratio test, it is tested which parameters are significant. For the Bayesian inequality constrained model, inequality constrained hypotheses are formulated and the posterior model probability is used to evaluate the support for the constrained hypotheses. The two research questions are (1) How satisfied are customers and does this relate to service level? and (2) How is satisfaction related to attitude toward the next purchase and how does this relate to service level?

Table 12.4. Percentages and odds ratios for the data in Table 12.1

	Satisfaction	Attitude	Satis * Att
Service level	% satisfied	% likely	Odds ratio
No account manager	76	53	1.2
Telephonic account manager	72	57	1.4
Account manager	73	57	2.1
Account team	70	53	1.8
Total	76	54	1.2

12.4.1 Levels of Satisfaction, the Log-linear Approach

Table 12.4 shows the observed percentages and odds ratios for the data in Table 12.1. The question of how satisfied are the customers can be answered informatively using the column containing the satisfaction percentages. As this is a sample of all customers, it is of interest to know whether it can be safely assumed that more customers are satisfied than nonsatisfied; that is, (a) Is the satisfaction percentage greater than 50%? (b) Is it true that *for each service level* the satisfaction percentage is greater than 50%?

To answer question a), the independence model (Service, Satisfaction, Attitude) is fit as explained in Section 12.2. Satisfaction has two levels; thus, there is one λ parameter to estimate. This parameter shows how much the observed counts differ from equally spread counts over the two categories; in other words, how far are the two categories away from .50/.50? A significant parameter indicates that indeed the counts are not equally spread over the categories. The estimate for the category "satisfied" equals 1.15 and is highly significant. It is concluded that customer satisfaction is different from 50%, and since the *observed* percentage is larger than 50%, it is also inferred that customer satisfaction is *larger* than 50%. Note that $\exp(1.15) = 3.15$ equals the ratio of .76 and .24 $(1 - .76)$. Note, furthermore, that the model as a whole does not fit the data ($LR(10) = 675, p = .000$); thus, although it can be concluded that customer satisfaction is different from 50%, the search for sources of dependencies has not yet finished.

To answer question b), the model (Service * Satisfaction, Attitude); that is, a model of type (AB,C) as discussed in Section 12.2 is fit. This model allows one parameter for satisfaction and for each service level a parameter that codes for the deviation from this satisfaction parameter. A likelihood ratio test for (Service, Satisfaction, Attitude) against (Service * Satisfaction, Attitude) shows a highly significant effect ($LR(3) = 132, p = .000$), which leads to the conclusion that the independence model does fit worse than the model that allows satisfaction to vary over the levels of service. In Table 12.5, the parameter estimates are displayed, along with their exponents and significance values. The "Account team" category of service level serves as reference category; thus, the value of satisfaction = "satisfied" can be filled in here. The

Table 12.5. Relevant parameters for model (Service * Satisfaction, Attitude)

Effect	λ	Exp(λ)	p
Satis="satisfied"	.86	2.4	.00
Satis="satisfied," Serv= No acc. man.	.31	1.4	.05
Satis="satisfied," Serv= Telephonic acc. man.	.10	1.1	.52
Satis="satisfied," Serv= Acc. man.	.12	1.1	.45
Satis="satisfied," Serv= Acc. team	–	–	–

exponentiated parameter for the reference category equals 2.4, indicating that there are 2.4 times as many observations in the category "Account team, satisfied" than in "Account team, unsatisfied," which can also be seen by the ratio of .70 and .30. The exponentiated parameter "No account manager, satisfied" equals 1.4, meaning there are 1.4 times more counts in the category "satisfied" than in the category "unsatisfied" *with respect to the reference category.* Thus, the category by itself contains 2.4 x 1.4 = 3.2 times more counts in the category "satisfied" than in "unsatisfied," which again can be verified by the ratio .76/.24. Note that also the significance should be interpreted as an effect with respect to the reference category. Thus, the reference category shows a highly significant effect, in the expected direction. All but the "No account manager, satisfied" effect are not significant, meaning that they do not differ with respect to "Account team, satisfied." The category "No account manager, satisfied" shows a marginally significant effect. Thus, the parameter for category "satisfied" is different from zero, and as it is observed greater than .5, and all deviations from this effect are not significant, it is concluded that for each category of service level the percentage "satisfied" is greater than the percentage "unsatisfied."

Since it is concluded that the model allowing for different satisfaction levels for each service level is a better model than the model with one main satisfaction level, it is of interest to test for a structure among these levels. In Table 12.4 it can be seen that the higher the service level, the lower the percentage satisfaction, with a small exception of the level "Account manager." It is of interest whether this trend can be shown to be significant. The model representing a linear trend is

$$\log(f_{ijk}) = \mu + \lambda_i^{satisfaction} + \lambda_j^{attitude} + \lambda_k^{service} + \beta u_i v_k,$$

where u_i and v_k are scores that refer to each category of the corresponding variable, as discussed in Section 12.2. To fit this model in standard software, a vector has to be created in the data matrix, containing the product of u_i and v_k, which are just the category labels $1, \ldots, I$ and $1, \ldots, K$. Furthermore, this vector has to be added as covariate to the independence model. In Table 12.6, the results for several likelihood ratio tests are displayed. Each cell in the table displays the resulting likelihood ratio test for the hypotheses in the respective row and column. First, there is no smaller model than (Service * Satisfaction,

Table 12.6. Likelihood ratio tests for the linear trend (lin)

	(Serv, Satis, Att, lin)	(Serv * Satis, Att)
(Serv, Satis, Att)	$LR(1) = 116, p = .000$	$LR(3) = 132, p = .000$
(Serv, Satis, Att, lin)	–	$LR(2) = 15,\ p = .000$

Table 12.7. Interpretation of the linear trend parameters

Effect	λ	$\exp(\lambda)$	p
Satis="satisfied'	1.67	3.2	.76
Satis="satisfied," Serv= No acc. man.	0	3.2	.76
Satis="satisfied," Serv= Telephonic acc. man.	−.16	2.8	.73
Satis="satisfied," Serv= Acc. man.	−.32	2.3	.70
Satis="satisfied," Serv= Acc. team	−.48	2.0	.67

Attitude) that still fits the data, as all significant effects indicate that the smaller models do fit worse. However, the distance between the independence model and model (Service * Satisfaction, Attitude) is 132 likelihood ratio points, whereas for the linear trend model, with only one parameter extra, this distance has decreased to 15 points, indicating that a linear trend improves the fit significantly ($LR(1) = 116, p = .000$).

Although the likelihood ratio test fails to recognize the linear trend model as a good model for the data, the linear trend parameter is highly significant (.000). The interpretation of this trend is discussed. The parameter for the linear trend has value −.16, or exponentiated .85, meaning that in each subsequent category of service level, the odds of being in the category "satisfied" decreases with factor .85. In the column λ in Table 12.7, the effect of the linear trend is displayed: Each step toward a higher service level, the linear trend parameter decreases with −.16. With respect to the baseline odds (3.2 for category "no account manager") the odds decreases with $\exp(-.16) = .85$. Finally, in the last column, the estimated probabilities, calculated by odds/(1+odds), under a linear trend are displayed; indeed, the probabilities are ordered and similar to those in Table 12.4.

It is concluded that the log-linear model does provide answers to the research questions at hand; however, the answers are not as straightforward as one would like.

12.4.2 Levels of Satisfaction, the Bayesian Approach

The Bayesian approach requires a set of hypotheses to be defined in advance of the model fitting. Table 12.8 displays these hypotheses. The cell probabilities are indexed such that i indicates the levels of satisfaction, j indicates the levels of attitude, and k indicates the levels of service level. Probability π_{1+1} is calculated by summing over attitude; that is, $\pi_{1+1} = \pi_{111} + \pi_{121}$.

Table 12.8. Constrained hypotheses for levels of satisfaction

$H_0 : \pi_{111}, \pi_{112}, \pi_{121}, ..., \pi_{222}$
$H_1 : \pi_{2++} > \pi_{1++}$
$H_2 : \{\pi_{2+1} > \pi_{1+1}\}, \{\pi_{2+2} > \pi_{1+2}\}, \{\pi_{2+3} > \pi_{1+3}\}, \{\pi_{2+4} > \pi_{1+4}\}$
$H_3 : \pi_{2+1}/\pi_{++1} > \pi_{2+2}/\pi_{++2} > \pi_{2+3}/\pi_{++3} > \pi_{2+4}/\pi_{++4}$
$H_4 : \pi_{2+1}/\pi_{1+1} > \pi_{2+2}/\pi_{1+2} > \pi_{2+3}/\pi_{1+3} > \pi_{2+4}/\pi_{1+4}$

Table 12.9. Results for the constrained hypotheses for levels of satisfaction

Hypothesis	PMP	BF_{t0}
H_1	.02	2.0
H_2	.13	15.7
H_3	.05	6.3
H_4	.80	98.0

Hypothesis H_0 is the unconstrained hypothesis. To be able to learn from the model a constrained hypothesis should at least fit better than H_0. Hypothesis H_1 states that, summed over all service levels and irrespective of attitude, the probability of falling into the category "satisfied" is larger than in the category "unsatisfied." Hypothesis H_2 specifies that for each category of service level, the probability of falling into the category "satisfied" is larger than in the category "unsatisfied." Note that if H_2 is true, also H_1 is true; however, the reverse is not necessarily true. Hypothesis H_3 specifies that the probability of being in the category "satisfied" is decreasing with service level. Note that each probability π_{2+k} is divided by π_{++k} to correct for the different number of observations in each service level. Hypothesis H_2 and H_3 have different character: Hypothesis H_2 is in terms of probabilities within one level and hypothesis H_3 orders probabilities among levels. If both hypotheses are supported by the data, it is of interest whether they are simultaneously supported by the data, hence hypothesis H_4, which is the combination of hypotheses H_2 and H_3. Hypothesis H_4 requires that in each category of service level, the probability of being satisfied is greater than being unsatisfied, and with increasing service level, the probability of being satisfied is restricted to decrease.

For each of the constrained hypotheses, the Bayes factor is calculated against the unconstrained hypothesis. This indicates whether the constrained hypotheses are supported by the data. A Bayes factor of one indicates that the unconstrained and the constrained hypothesis are equally supported. Next, the posterior model probabilities are calculated. These show which of the constrained hypotheses is mostly supported by the data. Note that the unconstrained hypothesis is not included in the set of hypotheses. The results are displayed in Table 12.9. First, for hypothesis H_1, it can be concluded that there is some evidence that the probability of being satisfied is larger than being unsatisfied; however, the PMP is not very large. This result occurs because the hypothesis does not restrict large parts of the parameter space; hence, when it is found to be true, this does not result in a large surprise. Hypothesis H_2 is much more restrictive, and as can be seen in the observed data (or the basic statistics in Table 12.4), the restrictions all conform to the data, which results in a large PMP. Hypothesis H_3 shows that the ordered probabilities hypothesis fits the data rather well. Hypothesis H_4, the combination of hypotheses H_2 and H_3, fits the data best (PMP = .80).

12.4.3 Comparison Between Log-linear and Bayesian Approach

As the Bayesian constrained hypotheses show, all hypotheses are directional (i.e., have a clear sign indicating in which direction the effect is expected). In the log-linear context, the parameter for satisfaction was found significantly different from zero, and then because the observed data showed it was larger than zero, it was automatically concluded that this significance could be interpreted as the parameter being significantly *larger* than zero. The latter step is not more than an eyeball test, which is rather subjective and informal. Although it is possible to do proper directional testing in a log-linear context, it requires a lot more effort – for example, setting up a Wald-test using specially crafted contrasts. Second, in the log-linear model, all hypothesis concerning satisfaction were tested in the larger model (Service * Satisfaction, Attitude), meaning that certain effects (e.g., the three-way interaction effect) are set to zero, whereas the Bayesian approach allows any effect, as long as it is satisfying the inequality constraints. The latter could also have been done in the log-linear context; however, it would have required a lot more interpretation of parameters (namely all effects higher than (Service * Satisfaction, Attitude)). Third, the log-linear hypotheses test whether the observed data could have occurred from a population where certain parameters are set to zero, thus only taking the deviation from the null hypothesis into account, whereas the Bayesian approach shows the evidence for both the null and alternative hypothesis. The latter is preferred, as when both the null and alternative hypotheses are not true, in a log-linear model, the null is rejected, and thus the alternative is accepted, whereas in the Bayesian approach, both hypotheses receive little support from the data. It can then correctly be decided that both the null and alternative hypothesis do not fit. Fourth, the linear trend hypothesis in the log-linear context requires that the differences between the adjacent categories are equal, whereas the inequality constrained approach only requires a decrease with increasing service level.

12.4.4 Association Between Satisfaction and Attitude, the Log-linear Approach

The question of interest in this section concerns the association between satisfaction and attitude. This breaks down in the following subquestions: (a) Is there a positive association between satisfaction and attitude? (b) Does it

Table 12.10. Relevant parameters for model (Service * Satisfaction * Attitude)

Effect	λ	$\exp(\lambda)$	p
Satis="satisfied" * Att="likely"	.60	1.8	.06
Satis="satisfied" * Att="likely," Serv= No acc. man.	−.41	1.2	.19
Satis="satisfied" * Att="likely," Serv= Telephonic acc. man.	−.26	1.4	.41
Satis="satisfied" * Att="likely," Serv= Acc. man.	.15	2.1	.65
Satis="satisfied" * Att="likely," Serv= Acc. team	–	–	–

differ for different service levels? (c) If so, how do the different service levels influence the association between satisfaction and attitude?

To answer question a), the model (Satisfaction, Attitude) is compared with the model (Satisfaction * Attitude). For this question, the variable service level has been excluded from the analysis, as the research question does not concern service level. This results in a highly significant p-value ($LR(1) = 382, p = .000$), and thus it is concluded that the independence model is not a good model for the data. The interaction parameter in the model (Satisfaction * Attitude) takes the value .198, which exponentiated indeed yields an odds ratio of 1.2 (see Table 12.4). Note that it is concluded that the odds ratio is different from one, and from the fact that the odds ratio takes the value 1.2, it is concluded that it is larger than one. Question b) is answered by investigating the model (Service, Satisfaction * Attitude) and compare it to (Service * Satisfaction * Attitude). In the latter model, the association between satisfaction and attitude is allowed to vary over different service levels. The likelihood ratio test shows that the smaller model is rejected ($LR(9) = 293, p = .000$). In Table 12.10, the relevant parameters are displayed. Note that the odds ratio for a specific three-way interaction is calculated by exponentiating the sum of the two-way effect (i.e., Satis = "satisfied" * Att = "likely") and the parameter for this specific three-way effect. For example, the odds ratio of satisfaction and attitude for the service level account team equals $\exp(.6) = 1.8$. This is because the service level account team is the reference category. To compute the odds ratio of satisfaction and attitude for the service level no account manager, the baseline parameter .6 is summed with the specific three-way effect −.41, which yields exponentiated an odds ratio of 1.2. Furthermore, note the discrepancy between the parameter tests and the likelihood ratio test: None of the parameter effects are significant, indicating that none of the associations in a specific service level is different from the association in the reference category ("the account team" level); however, the likelihood ratio test is significant, indicating that, jointly, there is an effect.

Question c) deals with finding structure in the interaction terms. One hypothesis is that the higher the service level, the stronger the association between satisfaction and attitude. This hypothesis can be represented by fit-

Table 12.11. Interpretation of the linear trend parameters

Effect	Linear effect	Odds
Satis="satisfied" * Att="likely"	.18	1.2
Satis="satisfied" * Att="likely," Serv= No acc. man.	–	1.2
Satis="satisfied" * Att="likely," Serv= Telephonic acc. man.	.20	1.5
Satis="satisfied" * Att="likely," Serv= Acc. man.	.39	1.8
Satis="satisfied" * Att="likely," Serv= Acc. team	.59	2.2

ting a linear term across the three-way interaction parameters. Formally, the model can be written as

$$\log(f_{ijk}) = \mu + \lambda_i^{satis} + \lambda_j^{att} + \lambda_k^{serv} + \lambda_{ij}^{satis*att} + \lambda_{ik}^{satis*serv} + \lambda_{jk}^{att*serv} + \beta u_i v_j w_k,$$

where u_i, v_j, and w_k are scores that refer to each category of the corresponding variable. Fitting this model in standard software requires a similar procedure as the linear trend model discussed in Section 12.4.1. The likelihood ratio test against the saturated model is not significant ($LR(2) = 4.6, p = .10$), indicating that the linear trend model is a good model for the data. The linear trend parameter β takes the value .195, which exponentiated equals 1.2. In Table 12.11, it can be seen that the ratio of each adjacent pair of odds equals 1.2, or in other words, in each higher service level, the association between satisfaction and attitude is increased with factor 1.2.

Finally, there exists the idea that up to the service level "account manager," the association between satisfaction and attitude is increasing, but one service level up, it is decreasing. The reason for this is that the personal relationship with the account manager plays an important role in placing orders, whereas placing orders in companies served by account teams is a matter of pricing due to competitive pressure. This hypothesis can also be tested using the linear trend model, and the categories "account team" and "account manager" are reversed. The observed data (see Table 12.4) indeed shows this ordering. The nonsignificant likelihood ratio test ($LR(2) = .490, p = .783$) shows that this is also a good model for the data. The question that remains is which of the linear trend models is best? This cannot be answered using the likelihood ratio test, as the number of parameters of both linear trend models is equal.

12.4.5 Association Between Satisfaction and Attitude, the Bayesian Approach

For the Bayesian scenario, the hypotheses are in an inequality constrained format as displayed in Table 12.12. Hypothesis H_0 is the unconstrained hypothesis. Hypothesis H_1 requires the odds ratio between satisfaction and attitude

Table 12.12. Constrained hypotheses for the association between satisfaction and attitude

$H_0 : \pi_{111}, \pi_{112}, \pi_{121}, ..., \pi_{222}$
$H_1 : (\pi_{11+} * \pi_{22+})/(\pi_{21+} * \pi_{12+}) > 1$
$H_2 : \theta_1 > 1, \theta_2 > 1, \theta_3 > 1, \theta_4 > 1$
$H_3 : \theta_1 < \theta_2 < \theta_3 < \theta_4$
$H_4 : \theta_1 < \theta_2 < \theta_3 > \theta_4$

Table 12.13. Results for the constrained hypotheses for association between satisfaction and attitude

Hypothesis	PMP	BF_{t0}
H_1	.07	2.0
H_2	.51	15.4
H_3	.25	7.5
H_4	.15	5.5

to be greater than one, when summed over all service levels. Hypothesis H_2 requires the association between satisfaction and attitude to be greater than one, but for each service level separately. Note that θ_k denotes the odds ratio of satisfaction and attitude at service level k. Hypothesis H_3 requires the odds ratios to be increasing with service level. Finally, H_4 reflects the hypothesis that up to the service level "account manager," the association is increasing, but for "account team," it is decreasing.

The results are displayed in Table 12.13. First, hypothesis H_1 is supported by the data, however, not convincing, mainly due to its little restrictive character. Hypothesis H_2 is strongly supported by the data, indicating that in each service level there is a positive association between satisfaction and attitude. The hypothesis of ordered odds ratios (H_3) is also supported by the data, and it is supported more than hypothesis H_4. Hypotheses H_3 and H_4 can be compared directly: Hypothesis H_3 is supported $7.5/5.5 = 1.36$ times more than H_4. With respect to the set of hypotheses, the posterior model probabilities show that hypothesis H_2 is mostly supported by the data (PMP = .51).

Since, both hypotheses H_2 and H_3 are supported by the data, it is of interest to investigate whether a combination of hypotheses H_2 and H_3 is also supported by the data. This hypothesis is given by $H_5 : 1 < \theta_1 < \theta_2 < \theta_3 < \theta_4$. It is very strongly supported by the data ($BF_{50} = 115$).

12.4.6 Comparison Between Log-linear and Bayesian Approach

Two differences with the log-linear analysis are discussed. First, in the log-linear context, apart from the linear trend hypotheses, all null hypotheses were rejected, indicating that the smaller model fitted worse than the satu-

Table 12.14. All log-linear models

Model	*LR*	*df*	*p*
satis, att, serv	675	10	.000
satis * att, serv	293	9	.000
satis, att * serv	574	7	.000
satis * serv, att	543	7	.000
linear(satis * serv), att	558	9	.000
satis * att, satis * serv	161	6	.000
satis * att, att * serv	192	6	.000
satis * serv, att * serv	442	4	.000
satis * att, satis * serv, att * serv	50	3	.000
satis * att, linear(satis * serv), att * serv, linear three-way	22	4	.000
satis * att, satis * serv, att * serv, linear three-way	4.6	2	.100
satis * att * serv	–	0	–

rated model, thus promoting the model with the least structure as the best model. Instead, the Bayesian approach showed more support for each of the constrained hypotheses than for the unconstrained hypothesis; thus, structure has been found, hence one learns something from the data. Second, in the log-linear context, it is not possible to compare two hypotheses with the same number of parameters. The idea that up to the service level "account manager," the association between satisfaction and attitude is increasing, but one service level up, it is decreasing could not be tested in a log-linear model (apart from the eyeball test); however, in the Bayesian analyses this was not a problem at all.

12.4.7 The Best Model

In the log-linear context, it is of interest to find the smallest model (i.e., fewest number of parameters that is not rejected against the saturated model). In Section 12.4.4, the linear trend model was found to be a good model for the data. This model included all possible effects (two-way and three-way interaction terms); however, the three-way interaction term was structured such that the model was not yet saturated. There still may be a more parsimonious model, and this is what is searched for in this section. Note that this is very explorative; however, this is the main method in log-linear models.

The likelihood ratio tests of all possible models of interest against the saturated model are displayed in Table 12.14. First, it can be seen that the fit greatly improves when a model includes the association between satisfaction and attitude. Next, the smallest (and only) model that is not rejected against the saturated model is the structured three-way interaction. Regarding the final model, all variables appear to be associated, whereby only the three-way interaction effect can be structured due to the ordinal nature of the variable service level.

Table 12.15. Combining Bayesian hypotheses

H_0 :	$\pi_{111}, \pi_{112}, \pi_{121}, ..., \pi_{222}$
H_1 :	$\pi_{2+1}/\pi_{1+1} > \pi_{2+2}/\pi_{1+2} > \pi_{2+3}/\pi_{1+3} > \pi_{2+4}/\pi_{1+4}$
H_2	$1 < \theta_1 < \theta_2 < \theta_3 < \theta_4$
H_3 :	$\pi_{2+1}/\pi_{1+1} > \pi_{2+2}/\pi_{1+2} > \pi_{2+3}/\pi_{1+3} > \pi_{2+4}/\pi_{1+4}$, $1 < \theta_1 < \theta_2 < \theta_3 < \theta_4$

Table 12.16. Results for the combined constrained hypotheses

Hypothesis	PMP	BF_{t0}
H_1	.01	98
H_2	.01	115
H_3	.98	11250

The Bayesian approach is more theory driven, and as such, there is no range of models to fit in search for structure. However, as is shown in the previous sections, supported hypotheses can be combined into a new hypothesis. In Section 12.4.2, the most supported hypothesis was the ordering of probabilities in combination with the restriction that the probability of falling into the category "satisfied" had to be greater than the probability of falling into the category "unsatisfied." This hypothesis is again displayed in Table 12.15 as hypothesis H_1. In Section 12.4.5, the most supported hypothesis was the ordered odds ratios in combination with all odds ratios being greater than one. This hypothesis is displayed as hypothesis H_2 in Table 12.15. Hypothesis H_3 is the combination of both hypotheses.

The results displayed in Table 12.16 show that the combined hypothesis H_3 is decisively supported by the data (PMP = .98). The conclusion for the Bayesian approach is that the satisfaction decreases with service level, and the association between service level and attitude is greater than 1 and increases with service level.

Finally, for the "final model," the (transformed) parameter estimates are displayed in Table 12.17, along with their observed counterparts. The column "Obs. % satis" displays the observed satisfaction per service level, whereas the column "Est. % satis." displays the estimated satisfaction per service level. The observed satisfaction is not strictly ordered with service level; however, the violation seems not large. The estimated satisfaction is strictly decreasing with service level, as the final hypothesis H_3 required. The column "Obs. OR" displays the observed odds ratio for the association between attitude and satisfaction for each service level and is not strictly ordered. The column "Est. OR" displays the estimated odds ratios. As can be seen, the estimates under the order restrictions do not differ largely from the observed quantities, hence the good fit.

Table 12.17. Customer satisfaction and Attitude by Service level

	Observed and Estimated quantities			
Service level	Obs. % satis.	Est. % satis.	Obs. OR	Est. OR
No acc. man.	76%	76%	1.2	1.2
Telephonic acc. man.	72%	73%	1.4	1.4
Acc. man.	73%	72%	2.1	2.0
Acc. team	70%	69%	1.8	2.5

Summarizing, the best model in the log-linear context showed a rich source of dependencies; however, they could not be reduced to clearly interpretable structure, apart from the linear three-way interaction effect. The Bayesian approach showed that both for the ordering of satisfaction and the association between satisfaction and attitude, the ordinal nature of service level played a role. Furthermore, the log-linear model agreed on counts that are not equally distributed over levels of satisfaction and showed a significant association between satisfaction and attitude. The Bayesian approach showed that there were indeed *more* counts in the category "satisfied" than in the category "unsatisfied" and showed a *positive* association between satisfaction and attitude.

12.5 Conclusion

In this chapter, a comparison was made between standard log-linear models and a Bayesian inequality constrained approach. It was shown how log-linear parameters are interpreted and how associations are tested. For the Bayesian approach, it was shown how the best of a set of hypotheses is selected using posterior model probabilities. In the example, it was shown how certain expectations about the data were translated in hypotheses, and the differences between the log-linear approach and the Bayesian approach were discussed. For an overview, Table 12.18 is presented to show the log-linear models needed to test the hypotheses together with their inequality constrained Bayesian counterparts.

Summarizing, the following differences were found. First, the interpretation of parameters in the log-linear context requires a detailed knowledge of the chosen parametrization and careful interpretation of the effects, as the exponentiated parameters are all multiplication factors with respect to a reference category. In the Bayesian approach, the hypotheses are formulated directly in terms of cell probabilities or odds ratios, which simplifies the interpretation. Second, the classical p-value takes only the null hypothesis into account, whereas the posterior model probability is both a measure of evidence for the null hypothesis and for the alternative hypothesis. The interpretation is straightforward: It is the probability that the hypothesis is true, given the

Table 12.18. Comparison of log-linear models and constrained hypotheses

Log-linear	Inequality constrained
satis * att	$(\pi_{11+} * \pi_{22+})/(\pi_{21+} * \pi_{12+}) > 1$
satis, att, serv	$\pi_{2++} > \pi_{1++}$
satis * serv, att	$\{\pi_{2+1} > \pi_{1+1}\},\{\pi_{2+2} > \pi_{1+2}\},$ $\{\pi_{2+3} > \pi_{1+3}\},\{\pi_{2+4} > \pi_{1+4}\}$
satis * att * serv	$\theta_1 > 1, \theta_2 > 1, \theta_3 > 1, \theta_4 > 1$
linear(satis * serv), att	$\pi_{2+1}/\pi_{++1} > \pi_{2+2}/\pi_{++2} > \pi_{2+3}/\pi_{++3} > \pi_{2+4}/\pi_{++4}$
all two-way, linear three-way	$\theta_1 < \theta_2 < \theta_3 < \theta_4$
satis * att * serv	$\pi_{111}, \pi_{112}, \pi_{121}, ..., \pi_{222}$

data and the set of hypotheses. Third, the use of the posterior model probability is not limited to two hypotheses, and the number of parameters does not need to differ (thus nested and non-nested hypotheses can be compared). Fourth, the Bayesian approach allows for directional testing, whereas in the log-linear approach the direction of effects is inferred from a combination of significant parameters and observed effects. The interested reader is referred to `http://www.fss.uu.nl/ms/informativehypotheses` for software for inequality constrained contingency table analysis.

References

[1] Agresti, A.: Categorical Data Analysis. New York, Wiley (1990)

[2] Agresti, A., Coull, B.A.: Order-restricted inference for monotone trend alternatives in contingency tables. Computational Statistics and Data Analysis, **28**, 139–155 (1998)

[3] Agresti, A., Coull B.A.: The analysis of contingency tables under inequality constraints. Journal of Statistical Planning and Inference, **107**, 45–73 (2002)

[4] Chib, S.: Marginal likelihood from the Gibbs output. Journal of the American Statistical Association, **90**, 1313–1321 (1995)

[5] Cowles, M.K., Carlin, B.P.: Markov chain monte carlo convergence diagnostics: A comparative review. Journal of the American Statistical Association, **91**, 883–904 (1996)

[6] Gelfand, A.E., Smith, A.F.M.: Sampling-based approaches to calculating marginal densities. Journal of the American Statistical Association, **85**, 398–409 (1990)

[7] Gelfand A.E., Smith A.F.M., Lee, T.M.: Bayesian analysis of constrained parameter and trucated data problems using Gibbs sampling. Journal of the American Statistical Association, **87**, 523–532 (1992)

[8] Gelman, A., Carlin J.B., Stern, H.S., Rubin, D.B.: Bayesian Data Analysis. London, Chapman & Hall (1995)

[9] Kass, R.E.: Bayes factors in practice. The Statistician, **42**, 551–560 (1993)

[10] Kass, R.E., Raftery, A.E.: Bayes factors. Journal of the American Statistical Association, **90**, 773–795 (1995)

[11] Klugkist, I., Kato, B., Hoijtink, H.: Bayesian model selection using encompassing priors. Statistica Neerlandica, **59**, 57–69 (2005)
[12] Laudy, O., Hoijtink, H.: Bayesian methods for the analysis of inequality constrained contingency tables. Statistical Methods in Medical Research, **16**, 123–138 (2007)
[13] Narayanan, A.: Computer generation of Dirichlet random vectors. Journal of Statistical Computation and Simulation, **36**, 19–30 (1990)
[14] Verdinelli, I., Wasserman, L.: Computing Bayes factors using a generalization of the Savage-Dickey density ratio. Journal of the American Statistical Association, **90**, 614–618 (1995)

13

Inequality Constrained Multilevel Models

Bernet Sekasanvu Kato[1] and Carel F.W. Peeters[2]

[1] Twin Research and Genetic Epidemiology Unit, St. Thomas' Hospital Campus, King's College London, Westminster Bridge Road, London SE1 7EH, United Kingdom bernet.kato@kcl.ac.uk

[2] Department of Methodology and Statistics, Utrecht University, P.O. Box 80140, 3508 TC Utrecht, the Netherlands c.f.w.peeters@uu.nl

13.1 Multilevel Models

13.1.1 Introduction

In many areas of research, datasets have a multilevel or hierarchical structure. By hierarchy we mean that units at a certain level are grouped or clustered into, or nested within, higher-level units. The "level" signifies the position of a unit or observation within the hierarchy. This implies that the data are collected in groups or clusters. Examples of clusters are families, schools, and firms. In each of these examples a cluster is a collection of units on which observations can be made. In the case of schools, we can have three levels in the hierarchy with pupils (level 1) within classes (level 2) within schools (level 3). The key thing that defines a variable as being a level is that its units can be regarded as a random sample from a wider population of units. For example, considering a multilevel data structure of pupils within classes within schools, the pupils are a random sample from a wider population of pupils and the classrooms are a random sample from a wider population of classrooms. Likewise the schools are a random sample from a wider population of schools. Data can then be collected at the pupil level (for example, a test score), at the classroom level (for example, teacher experience in years), and at the school level (for example, school's mean socioeconomic status). Variables like gender and social class are not levels. This is because they have a small fixed number of categories. For example, gender has only two categories, male and female. There is no wider population of gender categories that male and female are a random sample from. Another usual form of clustering arises when data are measured repeatedly on the same unit, for instance a patient. In this case the measurements from each patient would be at level 1 and the patients would be at level 2.

In all cases the elements of a cluster share some common characteristics. Therefore, the observations within a cluster tend to be more alike than observations from different clusters that is, they are correlated. For instance, in the pupils within classrooms example, pupils in the same classroom share some common characteristics (e.g., they have the same teachers); thus the test scores of pupils within a classroom will tend to be more alike than test scores from different classrooms. Multilevel data therefore have two sources of variation: In addition to the variation within clusters, the heterogeneity between clusters introduces an additional source of variation. Therefore, any analysis methods used should take the *within* cluster and *between* cluster variation into account. Because data can be clustered at more than a single level (e.g., pupils within classrooms within schools), data clustered at a single level (e.g., pupils within classrooms) are referred to as two-level data and the statistical models for the analyses are referred to as two-level models.

Multilevel or hierarchical data structures can occur in many areas of research, including economics, psychology, sociology, agriculture, medicine, and public health. Over the last 25 years, there has been increasing interest in developing suitable techniques for the statistical analysis of multilevel data, and this has resulted in a broad class of models known under the generic name of *multilevel* models. Generally, multilevel models are useful for exploring how relationships vary across higher-level units taking into account the *within* and *between* cluster variations. Considering an example of two-level data obtained on pupils within schools, there are two possible ways to deal with the data: either to focus separately on the pupils or on the schools. Focusing on the pupils by *pooling* together the data from all the schools ignores differences between schools and thus suppresses variation that can be important. Ignoring the clustering will generally cause standard errors of regression coefficients to be underestimated. On the other hand, focusing on schools by analyzing the data of each school separately ignores a lot of information and consequently renders low power for inferences. Multilevel modeling offers a compromise between these two extremes and enables researchers to obtain correct inferences.

13.1.2 The Multilevel Model

In this chapter we will confine ourselves to two-level models for continuous data, with one single outcome or response variable that has been measured at the lowest level and explanatory variables (or covariates) that have been measured at levels 1 and 2. For the sake of consistency, level 1 and level 2 units will be referred to as *individuals* and *groups*, respectively. Stated otherwise, *individuals* will be nested within *groups*.

To fix ideas, suppose we have J groups and N_j individuals in each group such that the total number of individuals is N. Furthermore, assume that one covariate a has been measured at the individual level and one covariate w has been measured at the group level and an outcome variable y has been measured on each individual. As an illustration suppose we have data on mathematics grades from N high school students from J classes as well as information on student socioeconomic background and teacher experience in years. In this case, each of the classrooms would be a *group* and the students would be the *individuals*. Furthermore, y would be the student level outcome variable "math grade," a would be student "socioeconomic status," and w would be "teacher experience" in years. Our interest is in modeling the outcome variable y in terms of the individual level variable a and the group level variable w using a multilevel model. At the individual level, for individual k (where $k = 1, \ldots, N_j$ for group j) within group j ($j = 1, \ldots, J$ groups in the sample) and $\sum_j N_j = N$, we have the following model:

$$y_{kj} = \pi_{1j} + \pi_{2j} a_{kj} + \varepsilon_{kj}. \tag{13.1}$$

In (13.1), π_{1j} is the intercept, π_{2j} is the regression coefficient for the covariate a, and ε is the residual error term. The residual errors ε_{kj} are assumed to have a normal distribution with mean 0 and variance σ^2. Model (13.1) implies that each group j has its own regression equation with an intercept π_{1j} and a slope π_{2j}. The next step in the modeling is to explain the variation of the regression coefficients π_{1j} and π_{2j} by introducing variables at group level:

$$\pi_{1j} = \beta_1 + \beta_2 w_j + u_{1j}, \tag{13.2}$$

$$\pi_{2j} = \beta_3 + \beta_4 w_j + u_{2j}, \tag{13.3}$$

where u_{1j} and u_{2j} are random residual error terms at group level. Note that in (13.2) and (13.3), the regression coefficients (β's) do not vary across groups and that is why they have no subscript j on them. Since they apply to all groups, they are sometimes referred to as *fixed* effects. Furthermore, all between group variation left in the π coefficients after predicting them with the group variable w_j is assumed to be random residual variation (at group level) which is captured by the terms u_{1j} and u_{2j}.

Substituting (13.2) and (13.3) into (13.1) renders the linear *two-level* regression model:

$$y_{kj} = \beta_1 + \beta_2 w_j + \beta_3 a_{kj} + \beta_4 a_{kj} w_j + u_{1j} + u_{2j} a_{kj} + \varepsilon_{kj}. \tag{13.4}$$

The right-hand side of model (13.4) has two parts to it: a *fixed* part $\beta_1 + \beta_2 w_j + \beta_3 a_{kj} + \beta_4 a_{kj} w_j$, where the coefficients are fixed, and a *random* part $u_{1j} + u_{2j} a_{kj} + \varepsilon_{kj}$. Note that in practice one can have several covariates measured at both individual and group level. Therefore, model (13.4) can be written in a slightly more general form using vector notation:

$$y_{kj} = \boldsymbol{x}_{kj} \boldsymbol{\beta}^T + \boldsymbol{z}_{kj} \boldsymbol{u}_j^T + \varepsilon_{kj}, \tag{13.5}$$

where $\boldsymbol{x}_{kj}$ is a vector of predictors (including main effects at levels 1 and 2 as well as interactions between level 1 and level 2 covariates) having coefficients $\boldsymbol{\beta}$. Furthermore, $\boldsymbol{z}_{kj}$ is a vector of predictors having random effects $\boldsymbol{u}_j$ at the group level and ε_{kj} is an error term. In the example above, $\boldsymbol{x}_{kj} = (1, w_j, a_{kj}, a_{kj} w_j)$, $\boldsymbol{z}_{kj} = (1, a_{kj})$, $\boldsymbol{\beta} = (\beta_1, \beta_2, \beta_3, \beta_4)$, and $\boldsymbol{u}_j = (u_{1j}, u_{2j})$. The vector of predictors $\boldsymbol{z}_{kj}$ will usually be a subset of the fixed-effects predictors $\boldsymbol{x}_{kj}$, although this is not a necessary requirement. The random terms

$\boldsymbol{u}_j = (u_{1j}, u_{2j})$ and ε_{kj} are assumed to be mutually independent and normally distributed:

$$\boldsymbol{u}^T \sim \mathcal{N}(\boldsymbol{0}, \boldsymbol{V}), \quad \varepsilon_{kj} \sim \mathcal{N}(0, \sigma^2), \tag{13.6}$$

where $\boldsymbol{V}$ is the variance-covariance matrix of the random effects and σ^2 is the residual variance. Thus, we can see that multilevel models provide a natural way to decompose complex patterns of variability associated with hierarchical structure.

In a frequentist analysis, estimation of parameters in the linear multilevel model is carried out by maximizing the likelihood function. To this end, direct maximization using the Newton-Raphson or Expectation-Maximization (EM) algorithm can be performed. For discussions on the methods, techniques, and issues involved in multilevel modeling in general, the interested reader is referred to [5, 11, 12, 14, 18, 24, 26]. This chapter is intended to illustrate model selection for inequality constrained two-level models. A Bayesian approach will be used for parameter estimation and model selection [15]. Bayesian estimation in multilevel models (without constraints on the model parameters) has also been implemented in the statistical package MLwiN [4].

13.2 Informative Inequality Constrained Hypotheses

Research scientists often have substantive theories in mind when evaluating data with statistical models. Substantive theories often involve inequality constraints among the parameters to translate a theory into a model; that is, a parameter or conjunction of parameters is expected to be larger or smaller than another parameter or conjunction of parameters. Stated otherwise and using β as a generic representation of a parameter, we have that $\beta_i > \beta_j$ or $\beta_i < \beta_j$ for some two parameters β_i and β_j. Additionally, inequality constraints also play a pivotal role when competing theories are presented as an expression of a multitude of initial plausible explanations regarding a certain phenomenon on which data are collected. Consider the following examples on two common multilevel models: school effects models and individual growth models. These examples will be the thrust of Sections 13.4 and 13.5.

Example 1: An educational researcher is interested in the effect of certain student and school level variables on mathematical achievement (*mathach*), and has obtained a dataset on students within schools. A students' ethnic background (*min*), student socioeconomic status (*ses*), a schools' average student socioeconomic status (*mses*), and the dichotomy between Catholic (*cat*) and public (*pub*) schools are hypothesized to be defining variables for the explanation of math achievement (cf. [2, 5, 7, 8, 23]). A possible formulation of the two-level model in the form (13.5) might be

$$\begin{aligned} mathach_{kj} = &\beta_1 cat_j + \beta_2 pub_j + \beta_3 mses_j + \beta_4 cat_j ses_{kj} \\ &+ \beta_5 pub_j ses_{kj} + \beta_6 mses_j ses_{kj} + \beta_7 min_{kj} \\ &+ u_{1j} + u_{2j} ses_{kj} + \varepsilon_{kj}. \end{aligned}$$

The reason for assigning an indicator variable to both the Catholic and public category of the constituent dichotomy is because this will enable one to estimate the regression coefficients corresponding to the covariates *cat* and *pub* and their interactions with other covariates rather than estimating contrasts.

The researcher can think of different plausible models regarding the direction and (relative) strength of the effects of the mentioned variables on the response math achievement. Subsequently, the researcher expresses the idea that students in Catholic schools have higher math achievement than those in public schools $\{\beta_1 > \beta_2\}$. Certain sociological work found that students belonging to a minority have lower math achievement than students not belonging to an ethnic minority $\{\beta_7 < 0\}$. Additionally, the researcher has the expectation that math achievement is positively related to socioeconomic status and that the effect of student socioeconomic status on mathematical achievement is more pronounced in public schools than in Catholic schools $\{\beta_4 < \beta_5\}$. These theories allow for several plausible models of differing complexity and with differing theoretical implications. The question of interest becomes: *Which of the plausible models best fits the data?*

Example 2: A researcher in child and adolescent psychology has obtained observational data on substance abuse collecting multiple waves of data on adolescents. This researcher sets out to assess the effects of alcoholic intake among peers (*peer*) and the fact that the adolescent has alcoholic (*coa*) or non-alcoholic (*ncoa*) parents on the development of adolescent alcohol use (*alcuse*) (cf. [6, 24]). The model can be formulated as

$$\begin{aligned} alcuse_{kj} = &\beta_1 coa_j + \beta_2 ncoa_j + \beta_3 peer_j + \beta_4 coa_j t_{kj} + \beta_5 ncoa_j t_{kj} \\ &+ \beta_6 peer_j t_{kj} + u_{1j} + u_{2j} t_{kj} + \varepsilon_{kj}, \end{aligned}$$

where t_{kj} is a time variable.

For these data, competing theories abound in the researchers' mind. A first plausible theory for him or her could be that adolescents with an alcoholic parent are more prone to have a higher alcoholic intake at baseline $\{\beta_1 > \beta_2\}$, as well as over time $\{\beta_4 > \beta_5\}$. A second plausible theory amends the first, with the additional expectation that for initial alcoholic intake, the effect of an alcoholic parent will be more influential that peer alcoholic intake $\{\beta_1 > \beta_3\}$, whereas for the time-dependent increase in alcoholic intake, peers will be more influential $\{\beta_4 < \beta_6\}$. The question of interest is: *Which of the theories best fits the data?*

The researchers' hypotheses are in fact informative, as they are hypotheses in which one explicitly defines direction or (relative) strength of relationships based on prior information for usage in confirmatory data analysis. Informative hypotheses have a direct connection to model translations of theory. For instance, the researcher from Example 2 would be interested in the following two hypotheses that have been arrived at by translating substantive theories via constraints on model parameters:

$$\begin{aligned} H_1 &: \{\beta_1 > \beta_2\}, \beta_3, \{\beta_4 > \beta_5\}, \beta_6 \\ \text{versus } H_2 &: \{\beta_1 > \beta_2\}, \{\beta_1 > \beta_3\}, \{\beta_6 > \beta_4 > \beta_5\}. \end{aligned}$$

The pertinent question is: Given H_1 and H_2, which of the two hypotheses has more support from the data?

A researcher might bring the classical or frequentist statistical viewpoint to bear on the central question of interest. One would then normally proceed to specify the traditional null hypothesis, which assumes that none of the covariate variables are associated with the response variable of interest against the alternative that at least one covariate variable is associated with the response variable:

$$H_0 : \text{all } \beta_i \text{ equal } 0 \quad \text{versus} \quad H_3 : \text{not all } \beta_i \text{ equal } 0.$$

There are several problems related to this procedure that leads one to infer little information regarding the actual hypotheses of interest, being H_1 and H_2. Generally, in the usual frequentist sharp null hypothesis test setting, the researcher often starts from the idea that H_3 holds and then tests H_0 using an appropriate test statistic. If we assume $\boldsymbol{\beta}$, the vector containing all β_i, is $\boldsymbol{\delta}$ away from the zero vector $\mathbf{0}$, with $\boldsymbol{\delta} > \mathbf{0}$ but very small, then by the consistency of the testing procedure, the rejection of H_0 becomes the sure event for N sufficiently large [21]. One could then actually choose N in accordance with the rejection of H_0. More specifically, if the null hypothesis is rejected, no information is gained regarding the fit of the inequality constrained hypothesis of interest. Note that the research questions of actual interest are not directly incorporated into the alternative hypothesis. Post hoc directional tests are then usually employed with certain corrections on the maintained significance level to assess the inequalities deemed interesting in the actual research hypothesis. If one considers H_1 above, these post hoc tests would amount to assessing:

$$\begin{aligned} H_{01} &: \beta_1 = \beta_2 \quad \text{versus} \quad H_{11} : \beta_1 - \beta_2 > 0 \\ \text{and } H_{02} &: \beta_4 = \beta_5 \quad \text{versus} \quad H_{12} : \beta_4 - \beta_5 > 0. \end{aligned} \tag{13.7}$$

The researcher is left with the situation in which several test results (those for the omnibus test and the post hoc tests) have to be combined to evaluate a single model translated theory. Such a situation may eventually force the researcher to make arbitrary choices. For example, how would one evaluate the situation where not all directional alternatives are accepted, or when the rather arbitrary significance threshold is surpassed by an arbitrarily small amount? Such problems abound especially in the social sciences where it is not uncommon to find situations where power is sufficient for obtaining significance somewhere while being insufficient to identify any specific effect [19]. The power gap between a single test and a collection of tests often renders the situation in which the omnibus test proves significant in the sense that the obtained p-value is smaller than or equal to the pre-specified significance level, while the individual post hoc tests lack power such that successive testing efforts may find erratic patterns of "significant" p-values.

If the null hypothesis is not rejected when testing H_0 against H_3, there is still a possibility that it could be rejected when testing it against the hypotheses of interest, namely H_1 and H_2. Inequality constraints contain information, in the form of truncations of the parameter space, and when properly incorporated, more efficient inferences can result. To gain power, one could therefore specify inequality constrained alternatives more in tune with substantive theoretical beliefs, instead of the traditional alternative H_3. This way the null hypothesis, if rejected, will be rejected in favor of the constrained alternative. Our researcher would then embark on testing

$$\begin{aligned} H_0 &: \beta_1 = \beta_2 = \beta_4 = \beta_5 = 0 \\ \text{versus } H_4 &: \beta_1 - \beta_2 \geqslant 0,\ \beta_4 - \beta_5 \geqslant 0, \text{ and} \\ &\quad \beta_1,\ \beta_2,\ \beta_4, \text{ and } \beta_5 \text{ do not all equal } 0 \end{aligned}$$

and (13.8)

$$\begin{aligned} H_0 &: \beta_1 = \beta_2 = \beta_3 = \beta_4 = \beta_5 = \beta_6 = 0 \\ \text{versus } H_5 &: \beta_1 - \beta_2 \geqslant 0,\ \beta_1 - \beta_3 \geqslant 0,\ \beta_6 - \beta_4 \geqslant 0,\ \beta_4 - \beta_5 \geqslant 0, \text{ and} \\ &\quad \beta_1,\ \beta_2,\ \beta_3,\ \beta_4,\ \beta_5 \text{ and } \beta_6 \text{ do not all equal } 0 \end{aligned}$$

respectively, in order to convey more information regarding the model translated theories of interest. Yet again, there are certain problems that render the information to be inferred from these omnibus tests to be limited.

First, there is an important difference between tests of the form (13.8) and tests of the form (13.7). The former states that a directional effect is present when the alternative is accepted, but it does not give which of the constituent directional effects is significant. For such an evaluation one needs to resort to tests of the latter form, which takes us back to the problems associated with combining several test results to evaluate a single model translated theory as discussed earlier. Moreover, for complex models and multivariate settings there may not generally be optimal solutions for frequentist inequality constrained testing alternatives such as those in (13.8). The interested reader is referred to [1, 22] for overviews on the possibilities of frequentist inequality constrained hypothesis testing. But even if these frequentist alternatives were available, the researcher would still run into a problem when wanting to evaluate which theory or plausible model fits the data best. One possibility is to test the null hypothesis against each of the theories in the form of inequality constrained alternatives. This would help one to obtain some evidence for the support for each of the separate theories, but it would still not answer the question concerning which theory is best. It is very well possible that in all of the tests the null hypothesis is rejected in favor of the inequality constrained alternative.

To assess the researchers' substantive theory in light of the available data, one needs to directly compare the constrained alternatives. This involves the simultaneous evaluation of multiple model translated theories, and for such an exercise, no frequentist possibilities are available. Therefore, Bayesian model selection is posed as an alternative to hypothesis testing. Posterior probabilities can be computed for all models under consideration, which enables the direct comparison of both nested and non-nested models. The incorporation of inequality constrained theory evaluation in a Bayesian computational framework has been formulated for multilevel models in [15]. In the next section it will be shown how the inequality constrained multilevel linear model can be given a Bayesian formulation, how the model parameters can be estimated using a so-called augmented Gibbs sampler, and how posterior probabilities can be computed to assist the researcher in model selection. Those wishing to skip this section may find general information regarding Bayesian estimation and model selection in Chapters 3 and 4. Subsequently, the two examples described above will be analyzed in the inequality constrained Bayesian framework to elaborate model selection among competing inequality constrained model translated theories. This will be done in Sections 13.4 and 13.5. The chapter will be concluded with a discussion in Section 13.6.

13.3 Bayesian Estimation and Model Selection

13.3.1 Introduction

In Bayesian analysis, model specification has two parts to it:

1. The likelihood function $f(\boldsymbol{D}|\boldsymbol{\theta})$, which defines the probability distribution of the observed data $\boldsymbol{D}$ conditional on the unknown (model) parameters $\boldsymbol{\theta}$.
2. The prior distribution $p(\boldsymbol{\theta})$ of the model parameters $\boldsymbol{\theta}$.

Bayesian inference proceeds via specification of a posterior distribution $p(\boldsymbol{\theta}|\boldsymbol{D})$ for $\boldsymbol{\theta}$, which is obtained by multiplying the likelihood and the prior distribution:

$$p(\boldsymbol{\theta}|\boldsymbol{D}) = \frac{f(\boldsymbol{D}|\boldsymbol{\theta})p(\boldsymbol{\theta})}{m(\boldsymbol{D})} \propto f(\boldsymbol{D}|\boldsymbol{\theta})p(\boldsymbol{\theta}), \tag{13.9}$$

where $m(\boldsymbol{D})$ is the marginal distribution of $\boldsymbol{D}$. The posterior distribution $p(\boldsymbol{\theta}|\boldsymbol{D})$ contains the state of knowledge about the model parameters given the observed data and the knowledge formalized in the prior distribution. Random draws from the posterior distribution are then used for inferences and predictions. In the sequel it will be explained how samples can be drawn from the posterior distribution.

For (13.5), the likelihood $f(\boldsymbol{D} \mid \boldsymbol{\theta})$ is

$$\prod_{j=1}^{J} \int_{\boldsymbol{u}_j} \left\{ \prod_{k=1}^{N_j} \frac{1}{\sqrt{2\pi}\sigma} \exp\left(-\frac{(y_{kj} - \boldsymbol{x}_{kj}\boldsymbol{\beta}^T - \boldsymbol{z}_{kj}\boldsymbol{u}_j^T)}{2\sigma^2} \right) \right\} p(\boldsymbol{u}_j \mid \boldsymbol{0}, \boldsymbol{V})\, d\boldsymbol{u}_j, \quad (13.10)$$

where $\boldsymbol{D} = (\boldsymbol{y}_{kj}, \boldsymbol{x}_{kj}, \boldsymbol{z}_{kj} : k = 1, \ldots, N_j;\ j = 1, \ldots, J)$, $\boldsymbol{\theta} = (\boldsymbol{\beta}, \boldsymbol{V}, \sigma^2)$, and $p(\boldsymbol{u_j} \mid \boldsymbol{0}, \boldsymbol{V})$ is a normal distribution with mean $\boldsymbol{0}$ and covariance matrix $\boldsymbol{V}$.

Suppose we have a total of S competing hypotheses or model translated theories H_s for $s = 1, \ldots, S$, where H_1 is the encompassing model (in the remainder of the text we will use the terms "hypothesis" and "model" interchangeably). The encompassing model is one where no constraints are put on the (model) parameters and therefore all other models are nested in H_1. If $p(\boldsymbol{\theta}|H_1)$ denotes the prior distribution of H_1, then it follows that the prior distribution of H_s for $s = 2, \ldots, S$ is

$$p(\boldsymbol{\theta}|H_s) = \frac{p(\boldsymbol{\theta}|H_1) I_{\boldsymbol{\theta} \in H_s}}{\int p(\boldsymbol{\theta}|H_1) I_{\boldsymbol{\theta} \in H_s} d\boldsymbol{\theta}}. \quad (13.11)$$

The indicator function $I_{\boldsymbol{\theta} \in H_s} = 1$ if the parameter values are in accordance with the restrictions imposed by model H_s, and 0 otherwise. Equation (13.11) indicates that for each model under investigation, the constraints imposed on the model parameters are accounted for in the prior distribution of the respective model. Using independent prior distributions for each of the model parameters, the prior distribution of the unconstrained encompassing model H_1 can be written as the product

$$p(\boldsymbol{\theta}|H_1) = p(\boldsymbol{\beta}) \times p(\boldsymbol{V}) \times p(\sigma^2), \quad (13.12)$$

where $p(\boldsymbol{\beta})$, $p(\boldsymbol{V})$, and $p(\sigma^2)$ are the prior distributions of $\boldsymbol{\beta}$, $\boldsymbol{V}$, and σ^2, respectively. In order to obtain a conjugate model specification, normal priors will be used for the fixed effects $\boldsymbol{\beta}$, a scaled inverse χ^2 prior for σ^2, and an inverse Wishart prior for $\boldsymbol{V}$. It follows that for the unconstrained encompassing model H_1, the posterior distribution of the parameters in $\boldsymbol{\theta}$ is proportional to the product of (13.10) and (13.12).

In what follows, it is explained how prior distributions for $\boldsymbol{\beta}$, $\boldsymbol{V}$, and σ^2 will be specified. As mentioned in Chapter 4 (see also [15, 17]), the encompassing prior should not favor the unconstrained or any of the constrained models. Because all constraints are on the parameters in the vector $\boldsymbol{\beta}$, each of the βs will be assigned the same prior distribution. In general, the estimate for the regression coefficient β_0 in a linear regression model with no covariates, $y = \beta_0 + \varepsilon$, where y is the dependent variable and ε is an error term, is the mean of y (i.e., $\widehat{\beta}_0 = E(y)$). Each of the parameters in $\boldsymbol{\beta}$ will therefore be assigned a normal distribution with mean equal to the mean of the response variable (from the data) and a large variance chosen so that the prior has minimal influence on the posterior distribution of the parameter. The prior distribution of σ^2 will also be data based – σ^2 will be assigned a scaled inverse χ^2-distribution with 1 degree of freedom and scale equal to the variance of the response variable. Lastly, $\boldsymbol{V}$ will be assigned an inverse Wishart prior distribution with $R+1$ degrees of freedom and as scale matrix the $R\times R$ identity matrix where R is the dimension of $\boldsymbol{V}$. Estimating the covariance matrix $\boldsymbol{V}$ is challenging especially when $R>2$. This is because each of the correlations (between the components of $\boldsymbol{u}$ in (13.6)) has to fall in the interval $[-1,1]$ and $\boldsymbol{V}$ must also be positive definite. Setting the degrees of freedom to $R+1$ ensures that each of the correlations has a uniform distribution on $[-1,1]$ ([11]). Although setting the degrees of freedom to $R+1$ ensures that the resulting model is reasonable for the correlations, it is quite constraining for the estimation of the variance terms in $\boldsymbol{V}$. Therefore, when $R>2$, it is recommended to model $\boldsymbol{V}$ using a scaled inverse Wishart distribution. The interested reader is referred to [11] for more details on the implementation.

13.3.2 Estimation

In this section it is explained how samples can be obtained from the posterior distribution of H_1 and how they can be used for inferences. With conjugate prior specifications, in (13.12), the full conditional distributions of $\boldsymbol{V}$ and σ^2 are inverse Wishart and scaled inverse χ^2 distributions, respectively, and the full conditional distribution of each parameter in the vector of fixed effects $\boldsymbol{\beta}$ is a normal distribution.

The Gibbs sampler (see, for example, [9, 15, 25]), which is an iterative procedure, can be used to sample from the conditional distribution of each model parameter – the set of unknown parameters is partitioned and then each parameter (or group of parameters) is estimated conditional on all the others. To sample from the posterior distribution of the encompassing model H_1 described in Section 13.3.1, first initial values are assigned to each of the model parameters. Next, Gibbs sampling proceeds in four steps, namely:

- Sample $\boldsymbol{u}_j$ for $j=1\ldots,J$ from $\mathcal{N}(\boldsymbol{\Phi}_j,\boldsymbol{\Sigma}_j)$ where

$$\boldsymbol{\Phi}_j=\frac{\boldsymbol{\Sigma}_j}{\sigma^2}\sum_{k=1}^{N_j}\boldsymbol{z}_{kj}^T(y_{kj}-\boldsymbol{x}_{kj}\boldsymbol{\beta}^T)$$

  and

$$\boldsymbol{\Sigma}_j=\left[\frac{\sum_{k=1}^{N_j}\boldsymbol{z}_{kj}^T\boldsymbol{z}_{kj}}{\sigma^2}+\boldsymbol{V}^{-1}\right]^{-1}.$$

- If the prior distribution $p(\sigma^2)$ of σ^2 is an inverse chi-square distribution with degrees of freedom γ and scale ω^2, then sample σ^2 from a scaled inverse χ^2-distribution with degrees of freedom $\gamma+\sum_{j=1}^{J}N_j$ and scale

$$\gamma\omega^2 + \sum_{j=1}^{J}\sum_{k=1}^{N_j}(y_{kj} - \boldsymbol{x}_{kj}\boldsymbol{\beta}^T - \boldsymbol{z}_{kj}\boldsymbol{u}_j^T)^2.$$

- If the prior distribution $p(\boldsymbol{V})$ of $\boldsymbol{V}$ is an inverse Wishart distribution with degrees of freedom λ and scale matrix $\boldsymbol{T}$, sample $\boldsymbol{V}$ from an inverse Wishart distribution with degrees of freedom $\lambda + J$ and scale matrix

$$\sum_{j=1}^{J}\boldsymbol{u}_j\boldsymbol{u}_j^T + \boldsymbol{T}.$$

- Let $\boldsymbol{\beta} = \{\beta_1, \ldots, \beta_p, \ldots, \beta_P\}$. If the prior distribution of β_p is a normal distribution with mean μ_p and variance τ_p^2, then sample β_p from a normal distribution with mean

$$\frac{\frac{\mu_p}{\tau_p^2} + \sigma^{-2}\sum_{j=1}^{J}\sum_{k=1}^{N_j}\left[y_{kj} - \sum_{\substack{i=1\\ i\neq p}}^{P}\beta_i x_{ikj} - \sum_{q=1}^{Q}u_{qj}z_{qkj}\right]x_{pkj}}{\tau_p^{-2} + \sigma^{-2}\sum_{j=1}^{J}\sum_{k=1}^{N_j}x_{pkj}^2}$$

and variance

$$\left[\frac{1}{\tau_p^2} + \frac{\sum_{j=1}^{J}\sum_{k=1}^{N_j}x_{pkj}^2}{\sigma^2}\right]^{-1}.$$

Effectively, the Gibbs sampler starts with initial values for all the parameters and then updates the parameters in turn by sampling from the conditional posterior distribution of each parameter. Iterating the above four steps produces a sequence of simulations $\boldsymbol{u}_1^{(1)}, \ldots, \boldsymbol{u}_J^{(1)}$, $\sigma^{2(1)}$, $\boldsymbol{V}^{(1)}$, $\beta_1^{(1)}, \ldots, \beta_P^{(1)}$; $\boldsymbol{u}_1^{(2)}, \ldots, \boldsymbol{u}_J^{(2)}$, $\sigma^{2(2)}$, $\boldsymbol{V}^{(2)}$, $\beta_1^{(2)}, \ldots, \beta_P^{(2)}$; $\boldsymbol{u}_1^{(3)}, \ldots, \boldsymbol{u}_J^{(3)}$, $\sigma^{2(3)}$, $\boldsymbol{V}^{(3)}$, $\beta_1^{(3)}, \ldots, \beta_P^{(3)}$; and so on until the sequence has converged. The first set of iterations, referred to as the *burn-in*, must be discarded since they depend on the arbitrary starting values. See Chapter 3 and references therein for more information on convergence diagnostics for the Gibbs sampler.

After convergence, samples drawn from the posterior distribution can be used to obtain parameter estimates, posterior standard deviations, and central credibility intervals. See, for example, [13]. To elaborate, suppose that $\boldsymbol{\beta} = (\beta_1, \beta_2)$ and we have a sample $(\beta_1^{(b)}, \beta_2^{(b)}), b = 1, \ldots, B$, from the posterior distribution. To estimate the posterior mean of β_1, a researcher would use

$$\frac{1}{B}\sum_{b=1}^{B}\beta_1^{(b)}, \tag{13.13}$$

and a 95% central credibility interval (CCI) for β_1 would be obtained by taking the empirical .025 and .975 quantiles of the sample of $\beta_1^{(b)}$ values. Furthermore, estimates of functions of parameters can also be obtained. For instance, suppose an estimate for the posterior mean of $\beta_1 - \beta_2$ and a credibility interval is required. This is easily obtained by taking the difference $\beta_1^{(b)} - \beta_2^{(b)}$, $b = 1, \ldots, B$, and using the computed values to obtain the posterior mean and credibility interval. Samples from the posterior distribution can also be used to draw histograms to display the distributions of parameters and functions of parameters.

13.3.3 Model Selection

If $p(H_s)$ and $m(\boldsymbol{D}|H_s)$ denote the prior probability and marginal likelihood of model H_s, respectively, then the posterior model probability (PMP) of H_s is

$$\mathrm{PMP}(H_s \mid \boldsymbol{D}) = \frac{m(\boldsymbol{D} \mid H_s)p(H_s)}{\sum_{s'=1}^{S} m(\boldsymbol{D} \mid H_{s'})p(H_{s'})}. \tag{13.14}$$

The method of encompassing priors (see [15, 17] and Chapter 4), can be used to obtain posterior probabilities for each model under investigation. If $1/c_s$ and $1/d_s$ are the proportions of the prior and posterior distributions of H_1 that are in agreement with the constraints imposed by model H_s, then the Bayes factor BF_{s1} comparing H_s to H_1 is the quantity c_s/d_s. Note that for each constrained model H_s, the quantities $1/c_s$ and $1/d_s$ provide information about the *complexity* ("size" of the parameter space) and *fit* of H_s, respectively. Subsequently, if H_1 is the encompassing model and assuming that each model H_s is a priori equally likely, it follows that

$$\mathrm{PMP}(H_s|\boldsymbol{D}) = \frac{BF_{s1}}{BF_{11} + BF_{21} + \cdots + BF_{S1}}, \tag{13.15}$$

for each $s = 1, \ldots, S$ and $BF_{11} = 1$. In practice, therefore, one only needs to specify the prior distribution and correspondingly the posterior distribution of the encompassing model. Next, samples are drawn from the specified prior and posterior distributions, which are then used to determine the quantities $1/c_s$ and $1/d_s$. Subsequently, posterior probabilities can be computed using (13.15) and the model with the highest posterior probability is considered to be the one that gets the highest support from the data. If the model with the highest posterior probability is one of the constrained models, then parameter estimates for the model can be obtained using the Gibbs sampling procedure presented in Section 13.3.2 with an extra step, namely that the β's are sampled from truncated normal distributions (see Chapter 3).

Note that if a diffuse encompassing prior is used, then for the class of models with strict inequality constraints, such as $\beta_1 > \beta_2 > \beta_3$ or $\beta_4 > 0$, the PMPs obtained will not be sensitive to the prior specification. However for models with equality constraints, such as $\beta_1 = \beta_2 = \beta_3$ or $\beta_4 = 0$, PMPs strongly depend on the actual specification of the encompassing prior. For details on this, the interested reader is referred to Chapter 4 and [15, 16, 17]. In this chapter, models with equality constraints are not considered, so sensitivity of PMPs to the choice of encompassing prior is not an issue.

13.4 School Effects Data Example

13.4.1 Data

The data used in this section are a subsample of the 1982 High School and Beyond Survey.[3] It includes information on 7,185 students nested within 160 schools. Data were obtained from: `http://www.ats.ucla.edu/stat/paper examples/singer/default.htm`.

The data set includes the following variables:

1. ***mathach***: The response variable, which is a standardized measure of mathematics achievement. The variable *mathach* has mean 12.75, standard deviation 6.88, and range −2.83 to 24.99.
2. ***ses***: A composite and centered indicator of student socioeconomic status. It was a composite of parental education, parental occupation, and income. The variable *ses* has mean 0.00014, standard deviation 0.78, and range −3.76 to 2.69.
3. ***minority***: A student level dummy variable that was coded as 1 if the student belonged to a minority and 0 otherwise. Numbers of minority and nonminority students were 1974 and 5211, respectively.
4. ***meanses***: School level variable indicating the average of student *ses* values within each school. As *ses* was centered around its mean a score of 0 can be interpreted as indicating a school with average (in fact average average) student *ses* values, whereas −1 and 1 indicate schools with below average and above average student *ses* values respectively. The variable *mses* has mean 0.0061, standard deviation 0.41, and range −1.88 to 0.83.
5. ***sector***: School level dichotomous variable where 1 indicates a Catholic school and 0 indicates a public school. Numbers of Catholic and public schools were 70 and 90, respectively.

Let $mathach_{kj}$ and ses_{kj} respectively represent the math achievement and student socioeconomic status for the kth ($k = 1, \ldots, 7185$) student in the jth school ($j = 1, \ldots, 160$). Let min_j be an indicator variable defined to be 1 if subject k in school j belongs to an ethnic minority, and 0 otherwise. Furthermore, let cat_j and pub_j be school level indicator variables defined to be 1 if a school is Catholic or public, respectively, and 0 otherwise. It should be noted that the variable *cat* is equivalent to the original variable *sector*. The reason for defining a new indicator variable *pub* is because in a regression model, this will make it possible to estimate the regression coefficients corresponding to the covariates *cat* and *pub* and their interactions with other covariates rather

[3] This data collection provides the second wave of data in a longitudinal, multi-cohort study of American youth conducted by the National Opinion Research Center on behalf of the National Center for Education Statistics. In the first wave, conducted in 1980, data were collected from 58,270 high school students and 1015 secondary schools by self-enumerated questionnaires, personal and telephone interviews, and mailback questionnaires.

than estimating contrasts. Furthermore, defining variables in this way enables one to put constraints on the model parameters. Finally, let $mses_j$ represent the continuous school level variable *meanses*.

13.4.2 Theory and Models

Research into child and adolescent mathematical achievement has spurred a vast stream of sociological, psychological, and educational literature; see, for example, [2, 5, 7, 8, 23]. Van den Berg, Van Eerde, and Klein [2] conducted research into the mathematical skills of ethnic minorities in the Dutch elementary school system. They concluded that children from ethnic minorities have less mathematical ability/maturity than children from the native Dutch population. These effects were, in their view, attributable to a language barrier and the differential use of educational skills between the home and the school environment. These effects are expected to persist throughout high school. Gamoran [7] found that Catholic schools produce higher overall math achievement in comparison to public schools. The (partial) explanation for this was found in the manner in which Catholic schools implement academic tracking. In addition, [5, 23] have indicated that higher math achievement occurs in schools where the average student socioeconomic status is higher. It is these expectations we want to express in a set of informative hypotheses.

Assuming a linear relationship between a student's mathematics achievement, *ses* and *min*, the relationship can be modeled using

$$mathach_{kj} = \pi_{1j} + \pi_{2j} ses_{kj} + \pi_{3j} min_{kj} + \varepsilon_{kj},$$

where

$$\begin{aligned} \pi_{1j} &= \beta_1 cat_j + \beta_2 pub_j + \beta_3 mses_j + u_{1j}, \\ \pi_{2j} &= \beta_4 cat_j + \beta_5 pub_j + \beta_6 mses_j + u_{2j}, \\ \pi_{3j} &= \beta_7, \end{aligned}$$

and with

$$\boldsymbol{u} = (u_{1j}, u_{2j})^T \sim \mathcal{N}(\boldsymbol{0}, \boldsymbol{V}), \quad \varepsilon_{kj} \sim \mathcal{N}(0, \sigma^2).$$

Thus, the school-specific intercepts (π_{1j}) and *ses* effects (π_{2j}) are related to the type of school and average socioeconomic status of the school. Note that the coefficient π_{3j} does not vary across schools. To keep things simple we are assuming it has the same value β_7 for each school ($j = 1, \ldots, 160$). Making the coefficient differ for each school, say by having $\pi_{3j} = \beta_7 + u_{3j}$, would give rise to a 3×3 covariance matrix $\boldsymbol{V}$ for $\boldsymbol{u} = (u_{1j}, u_{2j}, u_{3j})^T$. Effectively, the extra term u_{3j} introduces three new variance components, namely $\text{cov}(u_{1j}, u_{2j})$, $\text{cov}(u_{2j}, u_{3j})$, and $\text{var}(u_{3j})$ that have to be estimated from the data.

The following competing inequality constrained model translated theories will be compared:

$$
\begin{aligned}
&H_1 : \beta_1,\ \beta_2,\ \beta_3,\ \beta_4,\ \beta_5,\ \beta_6,\ \beta_7,\\
&H_2 : \{\beta_1 > \beta_2\},\ \beta_3,\ \beta_4,\ \beta_5,\ \beta_6,\ \beta_7,\\
&H_3 : \beta_1,\ \beta_2,\ \beta_3,\ \beta_4,\ \beta_5,\ \beta_6,\ \beta_7 < 0,\\
&H_4 : \{\beta_1 > \beta_2\},\ \beta_3,\ \beta_4,\ \beta_5,\ \beta_6,\ \beta_7 < 0,\\
&H_5 : \{\beta_1 > \beta_2\},\ \beta_3,\ \{\beta_4 < \beta_5\},\ \beta_6,\ \beta_7 < 0,\\
&H_6 : \{\beta_1 > \beta_2\},\ \beta_3,\ \{\beta_4 > \beta_5\},\ \beta_6,\ \beta_7 < 0.
\end{aligned}
$$

Model 1 is the unconstrained encompassing model. Model 2 expresses the idea that students in Catholic schools have higher math achievement than those in public schools $\{\beta_1 > \beta_2\}$. Model 3 expresses the viewpoint that students belonging to a minority will have lower math achievement than students not belonging to an ethnic minority. As min_j is an indicator variable defined to be 1 if subject k in school j belongs to an ethnic minority, the previous expectation means that β_7 should be negative, so that $\beta_7 < 0$. Model 4 combines the viewpoints in models 2 and 3, namely that student in Catholic schools perform better than those in public schools and that students belonging to ethnic minorities perform worse than those not belonging to an ethnic minority. Model 5 expresses the viewpoints of model 4, with the additional expectation that the slopes for *ses* are higher in public compared to Catholic schools $\{\beta_4 < \beta_5\}$. Lastly, model 6 expresses the viewpoints of model 4, with the additional expectation that the slopes for *ses* are higher in Catholic compared to public schools $\{\beta_4 > \beta_5\}$.

13.4.3 Results

As mentioned before, Bayesian analysis requires specification of prior distributions for all unknown parameters in the encompassing model (H_1). For all analyses diffuse priors were used. The regression coefficients $\beta_1, \ldots, \beta_7$ were each given normal prior distributions with mean 12.75 and variance 10^4 (that is, standard deviation 100). What this means is that each of the coefficients is expected to be in the range $(-87,\ 113)$, and if the estimates are in this range, the prior distribution is providing very little information in the inference. Because the outcome and all predictors have variation that is of the order of magnitude 1, we do not expect to obtain coefficients much bigger than 20, so prior distributions with standard deviation 100 are noninformative. The variance covariance matrix $\boldsymbol{V}$ was given an inverse Wishart prior distribution with 3 degrees of freedom and as scale matrix a 2×2 identity matrix. Lastly, σ^2 was given a scaled inverse χ^2 prior distribution with 1 degree of freedom and scale 47.

To obtain posterior model probabilities for the competing models, $200,000$ samples (after a burn-in of $10,000$) were drawn from the prior and posterior distributions of the encompassing model (H_1), respectively. For each of the constrained models $H_2, \ldots, H_6$, the proportion of samples from prior and posterior in agreement with the constraints on $\boldsymbol{\beta}$ were used to estimate the pos-

Table 13.1. Posterior model probabilities

Model	PMP
H_1	.059
H_2	.117
H_3	.118
H_4	.235
H_5	.471
H_6	.000

terior probabilities of each model. Table 13.1 shows the resulting estimated posterior probabilities, which express prior knowledge (model translated theories using inequality constraints) being brought up to date with empirical data. As can be seen in Table 13.1, H_5 gets most support from the data suggesting that, on average, students in Catholic schools have higher math achievement than those in public schools and that student level socioeconomic status is positively associated with mathematics achievement with public schools having higher slopes than Catholic schools. This is in line with the findings in [23]. Lastly, model 5 also suggests that students from an ethnic minority have lower math achievement than those who are not from a minority. These findings are similar to what was observed in a sample of children from the Netherlands [2]. It is worthwhile to note that models 2 and 3 are nested in model 5, implying that in a sense there is more evidence to support model 5 than just the PMP of 0.47. Stated otherwise, if models 2 and 3 were not part of the competing set of models, the PMP of model 5 would have been bigger than 0.47. Subsequently, estimates for parameters of model H_5 were obtained using constrained Gibbs sampling. Posterior distributions of the model parameters were monitored for 20,000 iterations after a burn-in of 10,000 and were summarized by posterior means, standard deviations, and 95% central credibility intervals. These are displayed in Table 13.2. Relating the estimates to the theories behind model H_5, it can be concluded that controlling for all other predictors in the model:

Table 13.2. Estimates for H_5

Parameter	Mean	*SD*	95% CCI
β_1	14.33	0.20	(13.93, 14.73)
β_2	12.67	0.19	(12.30, 13.03)
β_3	4.18	0.33	(3.53, 4.84)
β_4	1.16	0.18	(0.81, 1.51)
β_5	2.64	0.16	(2.32, 2.95)
β_6	0.98	0.30	(0.38, 1.57)
β_7	−2.76	0.19	(−3.14, −2.38)
$\mathrm{Var}(u_{1j})$	1.99	0.33	(1.42, 2.71)
$\mathrm{Cov}(u_{1j}, u_{2j})$	−0.04	0.19	(−0.01, 0.35)
$\mathrm{Var}(u_{2j})$	0.24	0.12	(0.09, 0.54)
σ^2	35.88	0.61	(34.71, 37.09)

1. Average predicted score for mathematics achievement is higher for Catholic than public schools. The average predicted mathematics achievement scores for students who are not minorities in schools with $meanses = 0$ are 14.33 and 12.67 for Catholic and public schools, respectively.
2. Students belonging to ethnic minorities have lower mathematics achievement than those who are not from minorities. The coefficient β_7 for min implies that the average predicted difference in mathematics achievement scores between students from minorities and nonminorities is 2.76.
3. Student level *ses* is positively associated with mathematics achievement with public schools having higher slopes than Catholic schools; for schools with average student *ses* values (i.e., $mses = 0$), each extra unit of *ses* corresponds to an increase of 2.64 and 1.16 in average mathematics achievement for public and Catholic schools, respectively. Furthermore, in both Catholic and public schools, the student level *ses* effect on math achievement increases with increasing *meanses*. Stated otherwise, the importance of *ses* as a predictor for math achievement is more pronounced for schools with higher values of *meanses*.

13.5 Individual Growth Data Example

13.5.1 Data

As part of a larger study regarding substance abuse, Curran, Stice, and Chassin [6] collected 3 waves of longitudinal data on 82 adolescents. Beginning at age 14, each year the adolescents completed a 4-item instrument that sought to assess their alcohol consumption during the previous year. Using an 8-point scale (ranging from 0 = "not at all", to 7 = "every day"), the adolescents described the frequency with which they (1) drank beer or wine, (2) drank hard liquor, (3) had 5 or more drinks in a row, and (4) got drunk. The data were obtained from URL: http://www.ats.ucla. edu/stat/examples/alda/.

The dataset includes the following variables:

1. ***alcuse***: The dependent variable. This (continuous) variable was generated by computing the square root of the mean of participants' responses across its constituent variables (the frequency with which the adolescents (1) drank beer or wine, (2) drank hard liquor, (3) had 5 or more drinks in a row, and (4) got drunk). The variable *alcuse* has mean 0.92 and standard deviation 1.06 (range 0 to 3.61).
2. ***age***: Variable indicating age of adolescent.
3. ***peer***: A measure of alcohol use among the adolescent's peers. This predictor was based on information gathered during the initial wave of data collection. Participants used a 6-point scale (ranging from 0 = "none", to 5 = "all") to estimate the proportion of their friends who (1) drank alcohol occasionally and (2) drank alcohol regularly. This continuous variable was generated by computing the square root of the mean of participants'

responses across its constituent variables. The variable *peer* has mean 1.02 and standard deviation 0.73 (range 0 to 2.53).

4. ***coa***: A dichotomous variable where a 1 indicates that an adolescent is a child of an alcoholic parent. Of the 246 adolescents, 111 are children of alcoholic parents and the rest are children of nonalcoholic parents.

Now let $alcuse_{kj}$ and age_{kj} be the response (alcohol use) and age, respectively, for the jth ($j = 1, ..., 82$) subject at age $k = 14, 15, 16$. Next, let $t_{kj} = (age_{kj} - 14)/(2 \times \text{std}(age))$, where std($age$) denotes the standard deviation of age. It follows that $t_{kj} = 0$ corresponds to the baseline age of 14. Also, let coa_j and $ncoa_j$ be indicator variables defined to be 1 if the subject is the child of an alcoholic or not the child of an alcoholic parent, respectively, and 0 otherwise. Additionally, let $speer_j$ be the centered and scaled measure of alcohol use among the adolescent's peers obtained by subtracting the mean and dividing by two standard deviations. In regression models that include both binary and continuous predictors, scaling the continuous predictors by dividing by 2 standard deviations rather than 1 standard deviation ensures comparability in the coefficients of the binary and continuous predictors [10, 11]. Note that for interactions between two continuous variables, say X_1 and X_2, each of the variables is scaled before taking their product; that is, the interaction term is not obtained by scaling $(X_1 \times X_2)$. It is the product of $(X_1 - \text{mean}(X_1))/(2 \times \text{std}(X_1))$ and $(X_2 - \text{mean}(X_2))/(2 \times \text{std}(X_2))$, where mean($X_r$) and std($X_r$) denote the mean and standard deviation of X_r, respectively.

13.5.2 Theory and Models

Previous longitudinal latent growth models have been used to examine the relation between changes in adolescent alcohol use and changes in peer alcohol use. Curran, Stice, and Chassin [6] found that peer alcohol use was predictive of increases in adolescent alcohol use. Furthermore, Singer and Willett [24] have shown that adolescents with an alcoholic parent tended to drink more alcohol as compared to those whose parents were not alcoholics. Additionally, it is expected that with regard to initial adolescent alcohol use, an alcoholic parent may be of more influence than peers, whereas for rate of change with regard to alcohol intake, peers may have more influence. It is these expectations we want to investigate in a model and accompanying informative hypotheses.

Assuming that the profiles of each subject can be represented by a linear function of time, the model can be written as

$$alcuse_{kj} = \pi_{1j} + \pi_{2j} t_{kj} + \varepsilon_{kj},$$

where

$$\pi_{1j} = \beta_1 coa_j + \beta_2 ncoa_j + \beta_3 speer_j + u_{1j},$$
$$\pi_{2j} = \beta_4 coa_j + \beta_5 ncoa_j + \beta_6 speer_j + u_{2j},$$

and

$$\boldsymbol{u} = (u_{1j}, u_{2j})' \sim \mathcal{N}(\mathbf{0}, \boldsymbol{V}), \quad \varepsilon_{kj} \sim \mathcal{N}(0, \sigma^2).$$

Thus, the subject-specific intercepts (π_{1j}) and time effects (π_{2j}) are related to peer alcohol use and whether parent(s) is/are alcoholic or not.

The following competing models will be compared:

$$\begin{aligned} &H_1 : \beta_1,\ \beta_2,\ \beta_3,\ \beta_4,\ \beta_5,\ \beta_6, \\ &H_2 : \{\beta_1 > \beta_2\},\ \beta_3,\ \beta_4,\ \beta_5,\ \beta_6, \\ &H_3 : \{\beta_1 > \beta_3\},\ \beta_2,\ \{\beta_4 < \beta_6\},\ \beta_5, \\ &H_4 : \{\beta_1 > \beta_2\},\ \beta_3,\ \{\beta_4 > \beta_5\},\ \beta_6. \end{aligned}$$

Model 1 is the unconstrained model. Model 2 expresses the theory that adolescents with an alcoholic parent are more prone to higher alcohol use at baseline $\{\beta_1 > \beta_2\}$. Model 3 expresses the theory that with regard to an adolescent's alcohol use, parents have more influence than peers at baseline $\{\beta_1 > \beta_3\}$, whereas over time peers have more influence $\{\beta_4 < \beta_6\}$. Model 4 expresses the theory that adolescents with an alcoholic parent are more prone to higher alcohol use at baseline $\{\beta_1 > \beta_2\}$, as well as over time $\{\beta_4 > \beta_5\}$.

13.5.3 Results

The prior distributions for the parameters in the encompassing model were specified as follows. The regression coefficients $\beta_1, \ldots, \beta_6$ were each given normal prior distributions with mean 0.92 and variance 10^4. The variance covariance matrix V was given an inverse Wishart prior distribution with 3 degrees of freedom and a 2×2 identity matrix as scale matrix. Turning to the prior on σ^2, we used a scaled inverse χ^2-distribution with 1 degree of freedom and scale 1.12. Subsequently, 200,000 samples (after a burn-in of 10,000) were drawn from the prior and the posterior distributions of the encompassing model, respectively. For each of the models H_2, H_3, and H_4, the proportion of samples from prior and posterior distribution of H_1 in agreement with the constraints on $\boldsymbol{\beta}$ were used to estimate the posterior probabilities of each model. These are displayed in Table 13.3.

The posterior probabilities suggest that the support in the data is highest for model H_2. Subsequently, estimates for parameters of model H_2 were obtained using constrained Gibbs sampling. Posterior distributions of the model

Table 13.3. Posterior model probabilities

Model	PMP
H_1	.208
H_2	.416
H_3	.000
H_4	.375

Table 13.4. Estimates for H_2

Parameter	Mean	*SD*	95% CCI
β_1	0.97	0.11	(0.75, 1.19)
β_2	0.39	0.10	(0.19, 0.59)
β_3	1.01	0.15	(0.70, 1.31)
β_4	0.43	0.15	(0.15, 0.72)
β_5	0.45	0.13	(0.19, 0.72)
β_6	−0.35	0.20	(−0.74, 0.04)
$\text{Var}(u_{1j})$	0.27	0.08	(0.14, 0.46)
$\text{Cov}(u_{1j}, u_{2j})$	−0.01	0.05	(−0.12, 0.07)
$\text{Var}(u_{2j})$	0.18	0.05	(0.09, 0.29)
σ^2	0.35	0.05	(0.26, 0.46)

parameters were monitored for 20,000 iterations after a burn-in of 10,000 and were summarized by posterior means, standard deviations, and 95% central credibility intervals, which are presented in Table 13.4. Looking at the PMPs for models 2 and 4 in Table 13.3 suggests that model 4 is not much worse than 2. In Table 13.4, the estimate for β_4 is less than that of β_5; this is opposite to the constraint $\beta_4 > \beta_5$ of model 4. This suggests that the reason why model 2 has a higher PMP than model 4 is because the constraint on the parameters β_4 and β_5 in model 4 is not in accordance with the data, whereas model 2 does not put any constraints on these parameters. Based on the estimates in Table 13.4, the following can be concluded:

1. Controlling for peer alcohol use, baseline (age = 14), adolescent alcohol use was higher in children of alcoholics than in children with nonalcoholic parents. The difference in average baseline alcohol use was $\beta_1 - \beta_2 = 0.58$ with 95% central credibility interval (0.28, 0.88).
2. Since β_3 is the coefficient for $speer = (peer - \text{mean}(peer))/(2 \times \text{std}(peer))$, it follows that the coefficient for the original variable $peer = 1.01/(2 \times \text{std}(peer)) = 0.69$. This implies that controlling for whether or not a parent is alcoholic, for every point difference in peer alcohol use, baseline adolescent alcohol use is 0.69 higher. Stated otherwise, teenagers whose peers drink more at age 14 also drink more at 14.
3. Adolescent alcohol use tended to increase over time at rates of $\beta_4 = 0.43$ and $\beta_5 = 0.45$ per year for children of alcoholics and nonalcoholics, respectively . However, there is no difference between the rates, $\beta_4 - \beta_5 = -0.02$ with 95% central credibility interval (−0.41, 0.37).
4. Since β_6 is the coefficient for the interaction between $t_{jk} = (age - 14)/(2 \times \text{std}(age))$ and $speer = (peer - \text{mean}(peer))/(2 \times \text{std}(peer))$, it follows that the coefficient for the interaction between *peer* and *age* is $-0.35/(4 \times \text{std}(peer) \times \text{std}(age)) = -0.15$. However, the CCI for β_6 contains 0, so there is no evidence to suggest that the coefficient is different from zero. This implies that *peer* alcohol use does not influence adoloscents' alcohol use over time.

13.6 Discussion

George Box is credited with the quote, "all models are wrong, but some are useful" [3]. A basic principle of scientific inference is that a good fit of a model to a set of data never proves the truth of the model. Indeed if one does find the best fitting model, it may not be theoretically plausible or represent the actual state of affairs. No (statistical) technique can prove that a model is correct; at best, we can give evidence that a certain model or set of models may or may not be a plausible representation of the unobservable forces that generated the dataset at hand.

There is, therefore, the possibility of the existence of unexplored models that may yield superior posterior probabilities compared to the set of models considered by a researcher. In practice, it would be possible to evaluate all possible combinations of constraints in the model set in order to obtain the best possible model given a certain index of model fit. This however takes us into the exploratory realm of data analysis, which may tempt us into hypothesizing after results are known and, as such, imposes physical as well as philosophical restrictions on a meaningful scientific method.

The crux of a meaningful scientific method is the exclusion of plausible alternatives. In the exploratory mode many models are included that may not be theoretically plausible or represent an approximation of the actual state of affairs, even when they report superior fit. Exploratory analysis in our view, as a tool of scientific advance, predates the scientific method in that it should be used for developing ideas about relationships when there is little or no previous knowledge. These ideas may then subsequently be tested in a confirmatory analysis thatadheres to the scientific method.

The inequality constrained Bayesian approach to analysis of multilevel linear models as advocated in this chapter explicitly encourages researchers to formulate plausible competing theories for confirmatory analysis and offers a framework in which one is able to simultaneously evaluate all possible alternative model translated theories with regard to model fit and complexity. As such it has a strong connection with the hypothetico-deductive scientific method and the concept of strong inference [20]. This method of scientific advance has, coupled to inequality constrained Bayesian confirmatory data analysis, the following form (also see [20]): (i) Devise on the basis of previous knowledge (such as a former exploratory data analysis on preliminary data, previous results, or expert opinion) alternative theories. These alternative theories will usually have inequality constraints among the parameters of its constituent hypotheses; (ii) devise a crucial experiment whose possible outcomes will be able to demarcate maximally the alternative theories or (when experiments are not possible) establish which observational data one would need to exclude one or more of the theories; (iii) perform the experiment or obtain the observational data and establish the "best model(s)" with the inequality constrained Bayesian confirmatory data analysis framework; (iv) repeat the cycle by refining the model(s) that remain(s) and/or by using the outcome as prior knowledge in a natural process of Bayesian updating.

In this chapter, we have considered only multilevel linear models. However, the ideas presented in this chapter can be extended and adapted to deal with multilevel logistic regression and other multilevel generalized linear models. In such settings extra complications are bound to arise because we are not dealing with continuous data.

Furthermore, in situations in which the posterior probabilities are similar or approximately equivalent for multiple models, the "best model" question may not be most appropriate and one then may want to embark on model averaging to take model uncertainty into account in a stricter manner. Such issues may be the topic of further research.

Acknowledgements. The authors would like to thank Judith Singer for indicating useful hierarchical datasets and the editors for useful comments that have improved this chapter.

References

[1] Barlow, R.E., Bartholomew, D.J., Bremner, J.M., Brunk, H.D.: Statistical Inference under Order Restrictions: The Theory and Application of Isotonic Regression. New York, Wiley (1972)

[2] Berg, W. van den, Eerde, H.A.A. van, Klein, A.S.: Proef op de som: Praktijk en resultaten van reken/wiskundeonderwijs aan allochtone leerlingen op de basisschool [Practice and Results of Education Arithmatics and Mathematics for Immigrant Children in Elementary School]. Rotterdam, RISBO (1993)

[3] Box, G.E.P., Draper, N.R.: Empirical Model-Building and Response Surfaces. New York, Wiley (1987)

[4] Browne, W.J.: MCMC Estimation in MLwiN (Version 2.0). London, Institute of Education University of London (2003)

[5] Bryk, A.S., Raudenbush, S.W.: Hierarchical Linear Models: Applications and Data Analysis Methods. London, Sage (1999)

[6] Curran, P.J., Stice, E., Chassin, L.: The relation between adolescent and peer alcohol use: A longitudinal random coefficients model. Journal of Consulting and Clinical Psychology, **65**, 130–140 (1997)

[7] Gamoran, A.: The variable effects of high school tracking. American Sociological Review, **57**, 812–828 (1992)

[8] Geary, D.C.: Children's Mathematical Development: Research and Practical Applications. Washington, DC, APA (1994)

[9] Gelfand, A.E., Smith, A.F.M., Lee, T.M.: Bayesian analysis of constrained parameter and truncated data problems using gibbs sampling. Journal of the American Statistical Association, **87**, 523–532 (1992)

[10] Gelman, A.: Scaling regression inputs by dividing by two standard deviations. Statistics in Medicine (in press)

[11] Gelman, A., Hill, J.: Data Analysis Using Regression and Multilevel/Hierarchical Models. Cambridge, Cambridge University Press (2007)

[12] Goldstein, H.: Multilevel Statistical Models (2nd edition). London, Edward Arnold (1995)
[13] Hoijtink, H.: Posterior inference in the random intercept model based on samples obtained with Markov chain Monte Carlo methods. Computational Statistics, **15**, 315–336 (2000)
[14] Hox, J.: Multilevel Analysis: Techniques and Applications. London, Lawrence Erlbaum Associates (2002)
[15] Kato, B.S., Hoijtink, H.: A Bayesian approach to inequality constrained linear mixed models: estimation and model selection. Statistical Modelling, **6**, 231–249 (2006)
[16] Klugkist, I., Hoijtink, H.: The Bayes factor for inequality and about equality constrained models. Computational Statistics & Data Analysis, **51**, 6367–6379 (2007)
[17] Klugkist, I., Kato, B., Hoijtink, H.: Bayesian model selection using encompassing priors. Statistica Neerlandica, **59**, 57–69 (2005)
[18] Longford, N.T.: Random Coefficient Models. London, Oxford University Press (1993)
[19] Maxwell, S.E.: The persistence of underpowered studies in psychological research: Causes, consequences, and remedies. Psychological Methods, **9**, 147–163 (2004)
[20] Platt, J.R.: Strong inference. Science, **146**, 347–353 (1964)
[21] Press, S.J.: Subjective and Objective Bayesian Statistics: Principles, Models, and Applications (2nd edition). New York, Wiley (2003)
[22] Silvapulle, M.J., Sen, P.K.: Constrained Statistical Inference: Inequality, Order and Shape Restrictions. Hoboken NJ, Wiley (2005)
[23] Singer, J.D.: Using SAS PROC MIXED to fit multilevel models, hierarchical models, and individual growth models. Journal of Educational and behavioral Statistics, **24**, 323–355 (1998)
[24] Singer, J.D., Willett, J.B.: Applied Longitudinal Data Analysis: Modeling Change and Event Occurrence. New York, Oxford University Press (2003)
[25] Smith, A.F.M., Roberts, G.O.: Bayesian computation via the Gibbs sampler and related Markov Chain Monte Carlo methods. Journal of the Royal Statistical Society, Series B, **55**, 3–23 (1993)
[26] Snijders, T., Bosker, R.: Multilevel Analysis: An Introduction to the Basic and Advanced Multilevel Modeling. London, Sage (1999)

Part IV

Evaluations

14

A Psychologist's View on Bayesian Evaluation of Informative Hypotheses

Marleen Rijkeboer and Marcel van den Hout

Department of Clinical and Health Psychology, Utrecht University, P.O. Box 80140, 3508 TC Utrecht, the Netherlands m.m.rijkeboer@uu.nl and m.vandenhout@uu.nl

14.1 Introduction

Psychologists, like other scientists, gather and analyse data to evaluate the explanatory power of theories. Typically they build on earlier studies, explicitly or implicitly formulating competing hypotheses and inferring different predictions about, for instance, the relative scores of different groups on an outcome measure in an experimental study. As a means to test their theories, psychologists are accustomed to the classical statistical tradition and most of them apply null hypothesis significance testing (NHST) that is dominant within this tradition. They are trained to use the Statistical Package for the Social Sciences (SPSS), which centers on NHST, and train their students to do the same. Yet several authors have noted that NHST is a suboptimal way to address the very questions that researchers are concerned with (cf. [13]). In principle, NHST provides researchers with "yes-or-no" answers about the tenability of H_0, in which two or more parameters are constrained to be equal. Rejection of H_0, however, does not render the alternative hypothesis to be relevant and the fact that deviations from H_0 do not give us the answer we need has not gone unnoticed. Within the realm of conventional statistics, a range of strategies have been suggested to make the latter more informative (e.g., the use of point estimates of effect sizes (ESs) and confidence intervals (CIs) [17]). Although these maneuvres denote an improvement, they remain adaptations to a statistical approach that is, according to its opponents, ultimately inadequate (e.g., [5, 8] and Chapter 9).

The Bayesian approach is logically opposed to NHST, one of the main differences between the two approaches being the basic question that is addressed. Let us suppose a researcher is interested in knowing whether treatment A is more effective than treatment B. NHST yields information about the probability of the data (or more extreme data) if H_0 were true; that is, both treatments are equally effective. The Bayesian approach, however, provides information on the probability that the alternative (or any other) hypothesis is true (e.g., treatment A is more effective than treatment B). This is the kind of information in which a researcher in psychology is actually interested [21]. It is intriguing, for that matter, that most students and even many researchers misinterpret NHST in a Bayesian way and believe that $p < .05$ means that the probability of this finding being coincidental is lower than 5%.

In this book convincing arguments are presented that Bayesian statistics in many respects are superior to significance testing when dealing with competing hypotheses. The Bayesian approach offers researchers an elegant and informative alternative for testing the explanatory strength of rivaling hypotheses. In this chapter, written by researchers in psychology, some considerations are formulated on shifting from NHST to the advocated Bayesian alternative.

14.2 Model Building and Model Selection

One of the major strengths of the Bayesian approach is its emphasis on prior knowledge and deductive reasoning. As was demonstrated in Chapter 2, complex psychological theories can often be converted into fine-grained models. In these models, divers and competing hypotheses are formulated that are characterized by a number of (in)equality constraints being imposed on relevant parameters (e.g., expected ordering of means, or ordering of differences between specific means). While NHST does not provide an incentive to devise well-specified hypotheses, Bayesian model building forces researchers *a priori* to explicitly state all expected relationships, preferably derived from "state of the art" theories within the specific research domain. So, whereas NHST contains explorative elements or is sometimes misused as a surrogate for theory [7], the Bayesian approach presented in this book is confirmative.

An illustration of the tendency of scientists, using the classical statistical approach, to rely on inductive data inspection instead of a priori theorizing, is the widespread use of exploratory factor analysis (EFA) in psychological research. Results obtained from exploratory techniques pose several inferential problems for researchers, as was discussed in Chapter 11 on latent class analysis. When applying these remarks to factor analysis, the following issues complicate the use of EFA: The number and the meaning of the factors has to be determined afterward, and the results may not be consistent with any of the theories researchers have in mind. See also [4] for a further elaboration of these matters.

The use of EFA, an analysis that is primarily data driven, is inevitable when no a priori information exists on the nature or dimensionality of the variables at hand. For example, Cattell performed an ambitious project to seek individual differences in personality by compiling a lexicon of trait-descriptive words. His assumption was that any language that has evolved over millennia includes words that describe qualities of personality [18]. He took a set of thousand English adjectives and removed obvious synonyms, after which a list of 171 trait names remained. Factor analysis of collected ratings on these words yielded 16 traits. In this case an explorative approach seems appropriate, as one has to rely on empirical data to reveal the underlying structure of traits. Yet this is rare. Generally, aspects of human behavior are operationalized by measures that are constructed on the base of theory and/or rich clinical experience. Therefore, in most cases it would be more appropriate to conduct confirmatory factor analysis (CFA), a less often used technique within the classical statistical tradition, which in contrast to EFA is theoretically grounded. Using CFA, psychologists are compelled to convert their often conflicting a priori held ideas regarding the relationships between the latent and observed variables into multiple alternative models. The relative fit of each model can be estimated, yielding a more clear-cut interpretation of the underlying structure. Moreover, an explicit test of measurement invariance *across* groups can be obtained [19], because multigroup CFA allows the researcher to simultaneously test factor models in relevant groups (e.g., men/women; clinical/nonclinical; African/Caucasian). Hence, CFA is useful for testing competing theories but also may provide valuable information to further develop these theories.

Structural Equation Modeling (SEM) allows a further elaboration of CFA models (cf. [1, 16]). Within the SEM framework, CFA models, path models, and other regression models can be combined, allowing researchers to test a wider variety of hypotheses than would be possible with traditional statistical techniques [3, 22]. All these models are built before the data are examined, so the approach is essentially confirmative in nature. Instead of absolute decisions as to either reject or accept a certain model, one gains information on the *relative* explanatory power of each model, whereby that model is selected that shows the best fit while being most parsimonious. When competing models are non-nested (i.e., not hierarchically related), using hypothesis testing to determine their relative value is not possible in SEM software. However, an alternative is the inspection of information criteria like Akaike's information criterion (AIC) and corrected AIC [2]. Both evaluate the value of a model combining its fit (in terms of likelihood) and penalize this with its size (in terms of the number of parameters needed to formulate a model). For SEM, user-friendly software is developed (e.g., AMOS, LISREL, M*plus*) and this approach, which is close to psychologists' statistical habits, is increasingly popular in the social sciences. So, SEM advantageously allows researchers to build competing models before they are fitted to the data, using easy and understandable software.

Although within the SEM framework both direct and indirect effects can be estimated simultaneously, no explicit test of the *order* of effects is provided. Hence, selection of models specified using inequality constraints on the parameters are, until now, not handled by SEM software. In the Bayesian approach, a straightforward evaluation of both nested and non-nested models is possible. Furthermore, as shown in this book, (in)equality or order constraints on the parameters can easily be handled when selecting the best of a set of competing models. This leads to more realistic models, as will be illustrated based on a CFA application.

Using standard procedures within SEM, a factor (e.g., Neuroticism) determines a specific set of indicators (i.e., items of the Neuroticism scale). Cross-loadings of other indicators within the model (items belonging to, e.g., Extraversion and Altruism) are not allowed and fixed at zero. In other words, the influence of a specific factor on the remaining indicators is assumed to be nonexistent. This, however, is not very realistic. In the process of instrument construction it is quit common that each item is constructed to tap a certain construct, but might reflect qualities embodied in other constructs as well. Therefore, it would be more informative to estimate the relative influence of each factor on all indicators. Bayesian methods allow for a test of these kind of models, in which several inequality constraints are imposed on the factor loadings of each set of indicators (see [9]). For example, the Neuroticism indicators (say items 1 to 5) should have substantial positive loadings on the factor Neuroticism (for instance, each loading should be larger than, e.g., .3), and small loadings on the factor Extraversion (e.g., loadings smaller than .3), whereas the Extraversion indicators (say items 6 to 10) should have low loadings on the factor Neuroticism and large loadings on the factor Extraversion.

14.3 Cumulative Knowledge: Prior and Posterior Distributions

At the heart of Bayesian statistics is the incorporation of prior knowledge about the distribution of the relevant parameters. Within the frequentist tradition, parameters are considered as fixed entities or truths one can get to know by an infinite replication of experiments (operationalized by point estimates). In Bayesian statistics, on the other hand, parameters are deemed random; that is, they are seen as unknown quantities that have a probability distribution. Parameters cannot be fixed entities, whereas knowledge is continuously changing. In this respect the Bayesian approach is fundamentally distinct from conventional approaches and may well serve psychological research in which multiple experiments are conducted or data of repeated measures are collected. Let us illustrate what this difference of both statistical accounts may imply for clinical trials.

Within the NHST tradition, data are approached as if no prior knowledge is available. Hence, in case a clinical trial is being replicated, no information on former results is being incorporated into the analyses, rendering the conclusions more or less independent from earlier knowledge. The only way prior information is used is in setting up additional studies, by carefully designing the study, and the calculation of the required sample size to reach adequate power, given the expected effect size and desired alpha-level. As a consequence, confidence intervals remain relatively large, leaving one uncertain what to decide in case results are not significant. Moreover, reliance on

NHST may not only lead to possible conflicting results *within* one study (e.g., in case of pairwise comparisons (see Chapter 2)) but also *between* studies. All kinds of moderators (e.g., sample characteristics like age, SES, comorbidity) might have caused these conflicting findings in the successively performed studies, but since no direct test of these interactions was employed, ad hoc inferences need to be made [20]. In the classical statistical tradition, therefore, the use of meta-analytic procedures is promoted, allowing one to accumulate data over studies. Meta-analysis, however, is useless for the many research questions that have *not* been addressed empirically in previous studies.

The Bayesian approach provides an alternative, in that it incorporates beliefs or knowledge on prior distributions of relevant parameters. Bayesians dictate explicit use of a priori information derived from theory, earlier research on the topic, and/or knowledge collected from experts in the field [21]. Thus, relative to classical approaches, the Bayesian approach allows for existing theoretical and empirical knowledge being accounted for in the analyses much more easily. As Krueger and Funder [14, p. 324] put it: "Bayesianism permits the integration of new evidence with theory and past research even at the level of the individual study."

To illustrate the possible inferential consequences of both statistical approaches, we adopt a practical example from Howard et al. [11, Study 1]. In this example actual data are analysed using NHST and Bayesian methods. In order to highlight the main disparities, a rather simplistic elaboration of the NHST approach is given. The authors presented a study in which the effect of a Psychology of Healthy Lifestyles course (PHL) on student's increase in alcohol consumption from senior year of high school to freshman year of college was evaluated. Students were randomly assigned to PHL or to other psychology seminars, the latter serving as a control condition. Changes were quantified by the difference in the mean number of drinks per week in the high school senior and college freshman year; $N = 49$; PHL: $M = 1.58$, $SD = 2.19$; control: $M = 2.21$, $SD = 2.98$. Although the students in the treatment condition seem better of compared to the control participants, a t-test for independent groups yielded a nonsignificant result: $t(47) = 0.70$, $p = .48$. As Howard et al. noted apart from taking into account the small sample size, a p-value as such might lead researchers to consider giving up on the PHL course, despite possible a priori (theoretical) ideas on the usefulness of the specific training. As an alternative, the Bayesian way of approaching the study and its resulting data was explicated. In the described example of Howard et al., before conducting the study, prior knowledge of experts (advanced graduate students and faculty) on drinking habits of freshmen was collected. The experts were asked to give information on the expected mean increase in number of drinks per week of freshmen in the absence of any intervention, and in case the PHL course was followed, and, finally, on how *certain* they were about their estimates given as a result of the first question. The assumptions on the effect of the PHL treatment and no treatment were quantified. Prior estimates of optimistic experts about the change in the mean number of drinks per week were 1.5 (SD = 0.5) for the PHL group and 5.5 (SD = 0.5) for the control group. Hereafter, this prior information was modified by the data that were gathered during the study, leading to the following posterior means: 1.53 (SD = 0.40) for the treatment group and 3.85 (SD = 0.35) for the control group. The difference between these two means is substantial and, as Howard et al. commented, is likely to encourage the researchers to continue their enterprise and to set up larger studies. Interestingly, in the example of Howard et al. two more studies with larger samples were performed. In the NHST approach, data of each study were analyzed in isolation, leading to conflicting results. In the Bayesian approach, however, the posterior distributions of each study served as priors for the newly collected data, which in the described study led to a convergence of estimates in favor of PHL.

Besides the incorporation of prior knowledge in the analyses to be executed, there are more grounds for Bayesian methods to provide an attractive alternative for the analysis of clinical trials (see also Chapter 9). One of the main advantages is that Bayesian inference permits unlimited inspection of the data as they accumulate. As [15, p. 1331] pointed to: "Unlike a classical trial, the number of patients to be enrolled in the trial or the timing of the interim analyses do not need to be predetermined. Other considerations, such as the rate of patient recruitment or funding constraints, can be used to determine the number and timing of the data analyses." Hence, Bayesian inference allows the researcher to constantly monitor the process, providing a means to stop the inclusion of more patients at the point that the experimental therapy appears to be ineffective or even harmful, or to quit the monitoring process when enough certainty about the treatment effect has been reached [21].

14.4 Equivocalness (or Reliability) of Results

Notwithstanding the qualities of Bayesian inference shown above, there do seem to be some problems around. The Achilles' heel of the Bayesian account, so it seems to us, is the choice of the prior distribution. The so-called "subjective" priors, or priors which are based on expert information, influence the outcome of the study. By way of illustration we return to the study as described by Howard et al. [11] in which, besides the information of the optimistic experts, the prognostic ideas of their pessimistic counterparts were gathered. When this latter prior information was incorporated into the data, in the first study far less favorable posterior probabilities were attained. Hence, the initially chosen priors determined the results yielded. However, after two replications of the study with larger samples this influence subsided and more or less comparable results as in the optimistic condition were found [11].

Objective Bayesians provide an alternative for the use of subjective priors (see Chapter 3; [10]) and have come up with ways to determine "objective" priors, like the encompassing prior described in Chapter 4 and the prior distributions described in Chapters 6, 7, and 8. However, in order for a technological innovation to become an attractive alternative for existing practices, the outcome of the novel approach should be reliable. Different users of the novel approach should end up with comparable conclusions. How do traditional NHST and the Bayesian alternative compare in this respect? In the present book the dataset of Huntjens et al. [12] served in a way as a testing case.

The study aimed at unraveling the nature of the interidentity amnesia as reported by patients suffering from Dissociative Identity Disorder (DID). For more details, see Chapter 2 and [12]. The dataset consisted of scores on a recognition test of four groups: DID-patients (*pat*), normal Controls (*con*), Simulators (*sim*), and True amnesiacs (*amn*). In the original article, scores of the DID-patients were compared to the scores of the other groups, using ANOVA with pairwise comparisons. Results obtained by Huntjens et al. suggest that DID-patients did not differ significantly from the Simulators, the latter ones being asked to deliberately simulate interidentity amnesia.

The classical ANOVA approach may have had its weaknesses, as was set out in the foregoing debate, but within the frequentist framework there is little room for doubt that the approach of Huntjens et al. was the right one. Moreover, colleagues, analyzing the same data set along conventional lines, would most likely choose the same procedure and come up with next to identical results. The ANOVA main effects and interactions may have been less informative than desirable, but we assume the interresearcher reliability would be pretty high. Of course, due to multiple pairwise comparisons, generally alpha corrections are performed. The options for these corrections are manifold, so at this point, differences in outcome might occur (although in the case of the Huntjens et al. data, the type of alpha correction did not influence the conclusions; Huntjens, personal communication). How about the "interevaluator reliability" of the data-analytic experts who reanalyzed the Huntjens et al. data along Bayesian lines?

The Bayesian statisticians who reanalyzed this data set all used "objective" priors, since they were solely equipped with material on the relevant theoretical stances to build their models and with the data collected in the study. All in the end dealt with the same question; that is, are DID-patients more similar in their recognition scores to Simulators than to True amnesiacs? Next to the null and the unconstrained model, the following two informative hypotheses were tested:

$$H_{1a} : \mu_{con} > \{\mu_{amn} = \mu_{did}\} > \mu_{sim}, \tag{14.1}$$

$$H_{1b} : \mu_{con} > \mu_{amn} > \{\mu_{did} = \mu_{sim}\}. \tag{14.2}$$

The performed Bayesian analyses, entailing the way the "objective" prior distributions have been determined (e.g., the encompassing prior) and the model selection criteria being used (e.g., the Bayes factor), differ with each contributor to this discussion (see Chapters 4, 6, 7, and 8). Irrespective of the applied "objective" prior type and selection criteria, all found the model that DID-patients resemble Simulators (H_{1b}) to be about three times more likely than the model in which DID-patients resemble True amnesiacs (H_{1a}). Although H_{1a} could not be ruled out completely, more support was yielded for H_{1b}. Thus, the various Bayesian statisticians drew comparable conclusions from the very same dataset.

Hence, in case "objective" priors are used when analysing inequality constrained models, there seems to be high "interevaluator reliability" of the divers Bayesian approaches. This, again, makes Bayesian statistics an appealing alternative for NHST in psychological research. Note, however, that the above presented findings resemble those by Huntjens et al., except for the fact that Bayesians are able to state on the relative account of each model. Of course, the latter is an enormous improvement. But might it be attractive enough for researchers accustomed to classical approaches, considering that Bayesian statistics are complicated to perform? To answer this question, the reader is referred back to Chapters 2, 4, and 5. There it is shown for two other datasets (the emotional reactivity data and the grief data) that classical approaches are not very useful for the evaluation of the informative hypotheses of interest, whereas this can straightforwardly be done with the Bayesian approach.

14.5 Dissemination of Bayesian Statistics

Before a major statistical reform will take place, some skepticism needs to be conquered. Resembling the history of psychology, classical and Bayesian approaches seem to be entangled in a battle of schools, each pointing to why the other approach is flawed. Clearly, the dominance of simplistic NHST for half a century in psychological research is at least problematic, but over the years many efforts have been undertaken to reform the classical approach. Most important is the discrediting of "mindless statistics" [8]. Researchers should be aware of the limitations of NHST and use additional analytic strategies. As Finch et al. [6] argued, this reform requires advocacy and support from many sources. The publication manual of the American Psychological Association has a vital role to play, but editors of peer reviewed journals should also clearly recommend proper statistical practice. Last but certainly not least, a lot can be won by changing the university curriculum in statistics.

Since the popularity of NHST will not diminish soon, some authors have proposed an integration of classical statistics with Bayesian concepts of hypothesis evaluation [14]. Clearly, the time is ripe for dissemination of the basic tenets of the Bayesian heritage. Less clear is how the dissemination will be most fruitful: by adapting and reforming the NHST from within, by piecemeal engineering, or by preparing an encompassing paradigm shift. The latter issue is not only a matter of statistics and logic but also a tactical question. It is not for us to provide an answer; it is up to the Bayesians.

References

[1] Bollen, K.A.: Structural Equations with Latent Variables. New York, Wiley (1989)

[2] Bozdogan, H.: Model selection and Akaike's information criteria (AIC): The general theory and its analytical extensions. Psychometrika, **52**, 345–370 (1987)

[3] Brown, T.A.: Advances in latent variable analysis: applications to clinical research. Behavior Therapy, **35**, 289–297 (2004)

[4] Byrne, B.M.: Factor analytic models: viewing the structure of an assessment instrument from three different perspectives. Journal of Personality Assessment, **85**, 17–32 (2005)

[5] Cohen, J.: The earth is round ($p <.05$). American Psychologist, **49**, 997–1003 (1994)

[6] Finch, S., Thomason, N., Gumming, G.: Past and future American Psychological Association guidelines for statistical practice. Theory and Psychology, **12**, 825–853 (2002)

[7] Gigerenzer, G.: Surrogates for theories. Theory and Psychology, **8**, 195–204 (1998)

[8] Gigerenzer, G.: Mindless statistics. Journal of Socio-Economics, **33**, 587–606 (2004)

[9] Hoijtink, H.: Beter een goed verhaal dan de hele waarheid [A good story is better than the whole truth]. In: Bronner, A.E., Dekker, P., de Leeuw, E., de Ruyter, K., Smidts, A., Wieringa, J.E. (eds) Ontwikkelingen in het Marktonderzoek [Developments in Market Research] (pp. 201–212). Haarlem, De Vrieseborch (2005)

[10] Hoijtink, H., Klugkist, I.: Comparison of hypothesis testing and Bayesian model selection. Quality and Quantity, **41**, 73–91 (2007)

[11] Howard, G.S., Maxwell, S.E., Fleming, K.J.: The proof of the pudding: An illustration of the relative strengths of null hypothesis, meta-analysis, and Bayesian analysis. Psychological Methods, **5**, 315–332 (2000)

[12] Huntjens, R.J.C., Peters, M.L., Woertman, L., Bovenschen, L.M., Martin, R.C., Postma, A.: Inter-identity amnesia in dissociative identity disorder: a simulated memory impairment? Psychological Medicine, **36**, 857–863 (2006)

[13] Krueger, J.I.: Null hypothesis significance testing. On the survival of a flawed method. American Psychologist, **56**, 16–26 (2001)

[14] Krueger, J.I., Funder, D.C.: Towards a balanced social psychology: Causes, consequences, and cures for the problem-seeking approach to social behavior and cognition. Behavioral and Brain Sciences, **27**, 313–376 (2004)

[15] Lewis, R.J., Wears, R.L.: An introduction to the Bayesian analysis of clinical trials. Annals of Emergency Medicine, **22**, 1328–1336 (1993)

[16] MacCallum, R.C., Austin, J.T.: Applications of structural equation modeling in psychological research. Annual Review of Psychology, **51**, 201–226 (2000)

[17] Masson, M.E.J., Loftus, G.R.: Using confidence intervals for graphically based data interpretation. Canadian Journal of Experimental Psychology, **57**, 203–220 (2003)

[18] Matthews, G., Deary, I.J., Whiteman, M.C.: Personality Traits (2nd ed.). Cambridge, UK, Cambridge University Press (2003)

[19] Raykov, T.: Behavioral scale reliability and measurement invariance evaluation using latent variable modelling. Behavior Therapy, **35**, 299–331 (2004)

[20] Schmidt, F.L.: Statistical significance testing and cumulative knowledge in psychology: Implications for training of researchers. Psychological Methods, **1**, 115–129 (1996)

[21] Spiegelhalter, D.J., Myles, J.P., Jones, D.R., Abrams, K.R.: An introduction to Bayesian methods in health technology assessment. British Medical Journal, **319**, 508–512 (1999)

[22] Tomarken, A.J., Baker, T.B.: Introduction to the special section on structural equation modeling. Journal of Abnormal Psychology, **112**, 523–525 (2003)

15

A Statistician's View on Bayesian Evaluation of Informative Hypotheses

Jay I. Myung[1], George Karabatsos[2], and Geoffrey J. Iverson[3]

[1] Department of Psychology, Ohio State University, 1835 Neil Avenue, Columbus, OH 43210 USA `myung.1@osu.edu`
[2] College of Education, University of Illinois–Chicago, 1040 W. Harrison Street, Chicago, IL 60607 USA `georgek@uic.edu`
[3] Department of Cognitive Sciences, University of California at Irvine, 3151 Social Science Plaza, Irvine, CA 92697 USA `giverson@uci.edu`

15.1 Introduction

Theory testing lies at the heart of the scientific process. This is especially true in psychology, where, typically, multiple theories are advanced to explain a given psychological phenomenon, such as a mental disorder or a perceptual process. It is therefore important to have a rigorous methodology available for the psychologist to evaluate the validity and viability of such theories, or models for that matter. However, it may be argued that the current practice of theory testing is not entirely satisfactory. Most often, data modeling and analysis are carried out with methods of null hypothesis significance testing (NHST). Problems with and deficiencies of NHST as a theory testing methodology have been well documented and widely discussed in the field, especially in the past few years (e.g., [42]). The reader is directed to Chapter 9 of this book for illuminating discussions of the issues. Below we highlight some of the main problems of NHST.

First of all, NHST does not allow one to address directly the questions she/he wants to answer: How does information in the data modify her or his initial beliefs about the underlying processes? How likely is it that a given theory or hypothesis provides an explanation for the data? – that is, one would like to compute Prob(*hypothesis*|*data*). Instead, the decision as to whether one should retain or reject a hypothesis is based on the probability of observing the current data given the assumption that the hypothesis is correct (i.e., Prob(*data*|*hypothesis*)). These two probabilities are generally not equal to each other and may even differ from each other by large amounts. Second, NHST is often conducted in a manner that it is the null hypothesis that is put to the test, not the hypothesis the researcher would like to test. The latter hypothesis called the alternative hypothesis does not get attended to unless the null hypothesis has been examined and rejected subsequently. In other words, there is an imbalance in weighing the null and alternative hypotheses against each other as an explanation of the data. Third, many NHST tests are loaded with simplifying assumptions, such as normality, linearity, and equality of variances, that are often violated by real-world data. Finally, the p-value, the yardstick of NHST, is prone to misuse and misinterpretation, and this occurs more often than one might suspect and is, in fact, commonplace (see, e.g., Chapter 9). For example, a p-value is often misinterpreted as an evidence measure of the probability that the null hypothesis is true.

These methodological problems go beyond just NHST and are intrinsic to any frequentist methodology. Consequently, they represent limitations and challenges for the frequentist approach to statistical inference. Are there any alternatives to NHST? Fortunately, there is one, namely the Bayesian approach to statistical inference, which is free of the problems we discussed earlier. Unlike NHST, in Bayesian inference, (1) one directly computes the probability of the hypothesis given the data, (2) two or more hypotheses are evaluated by weighing them *equally*, (3) any realistic set of assumptions about the underlying processes can easily be incorporated into a Bayesian model, and (4) interpretations of Bayesian results are intuitive and straightforward. What is apparent from the other chapters of the current book (see the evaluation given in Chapter 5) is the fact that it is much more straightforward to pose and test order restricted hypotheses with order constrains within the Bayesian framework, compared to the frequentist NHST approach to testing such hypotheses. By definition, an order restricted hypothesis is a hypothesis where a set of parameters are consistent with a particular order relation, and will be called an "informative hypothesis" in this chapter, so as to be consistent with the terminology used in the other chapters.

A purpose of this chapter is to present a review of recent efforts to develop Bayesian tools for evaluating order-constrained hypotheses for psychological data. In so doing, we provide our own critiques on some of the chapters in this book, discussing their strengths and weaknesses. Another purpose of writing this chapter is to present an example application of hierarchical Bayesian modeling for analyzing data with a structure that is ideal for an analysis of variance (ANOVA) and to compare performance of several Bayesian model comparison criteria proposed and discussed throughout the current book. We begin by reviewing the literature on Bayesian order restricted inference.

15.2 Bayesian Order Restricted Inference

Order restricted models (i.e., models with parameters subject to a set of order constraints) have long been considered in frequentist statistics (cf. [17, 18]). Isotonic regression exemplifies this approach, the theoretical foundations of which are summarized in [2, 35, 40]. It seems appropriate, then, to include a brief description of the frequentist approach to order restricted inference before discussing a Bayesian alternative.

For the purposes of testing order restricted hypotheses, the isotonic regression model leads to a special kind of likelihood-ratio test. Specifically, the test statistic in isotonic regression is the log-likelihood ratio of the maximum likelihood estimate of a reduced model with equal means to that of a full model with certain order constraints imposed on its means. Note that the former model is nested within the latter one. The sampling distribution of the test statistic is then sought under the null hypothesis that all means are equal against the alternative hypothesis that the means satisfy the order constraints. This turns out, however, to be a major hurdle to the method's widespread application in practice; there is no easy-to-compute, general solution for finding the sampling distribution for given forms of order constraints, unless the constraints belong to one of a few simplified forms.[4] Even if one is able to derive the desired sampling distribution, given the fact that isotonic regression is a NHST, the problems associated with the use of NHST and *p*-value for model evaluation are still at issue, as discussed at length in Chapter 9 and as critiqued by Kato and Hoijtink [23], who commented "Even though a great deal of frequentist literature exists on order restricted parameter problems, most of the attention is focused on estimation and hypothesis testing [as opposed to model evaluation and comparison]" (p. 1).

As an alternative to the frequentist framework, a Bayesian approach to order restricted inference was considered in the past (e.g., [39]). However, its application was limited due to the intractability of evaluating the posterior integral. This long-standing difficulty in Bayesian computation has been overcome in the 1990s with the introduction of general-purpose sampling algorithms collectively known as Markov chain Monte Carlo (MCMC cf. [9, 13, 34]). With MCMC, theoretical Bayes has become practical Bayes. In particular, Gelfand, Smith, and Lee [10] developed easily implementable MCMC methods for sampling from posterior distributions of model parameters under order constraints. Since then, a group of quantitative psychologists have demonstrated the application of the Bayesian framework on a wide range of order restricted inference problems in psychology, education and economics [16, 19, 20, 21, 24, 30]. This success prompted Hoijtink and his colleagues to organize the Utrecht Workshop in the summer of 2007, which subsequently led to the publication of the current book.

---

[4] As an alternative to the isotonic regression likelihood-ratio test, Geyer [12] proposed bootstrap tests in which one computes approximate *p*-values for the likelihood-ratio test by simulating the sampling distribution by an *iterated* parametric bootstrap procedure. One problem with the bootstrap, which may be easy to compute, is that it does not have finite sampling properties and, therefore, can give biased estimates of sampling distributions for finite samples [7]. Further, the bootstrap is a frequentist approach that is subject to the problems discussed earlier.

15.2.1 Why Bayesian?

Bayesian inference, at its core, is the process of updating one's initial belief (prior) about the state of a world in light of observations (evidence) with the use of Bayes' theorem, thereby forming a new belief (posterior). This way of making inferences is fundamentally different from that of frequentist inference. Among many differences between the two schools of statistics, Bayesian and frequentist, the most notable include the former's interpretation of probability as an individual's degree of belief as opposed to the long-run frequency ratio in the latter and also the Bayesian view of model parameters as random variables as opposed to fixed but unknown constants in frequentist statistics. For up-to-date and comprehensive treatments of Bayesian methods, the reader is directed to [11, 33].

Besides such theoretical and philosophical, differences between the two inference schemes, Bayesian inference offers many pragmatic advantages over its frequentist counterpart – in particular, in the context of evaluating informative hypotheses with parametric order constraints. The advantages may be termed directness of inference, automaticity, power of priors, and, finally, ease of computation. First, by directness of inference, we mean that the Bayesian inference process directly addresses the question the researcher wishes to answer – that is, how data modifies his or her belief about initial hypotheses. In contrast, frequentist inferences are based on the probability (i.e., p-value) of obtaining current data or more extreme data under the assumption that the researcher's initial hypothesis is correct, which seems awkward and even confusing. Second, Bayesian inference is automatic as there is just *one road* to data analysis: Each and every inference problem boils down to finding the posterior from the likelihood function and the prior by applying Bayes' theorem. Third, Bayesian statistics allows one to easily incorporate any available relevant information, other than observed data, into the inference process through priors. Being able to incorporate prior information into data modeling, which undoubtedly improves the quality of inferences, is indeed a powerful and uniquely Bayesian idea, with no counterpart in frequentist statistics. This is also one of the reasons Bayesian statistics has gained such popularity in fields dealing with practical problems of real-world significance such as biomedical sciences and engineering disciplines – one cannot afford to disregard potentially useful information that might help save lives or generate millions of dollars! Finally, as mentioned earlier, the recent breakthrough in Bayesian computation makes it routinely possible to make inferences about any given informative hypothesis. The necessary computations for any arbitrary form of order constraints can be performed via MCMC as easily as one is running simple simulations on computer.

In what follows, we provide a broad-brush overview of the Bayesian order restricted inference framework that is described and illustrated in greater detail by various authors of this book, with special attention given to the comparative review on the pros and cons of the Bayesian methods discussed in the various chapters.

15.2.2 The Specifics of the Bayesian Approach

The key idea of the Bayesian approach for testing and evaluating an informative hypothesis is to incorporate the order constraints specified by the hypothesis into the prior distribution. For example, for an informative hypothesis $H : \mu_1 < \mu_2 < \mu_3$ expressed in terms of means μ, the order constraint is represented by the following prior for the parameter vector $\boldsymbol{\theta} = (\mu_1, \mu_2, \mu_3)$:

$$p(\boldsymbol{\theta}) = \begin{cases} g(\boldsymbol{\theta}) & \text{if } \mu_1 < \mu_2 < \mu_3 \\ 0 & \text{otherwise} \end{cases} \tag{15.1}$$

for some probability measure function that integrates to 1. Given observed data $\boldsymbol{y} = (y_1, ..., y_n)$ and the likelihood $f(\boldsymbol{y}|\boldsymbol{\theta})$, the posterior is obtained from Bayes' rule as

$$p(\boldsymbol{\theta}|\boldsymbol{y}) = \frac{f(\boldsymbol{y}|\boldsymbol{\theta})p(\boldsymbol{\theta})}{\int f(\boldsymbol{y}|\boldsymbol{\theta})p(\boldsymbol{\theta})\, d\boldsymbol{\theta}}. \tag{15.2}$$

For a concrete example of the likelihood $f(\boldsymbol{y}|\boldsymbol{\theta})$ and prior distribution $p(\boldsymbol{\theta})$ the interested reader is referred to Section 15.3.2.

The posterior distribution in (15.2) represents a complete summary of information about the parameter $\boldsymbol{\theta}$ and is used to draw specific inferences about it. For instance, we may be interested in finding the posterior mean and Bayesian credible intervals. Each of these measures can be expressed as a posterior expectation. The trouble is that since the normalizing constant $\int f(\boldsymbol{y}|\boldsymbol{\theta})p(\boldsymbol{\theta})\, d\boldsymbol{\theta}$ in the denominator is commonly intractable for all but the simplest models, the posterior distribution is only known up to a proportionality constant. Even if the posterior is known in analytic form, finding its mean and credible intervals can be challenging. The next best thing, then, beyond knowing the exact expression of the posterior, is to generate a large number of samples that approximate the distribution and to use the samples to numerically estimate the expectation of interest. This is where MCMC comes in handy, as the technique allows us to draw samples from almost any form of posterior distribution without having to know its normalizing constant (i.e., the denominator in (15.2)).

When one entertains multiple hypotheses and wishes to compare them, this can be achieved using the Bayes factor (BF), which, for two hypotheses H_i and H_j, is defined as the ratio of their marginal likelihoods:

$$BF_{ij} = \frac{m(\boldsymbol{y}|H_i)}{m(\boldsymbol{y}|H_j)} = \frac{\int f(\boldsymbol{y}|\boldsymbol{\theta}, H_i)p(\boldsymbol{\theta}|H_i)\, d\boldsymbol{\theta}}{\int f(\boldsymbol{y}|\boldsymbol{\theta}, H_j)p(\boldsymbol{\theta}|H_j)\, d\boldsymbol{\theta}}, \tag{15.3}$$

where $m(\boldsymbol{y}|H_i)$ denotes the marginal likelihood under hypothesis H_i. The Bayes factor has several attractive features as a model selection measure. First, the Bayes factor is related to the posterior hypothesis probability: the probability of a hypothesis being true given observed data; that is, from a set of BFs computed for each pair of competing hypotheses, the posterior probability of hypothesis $p(H_i|\boldsymbol{y}), i = 1, ..., q$, is given as $p(H_i|\boldsymbol{y}) = BF_{ik}/\sum_{j=1}^{q} BF_{jk}, i = 1, ..., q$, for any choice of $k = 1, ..., q$, under the assumption of equal prior probabilities $p(H_i) = 1/q$ for all i's. Further, Bayes factor-based model selection automatically adjusts for model complexity and avoids overfitting, thereby representing a formal implementation of Occam's razor. What this means is that BF selects the one, among a set of competing hypotheses, that provides the simplest explanation of the data.

Another attractive feature of the Bayes factor, which is particularly fitting for evaluating order constrained hypotheses, is that the model selection measure is applicable for choosing between hypotheses that vary in the number of parameters but also, importantly, for comparing multiple informative hypotheses that posit different order constraints but share a common set of parameters. For example, consider the following three hypotheses: $H_1 : \mu_1, \mu_2, \mu_3$; $H_2 : \mu_1, \{\mu_2 < \mu_3\}$; and $H_3 : \mu_1 < \mu_2 < \mu_3$. It is worth noting here that commonly used selection criteria like the Akaike information criterion (AIC [1]) and the Bayesian information criterion (BIC [38]), which only consider the number of parameters in their complexity penalty term, are inappropriate in this case. This is because the two criteria treat the above three hypotheses equally complex (or flexible), which is obviously not the case.

Accompanying these desirable properties of the Bayes factor are some important caveats. First of all, the Bayes factor can be ill-defined and cannot be used under certain improper priors. An improper prior, by definition, does not integrate finitely so we will have $\int p(\boldsymbol{\theta})_{improper}\, d\boldsymbol{\theta} = \infty$. For example, the prior $p(\theta) \propto 1/\theta$ is improper over the parameter range $0 < \theta < \infty$ and so is the uniform prior $p(\theta) = c$ for an unspecified constant c over the same range of the parameter θ. To illustrate, suppose that each element of the data vector $\boldsymbol{y} = (y_1, ..., y_N)$ is an independent sample from a normal distribution $\mathcal{N}(\mu, \sigma^2)$ with unknown mean μ but known variance σ^2. In this case, the sample mean $\overline{y}$ is a sufficient statistic for parameter μ. The likelihood is then given by

$$f(\overline{y}|\mu) = \frac{1}{\sqrt{2\pi}\,(\sigma/\sqrt{N})} \exp\left(-\frac{1}{2\sigma^2/N}(\overline{y} - \mu)^2\right) \tag{15.4}$$

as a function of parameter μ. If we were to use the improper uniform prior $p(\mu) = c$ for $-\infty < \mu < \infty$, the marginal likelihood $m(\overline{y}) = \int_{-\infty}^{+\infty} f(\overline{y}|\mu)p(\mu)\, d\mu$ would contain the "unspecified constant" c, and as such, the Bayes factor value in (15.3) would be undetermined.[5] Interestingly however, for the present example, it is easy to see that the posterior distribution

[5] An exception to this "undetermined" Bayes factor case is when the marginal likelihood of the other hypothesis being compared against the current one also contains the same constant c so both "unspecified constants" do cancel each other out in the calculation of the ratio of the two marginal likelihoods.

$p(\mu|\overline{y})$ is proper with its finite normalizing constant. This is because the unspecified constant c "conveniently" cancels out in the application of Bayes' rule to find the posterior

$$p(\mu|\overline{y}) = \frac{f(\overline{y}|\mu)p(\mu)}{\int f(\overline{y}|\mu)p(\mu)\,d\mu} = \frac{f(\overline{y}|\mu)}{\int f(\overline{y}|\mu)\,d\mu}, \tag{15.5}$$

which integrates to one for $-\infty < \mu < \infty$. An important implication is that in a case like this, posterior-based inferences such as Bayesian confidence interval estimation and deviance information criterion (DIC)-based model selection [41] are welldefined and applicable, whereas the Bayes factor is not. We will come back to this later in this chapter.

Second, another caveat is about using the Bayes factor for the comparison of two nested models. It is well known that the Bayes factor can be highly sensitive to the choice of priors, especially under diffuse priors with relatively large variances. In other words, the Bayes factor value can fluctuate widely and nonsensically to incidental minor variations of the priors. This is connected to the Lindley's paradox (e.g., [31]). Therefore, for nested models, Bayes factors under diffuse priors must be interpreted with great care.

The last, and by no means least, challenge for the Bayes factor as a model selection measure is a heavy computational burden. The Bayes factor is nontrivial to compute. To date, there exists no general-purpose numerical method for routinely computing the required marginal likelihood, especially for nonlinear models with many parameters and nonconjugate priors.

Addressing these issues and challenges in Bayes factor calculations, Klugkist, Hoijtink, and their colleagues (see Chapter 4 and [24, 25]), have developed an elegant technique for estimating the Bayes factor from prior and posterior samples for order restricted hypotheses, without having to directly compute their marginal likelihoods. In following section, we provide a critical review of the essentials of the method, which may be called the encompassing prior Bayes factor approach, or the encompassing Bayes approach.

15.2.3 Encompassing Prior Bayes Factors

The encompassing Bayes approach has been developed specifically for model selection with informative hypotheses. Specifically, the approach requires the setting of two nested hypotheses, H_1 and H_2, that share the same set of parameters but differ from each other in the form of parametric constraints (e.g., $H_1 : \mu_1, \mu_2, \mu_3$ and $H_2 : \mu_1, \{\mu_2 < \mu_3\}$). For simplicity, in this section we assume that hypothesis H_2 is nested within hypothesis H_1. Another condition required for the application of the encompassing Bayes approach is that the prior distribution of the smaller hypothesis H_2 is obtained from the prior distribution of the larger hypothesis H_1 simply by restricting the parameter space of H_1 in accordance with the order constraints imposed by H_2. Formally, this condition can be stated as

$$p(\boldsymbol{\theta}|H_2) \propto \begin{cases} p(\boldsymbol{\theta}|H_1) & \text{if } \boldsymbol{\theta} \text{ is in agreement with } H_2 \\ 0 & \text{otherwise.} \end{cases} \tag{15.6}$$

With these two conditions met, it has been shown that the Bayes factor can be approximated as a ratio of two proportions (see [25] and Chapter 4):

$$BF_{21} \approx \frac{r_{post21}}{r_{pre21}}. \tag{15.7}$$

In the equation, r_{post21} denotes the proportion of samples from the posterior distribution of hypothesis H_1, $p(\boldsymbol{\theta}|\boldsymbol{y}, H_1)$, that satisfy the order constraints of hypothesis H_2. Similarly, r_{pre21} denotes the proportion of samples from the prior distribution $p(\boldsymbol{\theta}|H_1)$ that also satisfy the order constraints of hypothesis H_2. The beauty of the encompassing Bayes lies in that its implementation requires only the ability to sample from the prior and the posterior of the larger of the two hypotheses, without having to deal with their marginal likelihoods, which can be quite difficult to compute, as mentioned earlier.

The Bayes factor calculated using the computational "trick" in (15.7) may have large variances especially when the smaller hypothesis is too highly constrained to yield stable estimates of the proportions r_{post} and r_{pre}. In such cases, one may resort to the following more efficient estimation method. We first note that the Bayes factor for two nested hypotheses H_q and H_1, where $H_q \subset H_1$, can be rewritten in terms of a series of (artificial) Bayes factors corresponding to pairs of nested hypotheses created by recursively constraining the parameter space of H_1 as

$$BF_{q1} = BF_{q(q-1)} \cdot BF_{(q-1)(q-2)} \cdots BF_{21} \tag{15.8}$$

for $H_q \subset H_{q-1} \subset \cdots \subset H_2 \subset H_1$. Using this equality, one can then compute the desired BF_{q1} as a product of BF_{ij}'s, each of which is, in turn, estimated from an equation analogous to (15.7) using any standard MCMC algorithms or the ones that are specifically tailored to order constrained hypotheses (e.g., [10]). Equations similar to (15.8) are presented in Chapters 4 and 12.

The encompassing Bayes approach is quite an ingenious idea that allows one to routinely compute Bayes factors simply by sampling from prior and posterior distributions, thereby bypassing the potentially steep hurdle of computing the marginal likelihood. As demonstrated in various chapters of this book, the approach has been successfully applied to comparing order constrained hypotheses that arise in a wide range of data analysis problems, including analysis of variance, analysis of covariance, multilevel analysis, and analysis of contingency tables.

There is, however, one assumption of the encompassing Bayes approach that may limit its general application. This is the requirement that all hypotheses, constrained or unconstrained, be of the same dimension. To illustrate, consider the following two hypotheses:

$$\begin{aligned} H_1 &: \mu_1, \mu_2, \mu_3, \\ H_2 &: \mu_1 = \mu_2 < \mu_3. \end{aligned} \tag{15.9}$$

Note that H_1 has three free parameters, whereas H_2 has two. In this case, the Bayes factor in (15.7) is undefined, as both prior and posterior proportions is effectively equal to zero. Klugkist, in Chapter 4, outlined a heuristic procedure that may be employed to approximate the Bayes factor for equality constrained hypotheses. Briefly, according to the procedure, we first construct a series of "near equality" hypotheses of varying degrees,

$$H_2(\delta_i) : |\mu_1 - \mu_2| < \delta_i, \{\mu_1 < \mu_3\}, \{\mu_2 < \mu_3\} \quad (i = 1, 2, ..., q) \tag{15.10}$$

for $\delta_1 > \delta_2 > \cdots > \delta_q > 0$. We then estimate the Bayes factor using the formulation in (15.8) by letting $\delta_q \to 0$, provided that the estimate converges to a constant. This is quite an elegant trick, although a problem may arise if the estimate does not converge, meaning that the final estimate is highly dependent upon the particular choice of limiting sequences $\{\delta_1, \delta_2, ..., \delta_q\}$ and/or upon the choice of priors. Further theoretical work showing that this is not generally the case is clearly needed.

Continuing the discussion on the model selection problem with informative hypotheses involving equality constrained hypotheses, one can think of at least two alternative methods, other than the procedure of Klugkist described above.

The first is the completing and splitting method that is introduced in Chapter 7. To illustrate, consider again the two hypotheses $H_1 : \mu_1, \mu_2, \mu_3$ and $H_2 : \mu_1 = \mu_2 < \mu_3$. The basic idea of the completing and splitting method is to add a third "surrogate" hypothesis H_3 to the original two. The new hypothesis is constructed by removing the order constraint from H_2 but keeping the equality constraint (i.e., $H_3 : \{\mu_1 = \mu_2\}, \mu_3$). Note that H_3 is of the same dimension (i.e., 2) as H_2 so one can apply the encompassing Bayes approach to obtain the Bayes factor for these two hypotheses. Now, the desired Bayes factor BF_{21} we wanted to compute is then expressed in terms of the "surrogate" hypothesis H_3 as $BF_{21} = BF_{23} \cdot BF_{31}$. In this expression, the first factor BF_{23} on the right-hand side is calculated using the encompassing Bayes approach in (15.7). As for the second factor BF_{31} for two unconstrained hypotheses that differ in dimensions, this quantity may be computed by using an appropriate prior distribution with the usual Bayesian computational methods or, alternatively, with data-based prior methods such as the intrinsic Bayes factor [3] and the fractional Bayes factor [32]. Incidentally, it would be of interest to examine whether Klugkist's procedure would yield the same Bayes factor value as the completing and splitting method.

The second approach for dealing with equality hypotheses represents a departure from Bayes factor-based model selection. Model selection criteria proposed under this approach may be termed collectively posterior predictive selection methods and are discussed in great detail in Chapter 8. In the following section, we provide a critical review of these methods and their relations to Bayes factors.

15.2.4 Posterior Predictive Selection Criteria

The posterior predictive model selection criteria discussed in Chapter 8 are the L measure [4, 14], the DIC [11, 41], and the logarithm of the pseudomarginal likelihood (LPML [8, 15]). All three measures are defined with respect to the posterior predictive distribution (ppd) of future, yet-to-be-observed data $\boldsymbol{z}$:

$$f_{ppd}(\boldsymbol{z}|\boldsymbol{y}_{obs}) = \int f(\boldsymbol{z}|\boldsymbol{\theta})p(\boldsymbol{\theta}|\boldsymbol{y}_{obs})\, d\boldsymbol{\theta}, \tag{15.11}$$

where $\boldsymbol{y}_{obs} = (y_{1,obs}, ..., y_{N,obs})$ is the currently observed data. Samples from this predictive distribution represent predictions for future observations from the same process that has generated the observed data.

A posterior predictive criterion is designed to assess a model's or hypothesis's predictive accuracy for future samples. The above three criteria differ from one another in the form of the predictive accuracy measure employed

$$\begin{aligned} L \text{ measure} &= E(\boldsymbol{z} - \boldsymbol{y}_{obs})^2, \\ \text{DIC} &= E\left[-2\ln f\left(\boldsymbol{z}|\overline{\boldsymbol{\theta}}(\boldsymbol{y}_{obs})\right)\right], \\ \text{LPML} &= \sum_{i=1}^{N} \ln f_{ppd}\left(y_{i,obs}|\boldsymbol{y}_{obs}^{(-i)}\right), \end{aligned} \tag{15.12}$$

where $\overline{\boldsymbol{\theta}}$ denotes the posterior mean, $\boldsymbol{y}_{obs}^{(-i)}$ denotes $\boldsymbol{y}_{obs}$ with the i-th observation deleted, and, finally, all expectations $E(\cdot)$ are taken with respect to the posterior predictive distribution $f_{ppd}(\boldsymbol{z}|\boldsymbol{y}_{obs})$. Under suitable assumptions, each of the above "theoretical" measures is approximately estimated by the following "computable" expression

$$\begin{aligned} L \text{ measure} &= \sum_{i=1}^{N} \left(E_{\boldsymbol{\theta}|\boldsymbol{y}_{obs}}\left[E_{z_i|\boldsymbol{\theta}}(z_i^2|\boldsymbol{\theta})\right] - \mu_i^2\right) + \nu \sum_{i=1}^{N} \left(\mu_i - y_{i,obs}\right)^2, \\ \text{DIC} &= D(\overline{\boldsymbol{\theta}}) + 2p_D, \\ \text{LPML} &= \sum_{i=1}^{N} \ln E_{\boldsymbol{\theta}|\boldsymbol{y}_{obs}^{(-i)}}\left[f\left(y_{i,obs}|\boldsymbol{\theta}\right)\right]. \end{aligned} \tag{15.13}$$

In the first equation defining the L measure criterion, z_i is a future response with the sampling distribution $f(z_i|\boldsymbol{\theta})$, ν is a tuning parameter to be fixed between 0 and 1, and $\mu_i = E_{\boldsymbol{\theta}|\boldsymbol{y}_{obs}}\left[E_{z_i|\boldsymbol{\theta}}(z_i|\boldsymbol{\theta})\right]$, with the first expectation defined with respect to the posterior distribution $p(\boldsymbol{\theta}|\boldsymbol{y}_{obs})$ and the second expectation defined with respect to the sampling distribution $f(z_i|\boldsymbol{\theta})$. In the second expression defining DIC, $D(\boldsymbol{\theta})$ is the deviance function given data vector $\boldsymbol{y}_{obs}$ defined as $D(\boldsymbol{\theta}) = -2\ln f(\boldsymbol{y}_{obs}|\boldsymbol{\theta})$ (cf. [29]), $\overline{\boldsymbol{\theta}}$ denotes the mean of $\boldsymbol{\theta}$ with respect to the posterior distribution $p(\boldsymbol{\theta}|\boldsymbol{y}_{obs})$, and, finally, p_D is the effective number of model parameters, or a model complexity (flexibility) measure, defined as $p_D = \overline{D(\boldsymbol{\theta})} - D(\overline{\boldsymbol{\theta}})$. In the third expression regarding

LPML, the expectation is with regard to the posterior distribution $p(\boldsymbol{\theta}|\boldsymbol{y}_{obs}^{(-i)})$. For L measure and DIC, the smaller their value, the better the model. The opposite is true for LPML.

The three model selection criteria in (15.13) differ in at least two important ways from the Bayes factor. First, they are predictive measures, the goal of which is to pick a model or hypothesis that achieves best predictions for future data. In contrast, the goal of Bayes factor model selection is to find the model with the highest posterior model probability. Second, all three criteria are defined based on samples from the posterior distribution $p(\boldsymbol{\theta}|\boldsymbol{y}_{obs})$. As such, it is straightforward to compute the criteria with any standard MCMC methods for order constrained hypotheses and even for equality constrained hypotheses, which can be particularly thorny for Bayes factor computation.

Notwithstanding these attractive features of the predictive model selection criteria, one may object to them on the grounds that they may be intuitive but are based on arbitrary measures of predictive accuracy; that is, one may ask questions such as: Why the squared error loss function in L measure, or for that matter, the deviance function in DIC? Which of the three is the "best"? What should we do if their model choices disagree with one another? Further, regarding DIC, it is known to violate the reparameterization invariance rule [41]. Reparameterization invariance means that a model's data fitting capability does not change, as it should, when the model's equation is rewritten under a reparameterization. For instance, the model equation $y = \exp(-\theta x)$ can be re-expressed as $y = \eta^{-x}$ through the reparameterization $\eta = \exp(\theta)$. DIC is generally not reparameterization-invariant, as the posterior mean $\overline{\boldsymbol{\theta}}$ in the DIC equation (15.13) does change its value under reparameterization. In short, the reader should be aware of these issues and interpret the results from the application of the posterior predictive criteria with a grain of salt.

15.3 Hierarchical Bayes Order Constrained Analysis of Variance

In this section, we present and discuss an exemplary application of the Bayesian approach for analyzing ANOVA-like data. In particular, we implement and demonstrate a hierarchical Bayes framework. Also discussed in the example application is a comparison between the results from Bayes factor model selection and those from posterior predictive model selection using DIC.

15.3.1 Blood Pressure Data and Informative Hypotheses

We consider blood pressure data that are discussed in Maxwell and Delaney's book [28] on experimental designs. These are hypothetical data created to illustrate certain statistical ideas in their book. The data are imagined to be from an experiment in which a researcher wants to study the effectiveness of diet, drugs, and biofeeback for treating hypertension. The researcher designs a $2 \times 3 \times 2$ between-subjects factorial experiment in which the diet factor varies over two levels (absent and present), the drug factor over three levels (drug X, Y, and Z), and the biofeedback factor over two levels (absent and present). Blood pressure is measured for 6 individuals in each of 12 cells.

The full data are reported and summarized in Tables 8.12 and 8.13 of Maxwell and DeLaney [28]. Some general trends can be noticed from these tables. Both diet and biofeedback seem to be effective in lowering blood pressure. Also, among the three drugs, it appears that drug X is the most effective and that drug Z seems better than drug Y, although the latter differences may be due to sampling error. Results from an ANOVA applied to these data and reported in Table 8.14 of the book indicate that all three main effects are statistically significant, with each p-value being less than .001, and that one of the two-way interactions and the three-way interaction are marginally significant (i.e., $p = .06$ and $p = .04$, respectively).

Based on these ANOVA results, to illustrate a hierarchical Bayes order restricted inference framework, we consider five hypotheses. They include the null hypothesis, H_0, with no order constraints, and four informative hypotheses, $H_1 - H_4$, with varying degrees of order constraints on the population cell means:

$$
\begin{aligned}
H_0 &: \text{ Unconstrained } {\mu_{ijk}}'s \text{ for all } i, j, k, \\
H_1 &: \mu_{DB\bullet} < \{\mu_{D\underline{B}\bullet}, \mu_{\underline{D}B\bullet}\}; \{\mu_{D\underline{B}\bullet}, \mu_{\underline{D}B\bullet}\} < \mu_{\underline{DB}\bullet}, \\
H_2 &: \mu_{DB\bullet} < \mu_{D\underline{B}\bullet} < \mu_{\underline{D}B\bullet} < \mu_{\underline{DB}\bullet}, \\
H_3 &: \mu_{DBk} < \{\mu_{D\underline{B}k}, \mu_{\underline{D}Bk}\}; \{\mu_{D\underline{B}k}, \mu_{\underline{D}Bk}\} < \mu_{\underline{DB}k} \text{ for all } k, \\
&\quad \mu_{ijX} < \mu_{ijZ} < \mu_{ijY} \text{ for all } i, j, \\
H_4 &: \mu_{DBk} < \mu_{D\underline{B}k} < \mu_{\underline{D}Bk} < \mu_{\underline{DB}k} \text{ for all } k, \\
&\quad \mu_{ijX} < \mu_{ijZ} < \mu_{ijY} \text{ for all } i, j.
\end{aligned}
\tag{15.14}
$$

In the above equation the subscript i denotes the level of the diet factor (D: present; $\underline{D}$: absent), the subscript j denotes the level of the biofeedback factor (B: present; $\underline{B}$: absent), and, finally, the subscript k denotes the drug type (X, Y, or Z). The subscript $\bullet$ indicates that the result is averaged across all levels of the corresponding factor.

Shown in Figure 15.1 are the four informative hypotheses in graphical form. The data are found to violate none of the order constraints specified by hypothesis H_1 or by hypothesis H_2. In contrast, as marked by the asterisk symbol ($*$) in the figure, three violations of the order constraints under H_3 and four violations of the order constraints under H_4 are observed in the data. A question one might ask, then, would be: Are these violations "real" or just sampling errors? In the following section, we present a hierarchical Bayesian analysis that attempts to answer questions such as this.

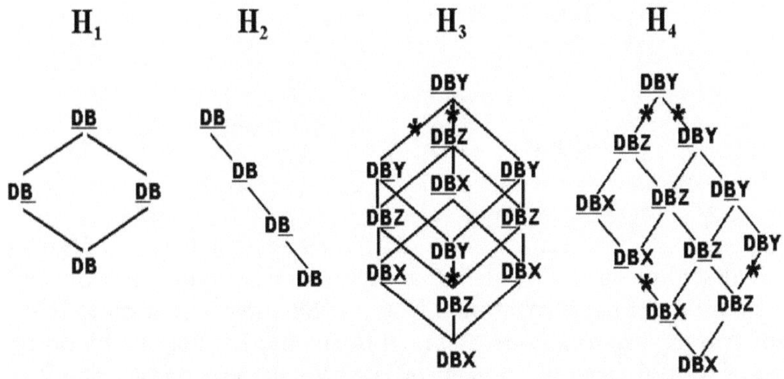

Fig. 15.1. The four informative hypotheses defined in (15.14). In each connected graph, for two treatment conditions that are connected to each other, the one that is positioned above the other has a higher population mean value and as such is *less effective* in treating high blood pressure than the other condition. The asterisk (∗) indicates a violation of the corresponding ordinal prediction in the data

15.3.2 Hierarchical Bayesian Analysis

Given the five hypotheses in (15.14), the model selection problem is to identify the hypothesis that best describes the blood pressure data. To this end, we present a hierarchical Bayesian framework and discuss results from its application to the data.

A defining feature of hierarchial Bayesian modeling is the setup of multilevel dependency relationships between model parameters such that lower-level parameters are specified probabilistically in terms of higher-level parameters, known as hyperparameters, which themselves may, in turn, be given another probabilistic specification in terms of even higher-level parameters, and so on [11]. The hierarchical modeling generally improves the robustness of the resulting Bayesian inferences with respect to prior specification [33]. Importantly, the hierarchical setup of parameters is particularly suitable for modeling various kinds of dependence structures that the data might exhibit, such as individual differences in response variables and trial-by-trial dependency of reaction times. Recently, the hierarchical Bayesian modeling has become increasingly popular in cognitive modeling, and its utility and success have been well demonstrated (cf. [26, 27, 36, 37]).

Using standard distributional notation, we now specify the hierarchical Bayesian framework for modeling the blood pressure data as

$$
\begin{aligned}
\text{Likelihood}: \quad & y_{ijkl} \sim \mathcal{N}(\mu_{ijk}, \sigma^2), \\
& \\
\text{Priors}: \quad & \mu_{ijk}|\eta, \tau^2 \sim \mathcal{N}(\eta, \tau^2), \\
& \eta|\psi^2 \sim \mathcal{N}(0, \psi^2), \\
& \tau^2|a, b \sim \mathcal{IG}(a, b), \\
& \sigma^2|c, d \sim \mathcal{IG}(c, d),
\end{aligned}
\tag{15.15}
$$

where $i = 1, ..., I$, $j = 1, ..., J$, $k = 1, ..., K$, and $l = 1, ..., N$, $\mathcal{N}(\cdot)$ denotes a normal distribution, $\mathcal{IG}(\cdot)$ denotes an inverse Gamma distribution,[6] and ψ^2, a, b, c, and d are fixed constants. Note in the above equation that η and τ^2 represent two hyperparameters assumed in the model. For the blood pressure data, there were 6 persons in each of the 12 cells created by the $2 \times 2 \times 3$ factorial design, and, as such, we have $I = 2$, $J = 2$, $K = 3$, and $N = 6$.

Let us define the data vector as $\boldsymbol{y} = (y_{1111}, ..., y_{IJKn})$ and the parameter vector as $\boldsymbol{\theta} = (\boldsymbol{\mu}, \eta, \tau^2, \sigma^2)$, where $\boldsymbol{\mu} = (\mu_{111}, ..., \mu_{IJK})$. The posterior density under the unconstrained hypothesis H_0 in (15.14) is then given by

$$
p(\boldsymbol{\theta}|\boldsymbol{y}) \propto f(\boldsymbol{y}|\boldsymbol{\mu}, \sigma^2)\, p(\boldsymbol{\mu}|\eta, \tau^2)\, p(\eta|\psi^2)\, p(\tau^2|a, b)\, p(\sigma^2|c, d), \tag{15.16}
$$

with the likelihood function of the following form:

$$
f(\boldsymbol{y}|\boldsymbol{\mu}, \sigma^2) = \prod_{i=1}^{I} \prod_{j=1}^{J} \prod_{k=1}^{K} \prod_{l=1}^{N} \frac{1}{\sqrt{2\pi}\,\sigma} \exp\left(-\frac{1}{2\sigma^2}(y_{ijkl} - \mu_{ijk})^2\right). \tag{15.17}
$$

From these expressions, one can easily derive the full conditional posterior distributions of various parameters as

$$
\begin{aligned}
p(\mu_{ijk}|\boldsymbol{y}, \boldsymbol{\mu}^{(-ijk)}, \eta, \tau^2, \sigma^2) &\sim \mathcal{N}\left(\frac{\frac{\sigma^2}{N}\eta + \tau^2 \frac{\sum_{l=1}^{N} y_{ijkl}}{N}}{\frac{\sigma^2}{N} + \tau^2}, \frac{\frac{\sigma^2}{N}\tau^2}{\frac{\sigma^2}{N} + \tau^2}\right), \\
p(\eta|\boldsymbol{y}, \boldsymbol{\mu}, \tau^2, \sigma^2) &\sim \mathcal{N}\left(\frac{\psi^2}{IJK\psi^2 + \tau^2} \sum_{i=1}^{I} \sum_{j=1}^{J} \sum_{k=1}^{K} \mu_{ijk}, \frac{\psi^2 \tau^2}{IJK\sigma^2 + \tau^2}\right), \\
p(\tau^2|\boldsymbol{y}, \boldsymbol{\mu}, \eta, \sigma^2) &\sim \mathcal{IG}(\left(a + \frac{IJK}{2}, \left[\frac{1}{b} + \frac{1}{2} \sum_{i=1}^{I} \sum_{j=1}^{J} \sum_{k=1}^{K} (\mu_{ijk} - \eta)^2\right]^{-1}\right), \\
p(\sigma^2|\boldsymbol{y}, \boldsymbol{\mu}, \eta, \tau^2) &\sim \mathcal{IG}(\left(c + \frac{IJKN}{2}, \left[\frac{1}{d} + \frac{1}{2} \sum_{i=1}^{I} \sum_{j=1}^{J} \sum_{k=1}^{K} \sum_{l=1}^{N} (y_{ijkl} - \mu_{ijk})^2\right]^{-1}\right).
\end{aligned}
\tag{15.18}
$$

From these full conditionals for the unconstrained hypothesis, a Gibbs sampler can be devised to draw posterior samples from an informative hypothesis with order constraints of the form $\alpha \le \theta_i \le \beta$, specifically, the

[6] The probability density function of the gamma and inverse-gamma distributions are defined as $\mathcal{G}(a, b) : f(x|a, b) = \frac{1}{\Gamma(a) b^a} x^{a-1} e^{-x/b} (a, b > 0;\ 0 < x < \infty)$ and $\mathcal{IG}(a, b) : f(x|a, b) = \frac{1}{\Gamma(a) b^a} x^{-a-1} e^{-1/bx} (a, b > 0;\ 0 < x < \infty)$, respectively. Note that $X \sim \mathcal{G}(a, b) \Longleftrightarrow 1/X \sim \mathcal{IG}(a, b)$.

following inverse probability sampling procedure [10]:

$$\theta_i = F_i^{-1}\left[F_i(\alpha) + U \cdot (F_i(\beta) - F_i(\alpha))\right], \tag{15.19}$$

where F_i is the cumulative full conditional distribution for θ_i of the unconstrained hypothesis, F_i^{-1} is its inverse, and U is a uniform random number on $[0,1]$. It should be noted that special care needs to be taken in applying this procedure for hierarchical models with constrained parameters. This is because the normalizing constants for lower-level parameters generally depend upon the values of higher-level parameters so the constants do not cancel one another out, thereby making the implementation of Gibbs sampling difficult, if not impossible. Chen and Shao [5] developed efficient Monte Carlo methods that address the problem. We implemented their methods in our application of the inverse probability sampling procedure.

From posterior samples, one can then compute the DIC criterion in (15.13) with the deviance function $D(\boldsymbol{\theta})$ for the data model in (15.15) expressed as

$$\begin{aligned} D(\boldsymbol{\theta}) &= -2\ln f(\boldsymbol{y}|\boldsymbol{\mu},\sigma^2) \\ &= \sum_{i=1}^{I}\sum_{j=1}^{J}\sum_{k=1}^{K}\frac{(\mu_{ijk}-\overline{y}_{ijk})^2}{\sigma^2/N} + IJK\cdot\ln\left(\frac{2\pi\sigma^2}{N}\right), \end{aligned} \tag{15.20}$$

where $\overline{y}_{ijk}$ represents the sample mean for cell ijk. The Bayes factors and the posterior model probabilities for the five hypotheses in (15.14) are estimated using the encompassing Bayes approach discussed earlier.

The model comparison results are presented in Table 15.1. The DIC results are based on the following parameter values for the hyperpriors: $a = 10$, $b = 0.01$, $c = 10$, $d = 0.01$, and $\psi = 4000$. For each hypothesis, the mean DIC value and the 95% confidence interval based on 10 independent runs of the inverse probability sampling procedure are shown. The encompassing prior Bayes factors are based on 30 million samples drawn from each of the prior and posterior distributions under the unconstrained hypothesis H_0. Shown in the second column of the table are the p_D values, which measure the effective number of parameters. All five hypotheses assume the same number of parameters (i.e., 15), including the two hyperparameters of η and τ^2, and yet, obviously they differ in model complexity (flexibility), as each imposes different degrees of order constraints upon the parameters. Note that the unconstrained hypothesis H_0 has the largest p_D value of 9.61 and then the complexity value decreases from top to bottom of the column. This pattern of result agrees with the intuitive notion that the more order constraints an informative hypothesis assumes, the less complexity the hypothesis presents. The DIC results shown on the third column indicate that among the five informative hypotheses, the simplest one, H_4, is the best predicting model from the posterior predictive standpoint.

The remaining columns of the table present the encompassing Bayes results. First of all, recall that the r_{preq0} and r_{postq0} values estimate the pro-

Table 15.1. Model comparison results for the five hypotheses in (15.14) and the blood pressure data in Maxwell and Delaney [28]

Hypothesis	p_D	DIC	r_{preq0}	r_{postq0}	BF_{q0}	$p(H_q\|\boldsymbol{y})$
H_0	9.61	37.06 ± 0.11	1.000	1.000	1.00	.0004
H_1	7.50	34.10 ± 0.52	.080	.570	7.15	.003
H_2	7.03	33.52 ± 1.42	.041	.49	12.0	.005
H_3	5.70	32.14 ± 1.57	5.0e-06	.0038	711	.31
H_4	5.03	30.96 ± 1.09	7.7e-07	.0012	1533	.67

portions of prior and posterior samples, respectively, drawn from the unconstrained hypothesis H_0 that satisfy the order constraints of an informative hypothesis. We note that both of these proportion values exhibit the same decreasing trend as the p_D values, although it is a much steeper for the r_{preq0} and r_{postq0} values. Next, the Bayes factor results, shown in the sixth column, clearly point to H_3 and H_4 as two "winners" in the model selection competition. Between these two, H_4 has a Bayes factor that is about double the corresponding factor for H_3. This result, taking into account the other Bayes factor values in the same column, translates into the posterior hypothesis probabilities of .67 and .31 for H_4 and H_3, respectively. So if we were to choose between these two informative hypotheses, it would then be H_4 as the one that is most likely to have generated the data. An implication of this conclusion is that the four violations in the data of the order constraints specified by H_4 (see Figure 15.1) are judged to be no more than sampling variations, and not due to systematic deviations of the underlying data-generating process from the said hypothesis.

To summarize, both DIC and Bayes factor-based selection criteria pick the hypothesis H_4 as the best model among the five competing hypotheses. Therefore, as far as the present data are concerned, the best predicting model turns out to be also the most likely model, which we find is often the case in practice.

15.4 Concluding Remarks

In this chapter we provided an overview of the recent developments in Bayesian order restricted inference that are well suited to theory testing in the psychological sciences. We also discussed an application of the Bayesian framework for hierarchical modeling. Fueled by a series of the computational breakthroughs in the early 1990s, Bayesian statistics has become increasingly popular in various scientific disciplines – in particular, in the biomedical and engineering sciences. We believe that it is the time for psychological researchers to take notice and reap the benefits of applying these powerful and versatile inference tools to advance our understanding of the mental and behavioral phenomena we are studying. We hope this chapter will serve as another example that demonstrates the power of the Bayesian approach.

We conclude the chapter by reiterating what we said earlier: The Bayesian methods developed over the past decade for testing informative hypotheses are quite impressive in their applicability and success across a wide array of data modeling problems, as illustrated in Chapters 2–5 and 10–13 of this book. The work is likely to be recognized in the years to come as a major contribution to the field of quantitative data modeling.

Acknowledgements. The authors are supported by United States National Science Foundation Grants SES-0241862 to JIM and SES-0242030 to GK.

References

[1] Akaike, H.: Information theory and an extension of the maximum likelihood principle. In: Petrox, B.N., Caski, F. (eds) Second International Symposium on Information Theory (pp. 267–281). Budapest, Academia Kiado (1973)

[2] Barlow, R.E., Bartholomew, D.J., Bremner, J.M., Bunk, H.D.: Statistical Inference under Order Restrictions. New York, Wiley (1972)

[3] Berger, J.O., Pericchi, L.R.: The intrinsic Bayes factor for model selection and prediction. Journal of the American Statistical Association, **91**, 109–122 (1996)

[4] Chen, M.-H., Dey, D. K., Ibrahim, J.G.: Bayesian criterion based model assessment for categorical data. Biometrika, **91**, 45–63 (2004)

[5] Chen, M.-H., Shao, Q.-M.: Monte Carlo methods on Bayesian analysis of constrained parameter problems. Biometrika, **85**, 73–87 (1998)

[6] Dunson, D.B., Neelon, B.: Bayesian inference on order-constrained parameters in generalized linear models. Biometrics, **59**, 286–295 (2003)

[7] Efron, B., Tibshirani, R.J.: An Introduction to the Bootstrap. New York, Chapman & Hall (1993)

[8] Gelfand, A. E., Dey, D. K.: Bayesian model choice: Asymptotics and exact calculations. Journal of the Royal Statistical Society, Series B, **56**, 501–514 (1994)

[9] Gelfand, A.E., Smith, A.F.M.: Sampling based approaches to calculating marginal densities. Journal of the American Statistical Association, **85**, 398–409 (1990)

[10] Gelfand, A. E., Smith, A.F.M., Lee, R.-M.: Bayesian analysis of constrained paramater and truncated data problems. Journal of the American Statistical Association, **87**, 523–532 (1992)

[11] Gelman, A., Carlin, J.B., Stern, H.S., Rubin, D.B.: Bayesian Data Analysis (2nd ed.). Boca Raton, FL, Chapman & Hall/CRC (2004)

[12] Geyer, C.J.: Constrained maximum likelihood exemplified by isotonic convex logistic regression. Journal of the American Statistical Association, **86**, 717–724 (1991)

[13] Gilks, W.R., Richardson, S., Spiegelhalter, D.J.: Markoc Chain Monte Carlo in Practice. Chapman & Hall, New York (1996)

[14] Ibrahim, J.G., Chen, M.-H.,, Sinha, D.: Criterion based methods for Bayesian model assessment. Statistical Sinica, **11**, 419–443 (2001)

[15] Ibrahim, J.G., Chen, M.-H., Sinha, D.: Bayesian Survival Analysis. New York, Springer-Verlag (2001)

[16] Iliopoulos, G., Kateri, M., Ntzoufras, I.: Bayesian estimation of unrestricted and order-restricted association models for a two-way contingency table. Computational Statistics & Data Analysis, **51**, 4643–4655 (2007)

[17] Iverson, G.J., Harp, S.A.: A conditional likelihood ratio test for order restrictions in exponential families. Mathematical Social Sciences, **14**, 141–159 (1987)

[18] Iverson, G.J.: Testing order in pair comparison data. Doctoral dissertation, Department of Psychology, New York University (1983)

[19] Johnson, V.E., Albert, J.H.: Ordinal Data Modeling. New York, Springer (1999)

[20] Karabatsos, G.: The Rasch model, additive conjoint measurement, and new models of probabilistic measurement theory. Journal of Applied Measurement, **2**, 389–423 (2001)

[21] Karabatsos, G., Sheu, C.F.: Bayesian order-constrained inference for dichotomous models of unidimensional non-parametric item response theory. Applied Psychological Measurement, **28**, 110–125 (2004)

[22] Kass, R.E., Raftery, A.E.: Bayes factor. Journal of the American Statistical Association, **90**, 773–795 (1995)

[23] Kato, B. S., Hoijtink, H.: A Bayesian approach to inequality constrained linear mixed models: estimation and model selection. Statistical Modelling, **6**, 1–19 (2006)

[24] Klugkist, I., Laudy, O., Hoijtink, H.: Inequality constrained analysis of variance: A Bayesian approach. Psychological Methods, **10**, 477–493 (2005)

[25] Klugkist, I., Kato, B., Hoijtink, H.: Bayesian model selection using encompassing priors. Statistic Neerlandica, **59**, 57–69 (2005)

[26] Lee, M.D.: A hierarchical Bayesian model of human decision-making on an optimal stopping problem. Cognitive Science, **30**, 1–26 (2006)

[27] Lee, M.D.: Three case studies in the Bayeian analysis of cognitive models. Psychonomic Bulletin & Review, **15**, 1–15 (2008)

[28] Maxwell, S.E., Delaney, H.D.: Designing Experiments and Analyzing Data: A Model Comparison Perspective (2nd ed.). Mahwah, NJ, Lawrence Erlbaum Associates (2004)

[29] McCullagh, P., Nelder, J.A.: Generalized Linear Models (2nd ed.). Boca Raton, FL, Chapman & Hall/CRC (1989)

[30] Myung, J.I., Karabatsos, G., Iverson, G.J.: A Bayesian approach to testing decision making axoims. Journal of Mathematical Psychology, **49**, 205–225 (2005)

[31] O'Hagan, A., Forster, J.: Kendall's Advanced Theory of Statistics Vol. 2B: Bayesian Inference (2nd ed.) (pp. 77–78). London, Arnold (2004)

[32] O'Hagan, A.: Fractional Bayes factors for model comparison. Journal of the Royal Statistical Society, Series B, **57**, 99–138 (1995)
[33] Robert, C.P.: The Bayesian Choice (second edition). New York, Springer (2001)
[34] Robert, C.P., Casella, G.: Monte Carlo Statistical Methods (second edition). New York, Springer (2004)
[35] Robertson, T., Wright, F.T., Dykstra, R.L.: Order Restricted Statistical Inference. New York, Wiley (1988)
[36] Rouder, J.N., Lu, J.: An introduction to Bayesian hierarchical models with an application in the theory of signal detection. Psychonomic Bulletin & Review, **12**, 573–604 (2005)
[37] Rouder, J.N., Lu, J., Speckman, P.L., Sun, D., Jiang, Y.: A hierarchical model for estimating response time distributions. Psychonomic Bulletin & Review, **12**, 195–223 (2005)
[38] Schwartz, G: Estimating the dimension of a model. The Annals of Statistics, **6**, 461–464 (1978)
[39] Sedransk, J., Monahan, J., Chiu, H.Y.: Bayesian estimation of finite population parameters in categorizal data models incorporating order restrictions. Journal of the Royal Statistical Society, Series B, **47**, 519–527 (1985)
[40] Silvapulle, M.J., Sen, P.K.: Constrained Statistical Inference: Inequality, Order, and Shape Restrictions. Hoboken, NJ, Wiley (2005)
[41] Spiegelhalter, D.J., Best, N.G., Carlin, B.P., Linde, A. van der: Bayesian measures of model complexity and fit. Journal of the Royal Statistical Society, Series B, **64**, 583–639 (2002)
[42] Wagenmakers, E.-J.: A practical solution to the pervasive problems of p values. Psychonomic Bulletin & Review, **14**, 779–804 (2007)

16

A Philosopher's View on Bayesian Evaluation of Informative Hypotheses

Jan-Willem Romeijn[1] and Rens van de Schoot[2]

[1] Department of Theoretical Philosophy, Groningen University, Oude Boteringestraat 52, 9712 GL, Groningen, the Netherlands j.w.romeijn@rug.nl
[2] Department of Methodology and Statistics, Utrecht University, P.O. Box 80140, 3508 TC Utrecht, the Netherlands a.g.j.vandeschoot@uu.nl

16.1 Bayesian Model Selection

This chapter provides an answer to the question: What it is, philosophically speaking, to choose a model in a statistical procedure, and what does this amounts to in the context of a Bayesian inference? Special attention is given to Bayesian model selection, specifically the choice between inequality constrained and unconstrained models based on their Bayes factors and posterior model probabilities.

Many of the foregoing chapters have provided examples of model selection by means of Bayes factors, and Chapter 4 has provided a thorough introduction to the subject. For the sake of completeness and in order to introduce some terminology that will be used in this chapter, we will briefly rehearse Bayesian model selection here. Say that we have some data E and that we think these data are sampled from a distribution $p_{\mu_p \mu_s}(E)$, characterized by two parameters $\mu_p \in [0,1]$ and $\mu_s \in [0,1]$. We say that each pair of values for μ_p and μ_s presents a specific hypothesis $H_{\mu_p \mu_s}$ concerning the data. By contrast, a statistical model consists of a set of hypotheses. One possible statistical model for the data allows for all possible values of both parameters; that is, $\langle \mu_p, \mu_s \rangle \in [0,1]^2$. Call this model $\mathcal{M}_0$; it consists of the entire range of hypotheses $H_{\mu_p \mu_s}$. Another possible model, $\mathcal{M}_1$, imposes the restriction that $\mu_p > \mu_s$; this model is restricted by an inequality constraint. Note that both models consist of a particular set of statistical hypotheses $H_{\mu_p \mu_s}$, each of which fixes a fully specified distribution for the data, $p(E|H_{\mu_p \mu_s}) = p_{\mu_p \mu_s}(E)$. Their difference is that the latter restrict the possible values for the parameters μ_p and μ_s. How can we compare these two models?

The Bayesian model selection procedure, as discussed in this book, presents an answer to the latter question. This answer employs the so-called marginal likelihoods of the models:

$$p(E|\mathcal{M}_j) = \int_0^1 \int_0^1 p(E|H_{\mu_p\mu_s}) p_j(H_{\mu_p\mu_s})\, d\mu_p\, d\mu_s$$
$$= \int_0^1 \int_0^1 p_{\mu_p\mu_s}(E) p_j(H_{\mu_p\mu_s})\, d\mu_p\, d\mu_s, \tag{16.1}$$

where $j = 0, 1$ indexes the models. Notice that for both models, the integration runs over the whole domain $[0,1]^2$. However, the prior for the two models is different: $p_0(H_{\mu_p\mu_s})d\mu_p\mu_s = 1$ and $p_1(H_{\mu_p\mu_s})d\mu_p\mu_s = 2$ if $\mu_p > \mu_s$ and $p_1(H_{\mu_p\mu_s})d\mu_p\mu_s = 0$ otherwise. Both these priors integrate to one, but the prior for $\mathcal{M}_1$ is such that only the distributions for which $\mu_p > \mu_s$ are included in the computation of the marginal likelihood. Finally, it must be emphasized that the marginal likelihood for a model is different from the ordinary likelihood of a hypothesis, although both are probabilities of the data E. The likelihood of a hypothesis is the well-known expression $p(E|H_{\mu_p\mu_s})$. The marginal likelihood of a model is essentially a mixture of the likelihoods of hypotheses that are included in the model, weighted with the probability of the hypotheses.

We may now use these expressions of the marginal likelihood to compute the Bayes factor for the models, $\mathcal{M}_0$ and $\mathcal{M}_1$:

$$BF_{01} = \frac{p(E|\mathcal{M}_0)}{p(E|\mathcal{M}_1)}. \tag{16.2}$$

It may then turn out that $BF_{01} \ll 1$, in which case the unrestricted model $\mathcal{M}_0$ seems strongly favored over the restricted model $\mathcal{M}_1$. But here we may wonder: What support exactly is provided by the high value of the Bayes factor? It must be emphasized that the comparison of two models (e.g., $\mathcal{M}_0$ versus $\mathcal{M}_1$), is not the same as a comparison between two rival hypotheses, for example $H_{1/2 1/2}$ versus $H_{1/3 2/3}$, because models are not themselves hypotheses, rather they are sets of hypotheses. In the case of hypotheses, a comparison by means of a Bayes factor makes perfect sense. But a Bayes factor may not be suitable for the comparison between models. This point is particularly pressing for comparisons of inequality constrained models, because they contain partially overlapping sets of hypotheses.

In this chapter we set out to investigate this latter question from a foundational perspective. We discuss what statistics was supposed to deliver in the first place and in what way Bayesian statistics delivers this. After we are clear on Bayesian statistics in its ordinary application, we can discuss the application of Bayesian statistics in the context of model selection, and in particular in the context of comparing models with different inequality constraints. It will be seen that this leads to a challenging question on the exact use, or function, of Bayes factors for models.

The chapter is set up as follows. In Section 16.2 we spell out the philosophical setting for statistical inference, dealing with the problem of induction and with the answers to this problem provided by Popper and Carnap. Section 16.3 presents a parallel between statistical inference and another system of reasoning: deductive logic. In Section 16.4, based on this parallel, we will describe the role of a statistical model in a Bayesian statistical inference as a specific type of premise in an inductive. We can thereby identify elements of the views of both Popper and Carnap in Bayesian statistical inference and extend Bayesian inference to model selection, in particular, the selection by means of Bayes factors. This leads to a discussion of some problematic aspects of Bayesian model selection procedures in Section 16.5. We will address two specific worries. First, a comparison of models in terms of their posterior model probabilities does not seem to make sense if the models overlap. We will remedy this by organizing the space on which the models are defined a bit differently. Second, and in view of this reorganization, we ask how we can interpret the probability assignments to hypotheses.

16.2 Statistics and the Problem of Induction

This section deals with statistics, its relation to the problem of induction, and the solutions that Popper and Carnap provided for this problem, drawing on standard textbooks in the philosophy of science such as Bird [2] and Curd and Cover [6]. We will see that these solutions, in this context termed inductivism and rationalism, are endpoints in a spectrum of positions and that, as such, they both miss out on an important aspect of statistical reasoning.

16.2.1 The Problem of Induction

Induction is a mode of inference that allows us to move from observed data to as yet unknown data elements and empirical generalizations. A typical example of an inductive inference is presented in Statements 1 and 2:

1. The sun has risen every morning up until now.
2. So, the sun will also rise tomorrow.
3. Even stronger, it will rise on all future days.
4. Alternatively, it will probably rise on all future days.
5. Or at least it will probably rise tomorrow.

Here the observed data is expressed in Statement 1, namely that the sun has always risen up until now. This observed data may be viewed as the sole premise of the inference. On the basis of it we may want to affirm several other statements, labeled 2 to 5, all of which can be viewed as conclusions of an inductive inference.

Premises and conclusions are statements, and as such they may be true or false. Of an inference, however, we cannot say that it is true or false. Rather we say that it is valid or invalid, where validity means that the inference provides a certain kind of guarantee: If the premisses are true and the inference from these premisses to a conclusion is valid, then we have the guarantee that the conclusion is true. When applied to the inductive inferences above, validity means the following: If the sun has indeed always risen up until now, and if the inductive inference is valid, then we can rely on the truth of the conclusion, namely that the sun will also rise tomorrow. The trouble with inductive inference, as presented above, is that its validity is very hard to establish. Nobody will seriously doubt the truth of the statement that the sun will rise tomorrow. But when asked whether on the basis of all its past risings we can validly infer, and hence whether we are justified to believe, that the sun will rise tomorrow, we are met with embarrassing difficulties.

Let us examine these difficulties in some more detail. David Hume asked himself the question of how we can derive new observations from observations that we have done in the past. He argued in An Enquiry Concerning Human Understanding [15]: "But why this experience should be extended to future times, and to other objects, which for aught we know, may be only in appearance similar; this is the main question on which I would insist" (pp. 33–34). In other words, inductive inferences seem to presuppose that a sequence of observations in the future will occur as it always has in the past. However, even very long series of the same observation are perfectly consistent with the next observation being quite different. The problem of induction is that no further basis can be found in the observations themselves for this presupposed constancy of the observations. To illustrate again with the example, we might conclude from the observations that the sun has risen every morning up until now that the next morning the sun will also rise. But what justification can there be for presupposing this constancy? In the next subsection we will try to provide some possible answers to this question, and we will show how these answers fail.

16.2.2 Uniformity Assumptions

A first possible answer is to justify the presupposition of constancy, and hence inductive inference, by using induction itself; that is, we could say that an inductive inference will work in the future because it has worked in the past. For example, we made many inductive inferences about many different topics (e.g.,that the sun has always risen), and until now all these inferences led to true conclusions. Or closer to scientific practice, we have often used a T-test successfully in the past, and so we may conclude that it will be a valuable method in the future as well. However, if we justify induction on the grounds that it has worked in the past, then we enter a vicious circle. The argument fails to prove anything, because it takes for granted what it is supposed to prove. We can therefore run the exact same criticism of induction again, this time on the level of the inferences. There is, again, no logical necessity that the previous success of the inferences guarantees future successes.

A second possible answer is to justify the constancy of observations by assuming an overall uniformity of nature. For example, we might say that the sun has always risen in the past and that since nature is uniform, this pattern will continue into the future. Note, however, that this is quite a strong assumption. It is not just questionable whether nature really is uniform, it is also dubitable whether we can apply this assumption to the natural and the human sciences alike. Uniformity could hold for the natural sciences, like the sunrise example, but whether it is also applicable to human science remains discussable. Is it, for example, true that the positive correlation between social isolation and aggression, as it might be established in psychology, continues to hold after the introduction of the Internet? To infer by induction, we have to assume a rather strong uniformity in nature.

Moreover, even while the assumption of uniformity must be very strong, one might argue that it is still not strong enough. Just stating that nature is uniform does not yet determine the exact patterns that will continue in future times. This problem is nicely brought out by Goodman's [11] so-called new riddle of induction, which we will present briefly. Say that the predicate Green belongs to emeralds, which appear to have the color green at any time. Suppose that up until the year 2008 we have observed many emeralds to be Green. We thus have evidence statements that emerald 1 is Green, emerald 2 is Green, etc. The standard inductive inference then is that all emeralds examined before the year 2008 were Green, so emeralds after that year will be Green as well. In this case we call the predicate Green projectable: Findings of the past can be projected unto the future. But now consider a somewhat different predicate: An object is Grue if either it has been observed before 2008 and it appeared green, or it has been observed after 2008 and appeared blue. Similarly, something is Bleen if observed before 2008 appearing blue, or after 2008 appearing green. We may redescribe what we observed until now as emerald 1 is Grue, emerald 2 is Grue, etc. So with the very same inductive inference just used on Green, but now taking Grue to be the projectable predicate, we might conclude that emeralds observed after 2010 will be Grue, so that we predict emeralds observed after 2010 will appear blue to us! It thus seems that simply assuming the uniformity of nature is not specific enough. If we are to apply the uniformity assumption, we must stipulate the exact predicates with respect to which nature is uniform.

In reaction to Goodman's riddle, we might argue that we can make a principled distinction between candidate predicates on grounds of their simplicity, defending induction by saying nature is uniform and simple. It seems that a model where emeralds are Green before time 2010 and are also Green after 2010 is simpler. However, we might also describe this model in a complicated way, saying that emeralds are Grue before time 2010 and are Bleen after 2010: In both cases, the result is that emeralds appear green throughout. Goodman points out [11, pp. 74–75] that predicates such as Grue and Bleen only appear to be more complex than the predicate Green or Blue. This is because we have defined Grue in terms of blue and green, whereas the predicate Green is only defined in term of the color green. In other words, the model we favor depends on which predicates are established in our language, leaving inductive inference relative to the language in which they are formulated. The ultimate question is therefore what predicates are considered the natural ones.

Hence, we cannot salvage inductive inference by imposing further simplicity constraints. We need a decision on the projectability of certain patterns or predicates.

This last remark concludes our discussion on the philosophical problem of induction. The fact that in inductive inference we must always make a choice for a specific projectable predicate will reappear in later sections.

16.2.3 Induction in Science

We will now explain the problem of induction and its relevance to scientific practice, by identifying inductive inference within a more scientific example. The first thing to note here is that, especially in social sciences, scientists make probabilistic inferences. In terms of the example of Section 16.2.1, from the data expressed in Statement 1, they generally derive statements like 4 and 5. This is because in social sciences, the data often show patterns that are not completely stable. However, we can still say that such probabilistic inferences are inductive.

To illustrate induction, we will use a rather simplified version of the example provided in Chapter 2 about amnesia in Dissociative Identity Disorder (DID). The study of Huntjens et al. [16] focuses on the question of whether DID-patients suffer from true amnesia or not. The design allowed the authors to compare the overall memory performance, called the Recognition Scores, between true DID-patients, controls, DID-simulators, and true amnesiacs. Let us say we are now only interested in the question of whether the memory performance of DID-patients differs from the performance of DID-simulators. If the performance of DID-patients is better than that of DID-simulators, we conclude that DID is not an iatrogenic disorder. To investigate this difference, the researchers selected a sample from a population of people diagnosed with DID and a sample of "normal" people who were asked to simulate DID. The memory performance of the two groups was observed in a number of trials and, based on the difference in the memory performance, a generalized statement was made about amnesia in DID.

With this scientific example of DID in place, we can restate the problem of induction. Suppose that the observations until now show that the entire group of DID-patients is better in memory performance than the group DID-simulators. By induction we might then infer that all DID-patients are better in memory performance than DID-simulators and, hence, that amnesia in DID is true amnesia rather than feigned. Or, alternatively, suppose that on average the DID-patients are better in memory performance than the DID-simulators. In that case we might infer, again by induction, that this average difference holds for the entire populations of DID-patients and DID-simulators and, hence, that a randomly chosen DID-patient can be expected to have better memory performance than a randomly chosen DID-simulator. This expectation is typically spelled out in terms of a probability assignment; in social sciences, such probabilistic conclusions are much more common than strict universal generalizations.

Because such general or predictive conclusions concerning DID-patients and DID-simulators are arrived at by induction, they are subject to the problem, sketched in the foregoing, that they are very hard to justify. More specifically, as the discussion of Goodman's riddle suggested, justifying such conclusions involves the explicit choice for predicates that are projectable. We will argue in the next two sections that the statistical justification of conclusions in the DID example requires such a choice. The predicates at issue are the test scores of the DID-patients and DID-simulators, respectively: If we want these test scores to be indicative of what is going on in the populations at large, we must somehow assume that they are based on, or refer to, some stable properties of the individuals in that population. As already announced, we return to this in later sections. In order to properly discuss the assumption, we first turn to two well-known responses to the problem of induction, from Carnap and Popper, respectively.

16.2.4 Carnap on the Problem of Induction

The philosophical discussion on the justification of induction is rich and multifaceted. In the following we will not provide an overview of this discussion, but rather we will present a specific take on it in order to portray statistics as a particular solution. For this we will first visit two important figures in the debate on induction: Karl Popper and Rudolf Carnap.

Carnap was one of the central figures of logical empiricism, a philosophical movement that dominated the philosophy of science in the first half of the twentieth century. In this movement, two discussions took center stage: One concerned the nature of science and its demarcation from pseudo-science and the other concerned the justification of science, which was intimately connected to the justification of conclusions arrived at by inductive inference. For the logical empiricists, as the name suggests, the main features of science were its firm foundation in primitive empirical fact and the further feature that more general scientific claims can be derived from these empirical facts by logical means. Hence, the logical empiricists faced a double challenge: To establish the firm foundations of science in primitive empirical fact and to provide a logical system that would allow us to derive more advanced scientific claims from these primitives.

Carnap's contribution to the second part of the logical empiricist program is also the salient part of the program for present purposes [4, 5]. Carnap tried to find the degree of confirmation that a given set of empirical evidence gives to some scientific hypothesis. To this aim he used both logic and probability theory. Both evidence and hypotheses were expressed in terms of a formal logical language, and the degree of confirmation was subsequently expressed in terms of a probability function over this language, the so-called confirmation function $c(H, E)$. The function $c(H, E)$ is the degree to which hypothesis H is supported by evidence E; in other words, c is the degree to which someone is rationally entitled to believe in the hypothesis H on the basis of full belief in the evidence E. The crucial ingredient in the determination of this function is Carnap's notion of logical probability: The probability assignment over the language in which H and E are sentences is fully determined by the structure of the language itself and symmetry requirements on the probability function with respect to the language. The confirmation function $c(H, E)$ can therefore be determined by a priori arguments from the language.

The main achievement of Carnap was that he managed to derive a general inductive rule on the basis of his concept of logical probability. This inductive rule allowed him to make justified predictions of future observations on the basis of a record of past observations. Say, for example, that we are given a record of the memory performance of n individuals, either DID-simulators or DID-patients, in which a number of n_0 people scored below guessing level and n_1 people scored above guessing level. We may denote each individual test result by Q_i^q, where $q \in \{0, 1\}$ and 0 means scoring below, and 1 means scoring above guessing level. The record of all n results is $E_n = \bigcap_{i=1}^n Q_i$. Carnap's c-function then gives the degree of confirmation for the next person passing the test, the event denoted by Q_{n+1}^1:

$$c(Q_{n+1}^1, E_n) = \frac{n_1 + \gamma\lambda}{n_0 + n_1 + \lambda}, \tag{16.3}$$

where γ is the initial probability for passing the test and λ is the firmness of that initial estimate. This degree of confirmation for Q_{n+1}^1 is the best guess we can make for the performance of the next individual; depending on the data, we may thus be able to conclude that the predictions for DID-patients and DID-simulators differ. Carnap maintained that in this way he solved the problem of induction. By casting the problem in a formal framework, defining a function that made explicit the degree to which we are rationally entitled to believe hypotheses on the basis of evidence, and by grounding this degree in the structure of the logical framework, he provided a logical system that allows us to derive predictions, albeit probabilistic ones, from the primitive empirical facts.

One of the weaknesses of Carnap's system is that it is fairly abstract and that it does not readily connect to the methods and statistical techniques used by scientists. For the purpose of this chapter, however, we would like to point to another set of related worries to do with language as a determining factor in the Carnapian system. Recall that the justification of the Carnapian inductive inferences rests on applying symmetry principles, as determined by the notion of logical probability, to some language. Moreover, following Goodman, we are stuck with an assumption on which predicates are projectable once the language is chosen. If the language adopts Grue and Bleen, then those are the predicates that will accumulate inductive confirmation or disconfirmation. The obvious question is: How do we determine the exact set of predicates to which the notion of logical probability can be applied? First of all, language in the Carnapian system is idealized and highly artificial, whereas most scientific theories are expressed in vague language, usually English. It is unclear how to isolate the salient predicates from the fluid scientific discourse. Related to this, to apply a Carnapian system we must hold this artificial language constant, and refuse new predicates to be introduced; otherwise we must accept that the degree of confirmation of scientific hypotheses will change whenever new predicates are introduced. But both options sit poorly with scientific method as we know it.

Finally, even if we accept the artificiality and the fixity of the language, we encounter a problem with its poverty, because the notion of a statistical or general hypothesis is virtually absent from it. The way Carnap has set up his inductive logic and the confirmation function $c(H, E)$ in it, both the evidence E and the hypothesis H must be finite expressions in a language that only has observations as primitive terms. Typically, the evidence and hypotheses are past and future observations, respectively, as in the example provided above. Now it must be admitted that this is largely due to a philosophical predisposition among the logical empiricists, namely to restrict scientific inference to the empirical realm. In principle, the formalism allows for extensions to general hypotheses, as attested by the inductive logical systems of Hintikka [12]. However, the inclusion of general hypotheses in Carnapian inductive logic remains very limited, and attempts to remedy that shortcoming have not exactly appealed to the general philosophical public.

We conclude that within Carnapian systems, we cannot formulate hypotheses on possible patterns in the data, let alone change or introduce them. In the following sections it will be seen that Bayesian statistics, as well as classical statistics, does better than the Carnapian inductive system on the count of both fixity and poverty.

16.2.5 Popper on the Problem of Induction

Before turning to statistics, we deal with another important contributor to the debate on inductive inference, Karl Popper [17] . Popper's views on induction can be explained most easily in conjunction with his position in the debate on the demarcation of science from pseudo-science. Popper rejected the view of the logical empiricists, who argued that science is defined by its roots in empirical fact and their logical implications, stating instead that falsifiability is the distinguishing feature of science. According to Popper, the hallmark of good science is that it puts itself at risk of being proven wrong. It generates distinct predictions that can be checked against the empirical facts and can subsequently be proven false. So, for example, the claim that the sun will rise tomorrow is scientific, because tomorrow we may find out that the sun has not risen, thus proving it wrong. The claim, on the other hand, that the sun will never rise anymore is not scientific, because at any point in time we must leave open the possibility of a future rising.

Popper's views on inductive inference can be seen as the continuation of this line of thought. From the view that claims are only scientific in virtue of their possible falsification, it is a small step to the view that the only claims that can be considered genuine scientific knowledge are those that result from falsification. So according to Popper, we cannot base any knowledge on inductive inference. In the example, we cannot conclude anything about future occasions of a rising sun from the fact that up until now the sun has always risen. As Popper would say, the theory that the sun will always rise is not yet disproved. But, at best, this motivates us to go on checking the claim that the sun will always rise. If, on the other hand, the sun does not rise tomorrow, science has truly advanced, because at that point we can be certain that the claim that the sun will always rise is false and, hence, that the claim that on some day the sun will not rise is true. In short, Popper argues that inductive inferences toward general claims cannot provide us with scientific knowledge but that deductive inferences toward the denial of general claims do provide knowledge. Deductive inference is valid, but inductive inference is not.

In our DID example, the question is how can we generalize toward a conclusion on the existence of DID, on the basis of observations of the memory performance of DID-patients and DID-simulators. Now, does Popper allow us to conclude that DID-patients are universally better in memory performance than DID-simulators and, therefore, that DID is a genuine disorder? Bypassing the further difficulty that in the DID example, the theory is cast in terms of probabilities and that probabilistic statements can strictly speaking never be proven false, Popper would argue there is never any positive evidence for such a general statement, let alone for concluding that amnesia in DID-patients is real rather than feigned amnesia. We can only conclude, by means of a single counterexample, that such a general statement is not true. So after our observation of a difference between DID-patients and DID-simulators, the theory that amnesia in DID is real is not disproved by the data and, therefore, the theory, for the time being, is not rejected. But it is not proven by the data either.

Admittedly, this is a rather critical view of inductive inference. Popper's position has aptly been named critical rationalism. But as the term rationalism suggests, the views of Popper also have a more positive part that is of interest to the present discussion. Whereas Carnap put the starting point of scientific knowledge in primitive empirical facts, as captured in a formal language, Popper put forward the view that science always starts with a hypothesis, some bold claim or general statement, that we may subsequently attempt to falsify. He referred to this as the searchlight theory of knowledge: The realm of empirical fact can provide some kind of knowledge, but the researcher has to provide a searchlight, more specifically a guiding hypothesis via which this realm can make itself known. Put differently, it is not the observations that come to us with their own message, rather we take the initiative to seek out the observations to meet our own interest. In comparison with the empiricist and inductivist views of Carnap, Popper's views show a marked rationalist tendency, in the fact that the mind rather than the world is the first cause in the production of knowledge.

Summing up, we have dealt with two very different views on the problem of induction in the foregoing. In the next two sections, we will argue that statistical inference occupies a middling position between the two views and that both Popper and Carnap fail to capture an important aspect of the solution inherent to statistical inference. On the one hand, statistical inference is inductivist, because it allows us to learn from the data. On the other hand, it is rationalist, because what is learned from the data is entirely determined by the statistical model that we choose.

16.3 Bayesian Inference as Deduction

The discussion on Carnap and Popper makes clear that the opinions on how to justify inductive inference diverge widely. Because Bayesian statistical inference is a way of dealing with inductive inference as well, the question of how it might be positioned relative to these diverging opinions arises. In the next two sections we will argue that Bayesian statistical inference contains both falsificationist and inductivist elements. More in detail, in this section we show that the methodology of Bayesian statistical inference can be spelled out by framing these inferences in a probabilistic logic, following ideas of Howson [13, 14] and Romeijn [18, 19]. It will become apparent that Bayesian inference is similar to deductive inferences. This will lead to a discussion of model selection procedures in the next section, which will reveal the position of Bayesian statistical inference in the spectrum between Carnap and Popper.

16.3.1 Deductive and Inductive Inference

Let us briefly compare deductive and inductive logic. Recall that in deductive logic, an argument is valid if the truth of its premises guarantees the truth of the conclusion. So a perfectly valid argument might lead to a false conclusion, on the grounds that one of its premises is false. Take, for example, the premises that all apples are fruit and that all fruit grows on bulldozers. By deductive inference, we therefore validly conclude that all apples grow on bulldozers, even though this is most certainly not true. Deduction serves to explain and rearrange our knowledge without adding to its content. Inductive inference, by contrast, seems to add to the content of our knowledge. We obtain observations and then amplify and generalize them to arrive at general conclusions. So an important difference between deduction and induction seems to be that whereas deduction is conceptually closed and only brings out the conclusions already present in the premises, induction adds to the content of the premises. As a result of this, conclusions obtained with inductive inferences do not necessarily have the same degree of certainty as the initial premises.

Nevertheless, in this section we will investigate the parallel between deductive and inductive inference. To do so, we will first study a specific deductive argument, and after that we will introduce an argument in Bayesian logic that can be seen as the inductive counterpart to the deductive argument. The example of deductive inference that we will study is the so-called proof by contraposition:

If H, then E (premise 1).
E is false (premise 2).
Therefore, H is false (conclusion).

To examine this inference in more detail, we will make use of the DID example we discussed earlier. The analogy between deductive and Bayesian inference suggests that, just like the deductive inference, Bayesian inference is valid.

16.3.2 Deduction in the DID Example

The full design of the study of Huntjes et al. [16] allowed the authors to compare estimations of memory performance of DID-simulators (μ_{sim}), DID-patients (μ_{pat}), true amnesiacs (μ_{amn}), and controls (μ_{con}). We can formulate many different general models concerning the memory performance of these groups:

$M_0 : \mu_{sim} < \mu_{pat} = \mu_{amn} < \mu_{con}$
$M_1 : \mu_{sim} = \mu_{pat} < \mu_{amn} < \mu_{con}$
$M_2 : \mu_{pat} = \mu_{con} = \mu_{sim} = \mu_{amn}$
$M_3 : \mu_{pat} > \mu_{con} > \mu_{sim} < \mu_{amn}$
...

Note that this is a list of models, not of general hypotheses. The statement that $\mu_{sim} < \mu_{pat} = \mu_{amn} < \mu_{con}$, for example, is consistent with a large number of different valuations of these parameters, and each of these valuations presents a separate hypothesis. So the statement concerns a set of hypotheses, or a model for short.

For convenience we will make the example of the present section a bit easier. First of all, we will abstract away from the parameters μ_{amn} concerning amnesiacs and μ_{con} concerning people from the control group. Second, in this section we will not deal with models but with specific hypotheses (e.g., specific valuations for the parameters μ_{pat} and μ_{sim}). Third, we are restricting attention to two hypotheses in particular, H_0 and H_1. For H_0 we choose particular values of the parameters such that $\mu_{pat} > \mu_{sim}$ and for H_1 we choose them such that $\mu_{pat} = \mu_{sim}$. Moreover, we assume for the time being that one of these two hypotheses is true and thus that all the other hypotheses are false, or in logical terms, $H_0 \vee H_1$, where the symbol $\vee$ can be read as "or." This expression is the first major premise in the deductive argument below. Note also that from their definitions, the hypotheses H_0 and H_1 are

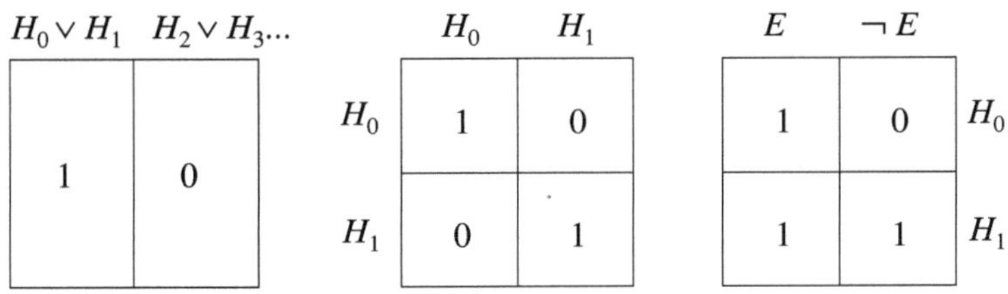

Fig. 16.1. These three squares summarize the premises of the logical argument. The leftmost square indicates that $H_0 \vee H_1$ and, thus, that all other hypotheses H_i for $i > 1$ are deemed false. The middle square indicates that $\neg(H_0 \wedge H_1)$, by setting the quadrants in which H_0 and H_1 overlap to 0. Finally, the rightmost square indicates that $H_0 \rightarrow E$, which is equivalent to $\neg(H_0 \wedge \neg E)$. The three quadrants labeled 1 in the rightmost square are the only logical possibilities consistent with the premises

mutually exclusive, so that $\neg(H_0 \wedge H_1)$, where $\neg$ means "not" and $\wedge$ can be read as "and." This will turn out to be convenient in the representation of the hypotheses below, but we will not use this premise in the argument.

So the inference concerns the two rival hypotheses H_0 and H_1. The empirical evidence, as, for instance, provided in the study of Huntjes et al. [16], is now used to adjudicate between these two hypotheses. First, we concentrate on a specific empirical difference between these two hypotheses, namely that according to H_0, DID-simulators have a worse memory performance than DID-patients, and according to H_1 the DID-simulators and true DID-patients have equal performance. Accordingly, the relevant observations are the scores of members of the two groups, patients and simulators, on some memory test. We might, for example, find that the difference of the scores of the two groups exceeds a certain threshhold, denoted E, or otherwise we might find that it does not exceed the threshhold, denoted $\neg E$. For the purpose of this example, we suppose that the test scores can tell the hypotheses apart unequivocally: If H_0 is true, then we are certain that the difference in scores on the memory test exceeds a certain threshhold, or in logical parlance, $H_0 \rightarrow E$.

We can specify so-called truth values for each combination of hypotheses and evidence, based on the premises of the above. It will be convenient and insightful to represent these premises as truth valuations over all the logical expressions that we can conceive; see the squares of Figure 16.1. As further explained in the caption, the truth values in the quadrants indicate whether the corresponding logical possibilities, or cells in the grid, are consistent with the premises. More specifically, given some truth valuation over the logical possibilities, we say that a proposition is true if and only if it is true in each of the cells that is assigned a 1. The premises $H_0 \vee H_1$ and $\neg(H_0 \wedge H_1)$ are worked out in the first two squares of Figure 16.1. They are, in a sense, implicit to the presentation of the truth valuations in the rightmost square of Figure 16.1, in which H_0 and H_1 are put side by side as mutually exclusive and jointly exhaustive possibilities.

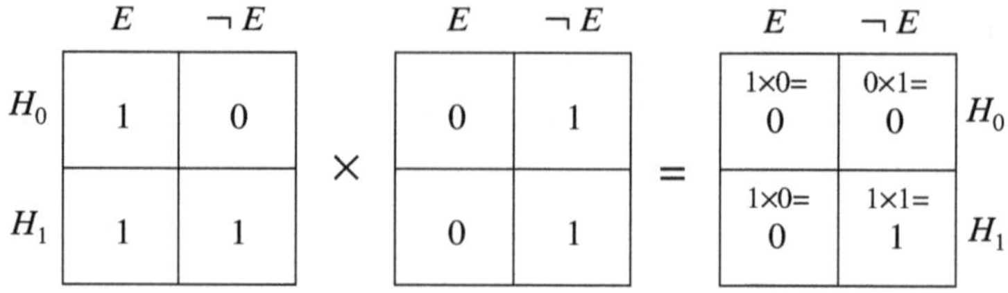

Fig. 16.2. This calculation with squares summarizes the logical argument that runs from the premises given previously, and the additional premise that $\neg E$, to the conclusion of H_1. The leftmost square is equivalent to the rightmost square of Figure 16.1. The middle square expresses the premise $\neg E$. The truth values in the rightmost square are obtained from the values in the other two squares by multiplying the values in each of the quadrants

The latter square also expresses how the hypotheses H_0 and H_1 relate to the data E. According to deductive logic, all the entailment $H_0 \rightarrow E$ says is that we cannot have the combination of H_0 being true yet E being false, so $H_0 \rightarrow E$ is equivalent to $\neg(H_0 \wedge \neg E)$. In sum, the three quadrants of the rightmost square that contain a 1 are the only logical possibilities consistent with the premises.

With this graphical representation of the premises in place, we can bring in the further premise presented by the observations. Say that we observed that the scores of the two groups on the memory test are slightly different, but that the difference does not exceed the given threshhold, so $\neg E$ receives a truth value of 1. In Figure 16.2, the corresponding truth values can be seen in the middle square. The observation itself does not involve the hypotheses and, therefore, $H_0 \wedge \neg E$ and $H_1 \wedge \neg E$ receive the truth value 1 and $H_0 \wedge E$ and $H_1 \wedge E$ receive the truth value 0. So the square on the left and in the middle of Figure 16.2 express the two main premises: One concerning the hypotheses, stemming from Figure 16.1, and one concerning the observations. The beauty of the graphical representation is that combining these premises is a straightforward operation on the truth valuations: We simply multiply the truth values of the two input premises, as expressed in the square on the right of Figure 16.2.

After combining the premises, we see that only $H_1 \wedge \neg E$ receives a truth value of 1. All the other cells have a truth value 0. We can therefore conclude all propositions that include the specific cell $H_1 \wedge \neg E$. Of course, we may conclude $\neg E$, but this is hardly surprising, because it was also one of the premises. However, we may also conclude H_1. Via $\neg E$ and $H_0 \rightarrow E$ we learn that H_0 cannot be true, so $\neg E$ falsifies H_0, and by $H_0 \vee H_1$ we can derive that H_1 must be true. We can conclude that the DID-simulators and DID-patients have equal capacities on memory performance.

16.3.3 Choosing a Model

In the previous subsection we used deductive inference to derive a conclusion from the premise concerning a finite set of hypotheses, the premise on how the hypotheses relate to evidence of the observed memory performance and, finally, a premise expressing what evidence we received. In this subsection and the next, we will use essentially the same premises, with a minor revision, as will be explained later, to derive a conclusion by means of Bayesian inference. The conversion has two aspects, namely the use of probabilistic valuations and of Bayes' theorem. In this subsection we will deal with the former.

Apart from providing us with a convenient way of representing the operation of combining premises, the graphical representation of Figures 16.1 and 16.2 can be used to illustrate the parallel between deductive and Bayesian logic, which we consider very telling. First, consider the graphical representation itself. As in the case of deductive inference, we take the logical possibilities provided by the hypotheses and the evidence as a starting point. We distinguish between E and $\neg E$, and, similarly, we consider hypotheses H_j with $j = 0, 1, 2, \ldots$. Now we want to connect these logical possibilities to probability theory, which is, according to the standard axiomatization, a function over sets, and, hence, we are taking the logical possibilities as sets as well. The logical possibility H_0 is the set of all those imaginable or possible worlds in which the hypothesis H_0 is true, and, similarly, E is the set of all those possible worlds in which the observation E occurred. Accordingly, instead of $H_0 \wedge E$, we will write $H_0 \cap E$; that is, instead of working with the logical operation $\wedge$, from now on we use the set-theoretical operation of intersection. Similarly, we will write $\neg E$ as $\bar{E}$, the set-theoretical complement of E.

Next consider the inference concerning the logical possibilities. Recall that the idea of deductive inference was to find a truth valuation of certain proposition, based on the truth valuations of a combination of premises. Again, Bayesian inference does roughly the same. The key difference between deductive and Bayesian logic is that Bayesian logic does not use truth values of 0 and 1, as does deductive logic. Rather it uses probabilistic valuations p, that is, valuations of logical possibilities within the interval $[0, 1]$ and satisfying the axioms of probability theory. So the cell $H_0 \cap E$ in the space of logical possibility receives some probability, $p(H_0 \cap E) = 2/5$ for instance. The probability values of all the cells must sum to 1. But apart from that difference in valuation function, the workings of Bayesian logic will turn out to be very similar to the workings of deductive logic. Just like deductive logic, Bayesian logic computes probabilistic conclusions on the basis of probability assignments over logical possibilities.

Let us have a look at the above deductive inference to make the above claims precise. The first premise in the foregoing is that we restrict ourselves to two hypotheses, H_0 and H_1. We assigned a truth value of 1 to $H_0 \vee H_1$, so that we ruled out all the H_j for $j > 1$. In Bayesian logic, we can do the same by assigning all probability to the hypotheses H_0 and H_1, $p(H_0 \cap H_1) = 1$; that is,

	E	$\neg E$
H_0	2/5	1/10
H_1	1/5	3/10

E	$\neg E$	
4	1	H_0
2	3	H_1

Fig. 16.3. The square on the left represents the probability assignment over the logical possibilities in terms of probability mass. The square on the right provides the same information in terms of odds

only these two hypotheses receive a probability and the remaining hypotheses $H_2, H_3, \ldots$ receive a probability of 0. But note that this probability assignment is not yet specific enough: We still have many ways of allocating the probability among the two hypotheses H_0 and H_1. On the basis of a symmetry argument, we might distribute the total probability evenly: $p(H_0) = p(H_1) = 1/2$.

We have now chosen the hypotheses, but we have not determined the probability assignment on the level of logical possibilities. Both the hypotheses H_0 and H_1 might allow for the occurrence of the observations E and $\neg E$, and we need to specify the probability valuations of these cells. Recall that in the deductive case we said that $H_0 \wedge \neg E$ was impossible. This was admittedly a rather strong assumption: Normally, test results cannot outright falsify any hypothesis, rather they make hypotheses more or less likely. By using the probability valuations, we can make such weak relations between observations and hypotheses precise. If H_0 is true, we think it is far more probable than not that the difference between the DID groups on memory performance exceeds the threshold, but this need not be strictly implied. So we might specify that conditional on H_0 being true, E is 4 times more likely than $\bar{E}$, so that $p(E|H_0) = 4/5$ and $p(\bar{E}|H_0) = 1/5$. Similarly, if H_1 is true, we might consider it somewhat less probable than not that the difference between the DID groups on memory performance exceeds the threshold, so we might specify $p(E|H_1) = 2/5$ and $p(\bar{E}|H_1) = 3/5$.

Together with the probability assignment over H_0 and H_1, we have thereby fixed the probability assignment for all the logical possibilities. We can compute $p(H_j \cap E) = p(H_j)p(E|H_j)$ and, similarly, $p(H_j \cap \bar{E}) = p(H_j)p(\bar{E}|H_j)$. This leads to the probability assignment over the logical possibility presented in the left square of Figure 16.3. The square on the right side of this figure effectively depicts the same probability assignment, but written down in terms of odds. The difference is that the odds do not have to add up to 1. Only their ratios matter. In the following we will only make use of the odds.

Finally, we want to point to the relation of the above with Bayesian statistics as we know it. In the foregoing we chose two hypotheses, defined the probabilities of the observations conditional on them, and chose the probabilities of the hypotheses themselves. In Bayesian statistics, this comes down to the choice of a model, or a set of possible statistical hypotheses, then the definition of a likelihood function for each of the hypotheses in the model, and the determination of so-called prior probabilities. Of course, statistical models are normally much more complicated and elaborate, but the general idea remains the same.

16.3.4 Bayesian Inference

As indicated, we are drawing an analogy between deductive inference and Bayesian inference. It will be clear that the determination of probabilities, or odds, over the logical possibilities in Figure 16.3 runs parallel to the first of the two main premises in the logical argument, as summarized in Figure 16.1. Now the second premise of the Bayesian inference is almost the same as the one we used for deductive inference. We observe $\bar{E}$, and in the deductive example, $\neg E$ therefore receives a truth valuation of 1. In Bayesian inference, as will be seen, we will say that the adapted probability for $\bar{E}$ must be 1. The question is how the addition of this premise reflects on the probability assignment over the logical possibilities, as given in Figure 16.3. In particular, how is the adapted probability distributed between the hypotheses H_0 and H_1?

Note first that the new premise is, strictly speaking, in contradiction with the probability assignment already given. We have $p(\bar{E}) = p(H_0 \cap \bar{E}) + p(H_1 \cap \bar{E})$ and hence $p(\bar{E}) = 2/5$. To express the probabilities after we observed $\bar{E}$, we must therefore make use of a so-called posterior probability assignment, which we will denote with $p_{\bar{E}}$. This is a new probability assignment, which is consistent with assigning $\bar{E}$ unit probability. To obtain the posterior probability assignment from the prior one, we can use the combination of Bayes' rule and Bayes' theorem:

$$p_{\bar{E}}(\cdot) = p(\cdot|\bar{E}) = p(\cdot)\frac{p(\bar{E}|\cdot)}{p(\bar{E})}. \tag{16.4}$$

Bayes' theorem is given by the second equality. It is a theorem of probability theory, and as such it is very hard to argue with. The interesting and contentious equality is the first one, which we might call Bayes' rule. Note that it is not a theorem of probability theory. Rather it relates two different probability functions, the prior distribution p and the posterior distribution $p_{\bar{E}}$, and thus expresses how we must adapt the probabilities if we add the further premise $\bar{E}$. In other words, Bayes' rule expresses how we can construct a new probability assignment $p_{\bar{E}}$ that incorporates the fact that we assign a probability 1 to the data $\bar{E}$, based on the old probability assignment p, in which the data $\bar{E}$ had a probability smaller than 1.

Now let us compute some posterior probabilities, based on the fact that we have $p_{\bar{E}}(\bar{E}) = 1$. By Bayes' rule, we can compute the posterior probability for the hypotheses H_0 and H_1 on the basis of the prior probability and the likelihoods. For H_1 we find

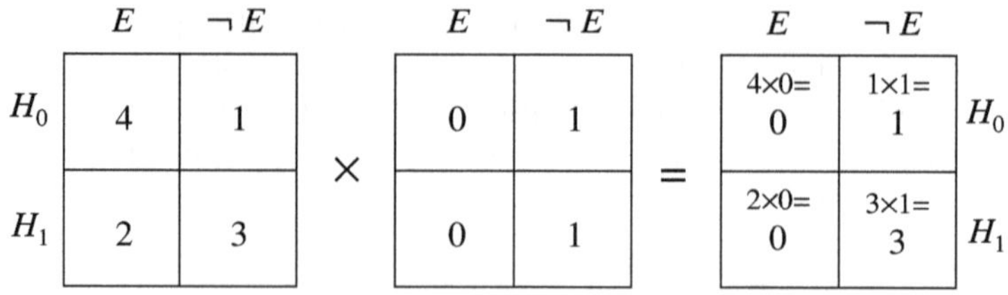

Fig. 16.4. This calculation with squares summarizes the Bayesian statistical inference. The leftmost square is equivalent to the square on the right-hand side of Figure 16.3. The middle square expresses the premise $\neg E$. The odds in the rightmost square are obtained from the values in the other two squares by multiplying the values in each of the quadrants

$$p_{\bar{E}}(H_1) = p(H_1|\bar{E}) = p(H_1)\frac{p(\bar{E}|H_1)}{p(\bar{E})} = \frac{1}{2} \times \frac{3/5}{2/5} = \frac{3}{4}; \tag{16.5}$$

in words, the observation that $\bar{E}$ leads to a posterior probability for H_1 that is higher than the prior probability. In this sense at least, Bayesian inference mimics the deductive inference, where $\bar{E}$ also favored H_1. But why are we to believe the posterior probabilities arrived at by means of Bayesian inference?

We will now argue that there is a much more genuine sense in which the Bayesian inference resembles the deductive inference. This resemblance provides us with a reason to believe that the posterior probabilities are in a sense the correct probabilities for the hypotheses after the observation of $\bar{E}$. As Figure 16.4 illustrates, if we represent the probability valuations as odds, we can combine the two main premises of the Bayesian inference in exactly the same way as in deductive inference.

It is not a coincidence that the results of this operation are the odds that correspond to the posterior probabilities arrived at by Bayesian inference. Changing the probability assignment in accordance with the observation $\bar{E}$, as laid down in Equation (16.4), is nothing but the rescaling of the probabilities to the proportions of the probabilities within $\bar{E}$. This is exactly what the formula does. Bayes' rule allows us to "zoom in" on the probability assignment over the hypotheses within $\bar{E}$.

Thus, Bayesian inference is like deductive inference in two important respects. First, they both make use of a valuation function over a set of elementary logical possibilities, although there are also differences here. As for these possibilities, in the case of deductive inference they are maximally specific propositions, and in the case of Bayesian inference they are sets of possible worlds. As for the valuations, in the case of deductive inference they are truth valuations, and in the case of Bayesian inference they are probabilities. Second, and most notably, the operation for combining a valuation with a further premise, in particular with an observation such as $\bar{E}$, is exactly the same. Rather suggestively, we might say that Bayesian inference is therefore valid in exactly the same way as that deductive inference is.

Although it has been noticed before, we want to emphasize again that the above example is nothing like a serious Bayesian statistical inference: Usually the model contains many more statistical hypotheses, and there are normally many more possible observations, or elements, in the sample space. However, the inferential steps are exactly the same. In Bayesian statistics we choose a model, fix the likelihoods of the hypotheses in the model, and, finally, determine a prior. Then we collect data and incorporate these data in the so-called posterior probability assignment over the model by means of a Bayesian update. We therefore maintain that the above example tells us something about Bayesian statistical inference in general.

16.3.5 Summing up

We have discussed how to derive a conclusion based on a set of premises, first by using deductive inference and then by using Bayesian inference. We have shown that Bayesian inference follows roughly the same procedure as deductive inference. This suggests that Bayesian inference, like deductive inference, is valid; that is, if the premises are true, then so is the conclusion. In the following we will elaborate how these ideas may be used to position Bayesian statistics in the philosophical debate over statistics and, in particular, how they can be applied to the Bayesian model selection described in Section 16.1.

16.4 Model Selection

We have seen that Bayesian statistics can be provided with a philosophical underpinning by portraying it as a logic. Against this backdrop we will now explain how statistics, and Bayesian statistics in particular, unites the views of Carnap and Popper on induction. This may well raise some eyebrows: In what sense do we do justice to Popper's views when we redistribute probability over a number of hypotheses in the light of data? Recall that an important aspect of Popper's view is falsificationism, which states that we can only learn from data if the data rule out some hypothesis. Bayesian statistical inference goes much further than that, because it allows us to learn positive facts from the data. Nevertheless, in the following we will argue that, in some important respect, Bayesian statistical inference retains the rationalist spirit.

16.4.1 Models as Uniformity Assumptions

The foregoing already indicated that in a Bayesian inference, the choice for a model can be understood as the choice of a certain kind of premise. We drew a parallel between, on the one hand, the choice for a prior restricted to $p(H_0) = p(H_1) = 1/2$ together with the likelihood functions of both hypotheses and, on the other hand, the choice for $H_0 \vee H_1$ together with $H_0 \rightarrow E$. In this subsection we will investigate this parallel further. In particular, we will relate the choice of a certain model, as elaborated in Section 16.3, with the choice of a certain set of projectable predicates, as discussed in Section 16.2.

Let us return to the nature of inductive inference as it was illustrated in the example of Section 16.2.1. It can be noted that the inference from Statement 1 to Statement 2, by itself, seems to miss a component. It is more or less implicit in the inference that what has happened in the past can be expected to happen in the future. As we discussed, one possible take on the problem of induction is that this component must be added to the inductive inference as an explicit premise. At first glance this premise might simply be that the world is a boring place and that the same events will keep repeating themselves. But it was easily seen that simply adding this premise cannot solve the problem: We ran into predicates like Grue. As exhibited clearly in the inductive logical systems devised by Carnap, if we want to infer anything inductively, we must choose the exact set of predicates with respect to which the world is boring, that is, the predicates that are supposed to stay constant. In philosophical parlance, we must select the projectable predicates.

There is a rather nice formal relation between the Carnapian systems and Bayesian statistical inference, which has an immediate bearing on this point. Note first that the c-function of Equation (16.3) only depends on the number of earlier results, n_0 and n_1, and not on the exact order in which these results were observed. Inductive logical systems with this property are called exchangeable. Famously, De Finetti [8] proved that any exchangeable inductive logical system can be represented as a Bayesian inference over a particular model, namely the model of binomial hypotheses, and furthermore that every prior over this model singles out a unique exchangeable system. As in the foregoing, we write Q^1_{n+1} for the result of person $n+1$ scoring above chance level in a memory test, meaning that this person scored better than the expected score of filling in the test randomly. We denote the binomial hypotheses with H_θ. These hypotheses have the following likelihoods:

$$p(Q^1_{n+1}|E_n \cap H_\theta) = \theta. \qquad (16.6)$$

This means that all test results are independent and identically distributed. The model of binomial hypotheses, which features in De Finetti's representation theorem, includes all these hypotheses: $\{H_\theta : \theta \in [0, 1]\}$. It can be proved that prior probability functions of the form $p(H_\theta) \sim \theta^{\gamma\lambda-1}(1-\theta)^{(1-\gamma)\lambda-1}$ lead to the Carnapian inductive systems of Equation (16.3); that is,

$$c(Q^1_{n+1}, E_n) = \int_0^1 p(Q^1_{n+1}|H_\theta \cap E_n)p(H_\theta|E_n)\, d\theta, \qquad (16.7)$$

in which $c(Q^1_{n+1}, E_n)$ is the expected value of the response of subject $n+1$ given earlier responses E_n, $p(Q^1_{n+1}|H_\theta \cap E_n)$ is the likelihood of the hypothesis

H_θ for the event of this subject scoring above chance level, and $p(H_\theta|E_n)$ is the posterior probability over all hypotheses H_θ in the model of binomial hypotheses, given the earlier responses E_n. The interested reader may consult Festa [9] for further details on this. For present purposes it is only important to remember that Carnapian inductive systems can be replicated in a Bayesian inference.

This mathematical fact provides us with crucial insight into the nature of choosing a model. Recall that in the Carnapian system, the choice of the predicates, in this case scoring above, on, or below chance level in the memory test, effectively determined the projectable or stable pattern in the data: The observed relative frequencies of scoring were supposed to be indicative of the scoring of future subjects. But we can identify exactly these projectable patterns in the statistical model that, according to the representation theorem, underpins the Carnapian system. For each of the binomial hypotheses, the probability of scoring above chance level is stable and constant over time. The choice for this specific set of hypotheses, or this statistical model for short, is effectively the choice for a set of projectable predicates, namely the chance for scoring over a certain level is stable and constant over time. In our view this is exactly the function of choosing a model as part of a Bayesian statistical inference: to fix the starting point, namely the set of hypotheses and the associated probabilistic patterns, so that the data are allowed to select the most fitting one.

The choice for a specific model, or for specific hypotheses to be part of the model, reflects the interest and often the background knowledge of the researcher. But this also means that a researcher can help herself to more informative conclusions by choosing her hypotheses well and, similarly, that she can ruin it by choosing her model poorly. For instance, she might choose for the gruesome variants of the binomial hypotheses introduced in the above:

$$p(Q^1_{n+1} = 1|E_n \cap G_{N\theta}) = \begin{cases} \theta & \text{if } n < N \\ (1-\theta) & \text{if } n \geq N; \end{cases} \tag{16.8}$$

in words, the hypotheses $G_{N\theta}$ dictate that up until the Nth observation Q^q_n for $n < N$ the probability for $q = 1$ is θ, but that for $n \geq N$ the probability for $q = 1$ is $(1 - \theta)$. We might take the model $\{G_{N\theta} : \theta \in [0, 1]\}$ for some large N, choose a uniform prior $p(G_{N\theta})d\theta = 1$, and then start updating with observations of subjects doing the memory test. For values of $n + 1 < N$ the choice of this model leads straightforwardly to the Carnapian prediction rule of Equation (16.3), with $\gamma = 1/2$ and $\lambda = 2$. Now say that, by far, most subjects $i < N$ pass the test, so that $n_1 >> n_0$. Using the Carnapian system and assuming that $n < N$, we have $p(Q^1_{n+1}|E_n) \gg p(Q^0_{n+1}|E_n)$. But what can we predict for the subject indexed i with $i > N$ on the basis of E_n? Because of the sudden reversal in the likelihood functions of the hypotheses, we effectively swap the places of scoring on or above chance level in the prediction, so on the basis of a large majority of people exceeding chance level in E_n, we predict that subjects $i > N$ will most likely fail! Or in mathematical words, $p(Q_i^1|E_n) \ll p(Q_i^0|E_n)$ for $i \geq N$.

We saw in the example on gruesome predicates of Section 16.2.2 that the wrong predicate choice may lead to useless predictions, and we have here seen that the same holds for the choice of models, thus indicating how the choice of a certain model resembles the choice of a projectable predicate. Bayesian statistical inference therefore has, at least, this one distinct Carnapian streak: It allows for inductive inference on the basis of a specific uniformity assumption.

16.4.2 Models as Searchlight

In the foregoing we claimed that Bayesian statistical inference occupies a middle position between Carnap and Popper. Partly the link with Carnap has now been made clear, and so we turn to the relation with Popper, in particular with his searchlight theory of knowledge alluded to in Section 16.2.5.

We first identify this searchlight theory in Bayesian statistical inference, which will point us to an important difference between Carnapian inductive systems and Bayesian statistical inference. We have already seen how both make use of specific uniformity assumptions. However, in the case of Carnapian systems, there seems to be very little by way of actively choosing, let alone comparing the assumptions. In the views of Carnap, the choice for a language, and thus the uniformity assumptions inherent to it, is a precondition for dealing with the problem of induction in terms of a logic. In fact, according to Carnap [3], it is a precondition for dealing with philosophical problems in general. So it seems that for Carnapian systems, the choice for a specific uniformity assumption is beyond the reach of logical analysis. By contrast, in Bayesian statistical inference the choice for a uniformity assumption, by choosing a model, is an explicit part of the logical account. As also argued in the foregoing, the choice of a model determines the type of probabilistic pattern that we can identify in the data. In other words, it provides us with a searchlight looking at the data. The explicit choice for a model signals a rationalist tendency in Bayesian statistical inference. The origin of empirical knowledge is not naked observation, but observation within the context of a theoretical starting point, namely a model.

We may wonder whether we can extend the parallel between the Popperian view on induction and Bayesian statistics, in particular whether Bayesian statistics presents us with a notion of falsification. To answer this, consider the probabilistic inference in the example of Section 16.3.4. We might argue that this Bayesian inference already exhibits a weak form of falsification: The hypothesis H_0 is proved unlikely by the data, and so we may decide to discard that hypothesis, or at least not use it in predictions or decision making. However, apart from the fact that low probability is not the same as logical impossibility, the use of the specific model $\{H_0, H_1\}$ determines that either one of them will accumulate most probability in the light of the data. So by discarding H_0 we can infer H_1. Therefore, discarding H_0 is not a falsification of the starting point of the inference, namely the model $\{H_0, H_1\}$.

In the DID example, it may happen that we find further data E' for which $p(E'|H_1)$ is very small, so that H_1 fits poorly with the data as well. In such a case the whole model fits poorly with the data. Similarly, in the example concerning the hypotheses $G_{N\theta}$ of the preceding section, we may observe further subjects $i \geq N$ doing the memory test. On the basis of our model choice and the fact that subjects $i < N$ performed very well, we expect these new subjects to perform very poorly. But it may certainly happen that the subjects $i \geq N$ perform very well. We then want to conclude that something was amiss with the model choice, (e.g., that the true hypothesis is not to be found among the hypotheses in the model). Note also that such cases do not allow us to draw any positive conclusions: We just conclude that none of the hypotheses in the model is any good and that some unspecified other hypothesis would have been better. Such cases of poor model fit come a bit closer to the idea of falsification in Popper. Now we want to emphasize immediately that finding a poor model fit is not the same as definitively falsifying the model, in the same way as that finding a low probability for H_0 is nothing like logically deriving the falsehood of H_0. Low probability, or even zero probability for that matter, is entailed by but does not entail logical impossibility. Still, the closest we can get within Bayesian statistical inference to the idea of falsification is the idea of poor model fit.

However, the falsification of a model is not an integral part of the Bayesian inference machinery. The model can be chosen explicitly in the Bayesian inference, by distributing prior probability over a restricted set of hypotheses. But the tools of Bayesian inference do not allow for changes to that initial choice for a model. In the words of Dawid [7], the Bayesian is "well-calibrated": Inherent to the choice of set of hypotheses (i.e., a model), is the assumption that the true hypothesis is among them. It is impossible to change this assumption without after the fact changing the prior probability, which is a non-Bayesian move. Of course we can change a statistical model in a controlled and rational way, by turning to model selection techniques [1]. There are various criteria for model fit and various ways of off-setting model fit against the complexity of models. But with the exception of the Bayesian information criterion, the standard model selection techniques do not take the explicit form of a Bayesian inference. Even the Bayesian information criterion only employs an approximation of posterior model probabilities.

16.4.3 Bayesian Model Selection

We are now ready to present Bayesian model selection, as it was presented in Section 16.1, against the philosophical background of Bayesian statistics. Concerning this philosophical background, we argued that it combines the inductivist view of Carnap with the falsificationist view of Popper. As in the work of Carnap, Bayesian statistics allows us to reason inductively from the data by assuming that certain data patterns, summarized in a model, are invariant. But this is only possible once we have made a specific selection of hypotheses to begin with, and in this sense, Bayesian statistics also have a marked Popperian component. In the same line, the assessment of a model against the data runs parallel to falsification in the view of Popper.

How does Bayesian model selection fit into this background? It is important to keep clear on the roles of models and hypotheses here. Bayesian model selection deals with the assessment of model fit (i.e., with the fit of a collection of statistical hypotheses). It therefore extends the reach of standard Bayesian statistical inference, which concerns the fit of specific statistical hypotheses once the model is given. On the other hand, in Bayesian model selection the rival models are understood as statistical hypotheses themselves, that is, they are somehow understood as claims about patterns in the data, as expressed in a likelihood function. These likelihood functions are not straightforwardly defined, as they are in the case of a normal Bayesian statistical inference. They are so-called marginal likelihoods, because they involve the likelihoods of the hypotheses inside the rival models. Bayesian model selection is thus similar to standard Bayesian statistical inference, in the sense that rival models are treated as if they were normal statistical hypotheses. This makes Bayesian model selection very attractive: It benefits from all the arguments standardly given to support Bayesian statistical inference. However, the key difference also leads to some problematic aspects, to which we will now turn.

16.5 A Challenge for Bayesian Model Selection

This section discusses some problematic aspects of applying Bayesian inference to models. These aspects relate directly to the philosophical background for Bayesian statistical inference, as provided in the preceding sections. First, we take a closer look at the fact that in Bayesian model selection, models are conceived as hypotheses. Second, we discuss how to understand the probability assignments over models. First, we provide a tentative solution, but it will be seen that this solution puts more weight on the second problem. The section ends with a challenge to the proponents of Bayesian model selection.

16.5.1 Models as Hypotheses?

To illustrate the first of our two concerns, it is useful to recollect a well-known finding from the psychology of reasoning, concerning the so-called conjunction fallacy. In an experiment done by Tversky and Kahneman [20], subjects were presented with the following story:

> Linda is 31 years old, single, outspoken, and very bright. She majored in philosophy. As a student, she was deeply concerned with issues of discrimination and social justice, and also participated in anti-nuclear demonstrations. Which is more likely?

1. Linda is a bank teller.
2. Linda is a bank teller and is active in the feminist movement.

Rather surprisingly, a majority of normal subjects think the second fact to be the more likely one. This is odd, because the axioms of probability do not allow a conjunction to be more probable than either of its conjuncts: It is a theorem that $p(E \wedge E') \leq p(E)$ for any pair of events or facts E and E'. Clearly, people do not follow the axioms of probability in their intuitive judgments of likeliness.

Next, consider the example of Bayesian model selection in Section 16.1, and in particular the two models that are being compared: $\mathcal{M}_0$ and $\mathcal{M}_1$. Recall that both models consisted of the same hypotheses $H_{\mu_{pat}\mu_{sim}}$, that $\mathcal{M}_0$ contained all these hypotheses, and that the model $\mathcal{M}_1$ was subject to the further constraints that $\mu_{pat} < \mu_{sim}$. At first sight, this situation is completely identical to the situation with Linda the bank teller. We may write the model $\mathcal{M}_1$ as a conjunction of facts, namely the model $\mathcal{M}_0$ and the further fact that $\mu_{pat} < \mu_{sim}$. This fits well with the fact that the set of hypotheses associated with $\mathcal{M}_1$ is strictly included in the set of hypotheses associated with $\mathcal{M}_0$. It is, under closer scrutiny, truly remarkable that a set that is strictly included in another set can nevertheless have a larger probability. Is Bayesian model selection implicitly violating the axioms of probability?

The reader will be relieved to find that the answer to this question is negative. To explain this, we simply need to cast the comparison of both models and hypotheses in a different set-theoretical framework, as illustrated in Figure 16.5. As we have conceptualized the two models $\mathcal{M}_0$ and $\mathcal{M}_1$ in the above, they are overlapping sets. Even stronger, all elements $H_{\mu_{pat}\mu_{sim}}$ in $\mathcal{M}_1$ are also a member of $\mathcal{M}_0$. However, nothing prevents us from using two distinct sets of hypotheses, labeled $H0_{\mu_{pat}\mu_{sim}}$ and $H1_{\mu_{pat}\mu_{sim}}$, which are different from a set-theoretical point of view by virtue of being labeled differently, even while they have exactly the same likelihood functions over the data. The model $\mathcal{M}_0$ consists of the hypotheses $H0_{\mu_{pat}\mu_{sim}}$, whereas the model $\mathcal{M}_1$ consists of the different hypotheses $H1_{\mu_{pat}\mu_{sim}}$. The model $\mathcal{M}_1$ is further restricted by the fact that $p(H1_{\mu_{pat}\mu_{sim}}) = 0$ if $\mu_{pat} \geq \mu_{sim}$. In this

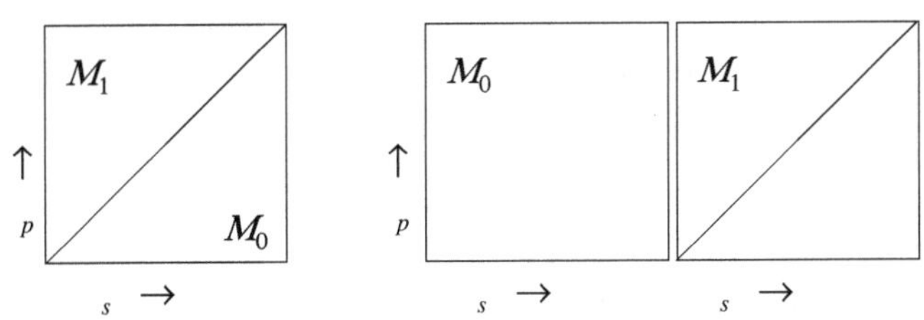

Fig. 16.5. The leftmost square shows the two models as nested sets of statistical hypotheses. On the right side, the two models are disjunct sets of statistical hypotheses, but these hypotheses have identical likelihood functions

framework, Bayesian model selection is not presenting a blatant violation of the axioms of probability. However, we may now argue that something else is wrong.

The empirical content of ordinary statistical hypotheses is in their likelihood function, that is, statistical hypotheses can in a sense be told apart by the data, even though they are distinguishable only in the limit. Consider, for example, the hypotheses of Section 16.4.1, as defined in Equation (16.6). It is logically possible that the hypothesis H_θ with $\theta = 1/2$ is true and that nevertheless the limiting relative frequency of passings in an infinitely long sequence of test results is equal to some other fraction, such as $3/4$; the interested reader may consult Gaifman and Snir [10]. But the probability of this happening is 0. In close connection to this, there are the so-called convergence theorems of Bayesian statistical inference, which show, in general, that if the hypothesis H_θ is true, the posterior probability $p(H_\theta|E_n)$ will tend to 1 in the limit of larger and larger datasets E_n. In this particular sense we can say that ordinary statistical hypotheses can be told apart by the data.

With this notion of empirical content in place, consider the two statistical models $\mathcal{M}_0$ and $\mathcal{M}_1$ of the DID example, which consist in part of statistical hypotheses that have identical likelihood functions. Can they be told apart by the data in the limit? Of course, if the true hypothesis does not satisfy the restriction imposed by the model $\mathcal{M}_1$, namely that $\mu_{sim} < \mu_{pat}$, then given sufficient data, the posterior probability of model $\mathcal{M}_0$ will tend to 1. However, if the true hypothesis does satisfy the restriction imposed by the model $\mathcal{M}_1$, then there is no such limiting behavior. In that case there are two hypotheses with correct values for μ_{pat} and μ_{sim}, namely $H_{0\mu_{pat}\mu_{sim}}$ and $H_{1\mu_{pat}\mu_{sim}}$. These two hypotheses have exactly the same likelihood function, hence there can never be any piece of data that tells against the one and in favor of the other. Admittedly, within the two models $\mathcal{M}_0$ and $\mathcal{M}_1$ separately, the convergence theorems alluded to in the foregoing take care that the hypotheses $H_{0\mu_{pat}\mu_{sim}}$ and $H_{1\mu_{pat}\mu_{sim}}$ will both attract all the probability. But exactly because $H_{0\mu_{pat}\mu_{sim}}$ and $H_{1\mu_{pat}\mu_{sim}}$ will in the limit attract all probability within their respective models, the initial probability ratio between the two hypotheses $H_{0\mu_{pat}\mu_{sim}}$ and $H_{1\mu_{pat}\mu_{sim}}$ will be retained. To be precise, we have $p(H_{0\mu_{pat}\mu_{sim}})d\mu_{pat}\mu_{sim} = 1/2$ and $p(H_{1\mu_{pat}\mu_{sim}})d\mu_{pat}\mu_{sim} = 1$, because the prior over models is $p(\mathcal{M}_0) = p(\mathcal{M}_1) = 1/2$, where within the two models the prior is uniform, and thus $p(\mathcal{M}_0|E_n) = 1/3$ and $p(\mathcal{M}_1|E_n) = 2/3$ for $n \to \infty$. For a more detailed discussion of this effect in the DID example, we refer to Chapter 4.

Summing up, it seems that we can avoid a violation of the axioms of probability in Bayesian model selection. We can do so by reconceptualizing the models involved in the selection. However, understanding models in this way may leave us with an identifiability problem: If the true parameter values satisfy the restriction at issue, the data do not single out a unique statistical hypothesis, or a single model for that matter. Instead we retain the difference between the hypotheses and models that we have ourselves imposed at the onset. We may argue that this is not a big deal. After all, once we have gained access to true parameter values, the distinction between $H_{0\mu_{pat}\mu_{sim}}$ and $H_{1\mu_{pat}\mu_{sim}}$, or between $\mathcal{M}_0$ and $\mathcal{M}_1$, may be inessential. This reaction leads us to consider the following question: How can we interpret the intermittent probability assignments over the two models, as long as we do not have the true parameter values? What, if we are eventually interested in the true parameter values, are these probability assignments about?

16.5.2 The Probability of a Model

Unfortunately, these questions are not a cliffhanger, or some other rhetorical device. By way of an answer we only have some suggestions to offer. However, we do feel that these suggestions invite further research, and we are confident that such research will not be in vain.

One rather natural answer to the above questions is that the probability of the model presents us with a specific trade-off between two different aspects of model selection. On the one hand, the probability of the models measures model fit: The better the hypotheses within a model fit the data, the higher the marginal likelihood of the model, and hence the higher the posterior model probability. On the other hand, the probability of the model reflects the simplicity of the model. The number of inequality restrictions in a model is directly related to the value of the probability density function within the model. For example, as indicated in the foregoing, the hypotheses in $\mathcal{M}_1$ have a probability that is twice as large as that of their empirically equivalent counterparts in $\mathcal{M}_0$, because in an intuitive sense the space occupied by $\mathcal{M}_1$ is half of that occupied by $\mathcal{M}_0$. The probability density over the restricted model is therefore twice as large as the probability density over the unrestricted model. Hypotheses in a restricted and hence simpler model are thus given a head start via the prior. This is reminiscent of the standard situation in model selection, in which typically the more complex model has more parameters and hence occupies a larger space as well.

This view on Bayesian model selection invites a host of further questions. One question is whether we have any reason for choosing this specific trade-off between simplicity and model fit. It is as yet unclear whether the bonus for simplicity that is implicit in Bayesian model selection always latches onto our intuitive or independently motivated criteria for the model selection at hand. If this is not the case, we may tweak the priors over the models, as they can be used as an independent component in Bayesian model selection. Another question is how the trade-off between simplicity and fit fares in cases in which the two models are of different dimensionality, for example if we compare the model $\mathcal{M}_0$ to a third model, $\mathcal{M}_2$, which has the restriction that $\mu_{pat} = \mu_{sim}$. In such cases of differing dimensionality, we may also ask how Bayesian model selection relates to other ways of trading off simplicity and fit (e.g., Aikaike's criterion), which concerns differing dimensionality as well. These are all legitimate research questions. We expect that a study into the relation between Bayesian model selection and complexity will therefore be very fruitful.

Apart from weighing simplicity and fit against each other, we can conceive of another function for comparing models in a Bayesian model selection procedure. It may be that eventually the interest of a Bayesian statistical inference lies in determining the values of the parameters in a statistical model. The employment of several models in a Bayesian model selection procedure may be a way of finding the best estimate for some parameter efficiently. This view on the use of several models leads us to consider an interpretation of the posterior model probabilities of an entirely different nature, namely as a clever means to enhance the convergence properties of the Bayesian inference. But before we wholeheartedly adopt this view, it will be wise to investigate the convergence properties of Bayesian statistical inference using multiple models in more detail.

Whatever the exact results of either of the two research lines suggested in the foregoing, we feel that we have already taken one step forward. By describing Bayesian model selection as the continuation of Bayesian statistical inference and by describing the latter as the continuation of deductive inference, we have provided a context for understanding Bayesian model selection in a philosophical way. We hope that the groundwork is laid and that any further investigations into understanding posterior model probabilities and Bayes' factors do not have to start at square one.

References

[1] Barnet, V.: Comparative Statistical Inference. New York, Wiley (1999)
[2] Bird, A.: Philosophy of Science. Montreal, McGill-Queen's University Press (1998)
[3] Carnap, R.: Scheinprobleme in der Philosophie. Berlin, Weltkreis-Verlag (1928)
[4] Carnap, R.: The Foundations of Probability. Chicago, University of Chicago Press (1950)
[5] Carnap, R.: The Continuum of Inductive Methods. Chicago, University of Chicago Press (1952)
[6] Cover, J.A., Curd, M.: Philosophy of Science: The Central Issues. New York , Norton and Co. (1998)
[7] Dawid, A.P.: The well–calibrated Bayesian. Journal of the American Statistical Association, **77**, 605–613 (1982)
[8] De Finetti, B.: Probability, Induction and Statistics. New York, Wiley (1972)
[9] Festa, R.: Optimum Inductive Methods. Dordrecht, Kluwer (1993)
[10] Gaifman, H., Snir, M.: Probabilities over rich languages. Journal of Symbolic Logic **47**, 495–548 (1982)
[11] Goodman, N.: Fact, Fiction, and Forecast. Cambridge (MA), Harvard University Press (1955)

[12] Hintikka, J.: A Two–dimensional Continuum of Inductive Methods. In: Hintikka, J., Suppes, P. (eds) Aspects of Inductive Logic. Amsterdam, North Holland (1966)

[13] Howson, C.: A logic of induction. Philosophy of Science, **64**, 268–90 (1997)

[14] Howson, C.: Hume's Problem. Oxford, Clarendon Press (2000)

[15] Hume, D.: An Enquiry concerning human understanding (1748). Tom Beauchamp (ed), Oxford, Oxford University Press (1999)

[16] Huntjens, R.J.C., Peters, M.L., Woertman, L., Bovenschen, L.M., Martin, R.C., Postma, A.: Inter-identity amnesia in dissociative identity disorder: A simulated memory impairment? Psychological Medicine, **36**, 857–863 (2006)

[17] Popper, K.: The Logic of Scientific Discovery. London, Hutchinson (1959)

[18] Romeijn, J.W.: Hypotheses and inductive predictions. Synthese, **141**, 333–364 (2004)

[19] Romeijn, J.W.: Bayesian inductive logic. Ph.D. thesis, University of Groningen (2005)

[20] Tversky, A., Kahneman, D.: Extension versus intuitive reasoning: The conjunction fallacy in probability judgment. Psychological Review, **90**, 293-315 (1983)

Index

Springer
the language of science

springer.com

Statistical Learning from a Regression Perspective

Richard A. Berk

This book considers statistical learning applications when interest centers on the conditional distribution of the response variable, given a set of predictors, and when it is important to characterize how the predictors are related to the response. As a first approximation, this is can be seen as an extension of nonparametric regression. Among the statistical learning procedures examined are bagging, random forests, boosting, and support vector machines. Response variables may be quantitative or categorical.

2008. Approx. 370 pp. (Springer Series in Statistics) Hardcover
ISBN 978-0-387-77500-5

Modern Multivariate Statistical Techniques

Allen Julian Izenman

This book is for advanced undergraduate students, graduate students, and researchers in statistics, computer science, artificial intelligence, psychology, cognitive sciences, business, medicine, bioinformatics, and engineering. Familiarity with multivariable calculus, linear algebra, and probability and statistics is required. The book presents a carefully-integrated mixture of theory and applications, and of classical and modern multivariate statistical techniques, including Bayesian methods.

2008. Approx 760 pp. (Springer Texts in Statistics) Hardcover
ISBN 978-0-387-78188-4

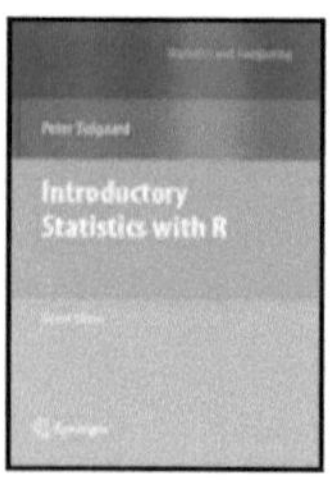

Introductory Statistics with R
Second Edition

Peter Dalgaard

This book provides an elementary-level introduction to R, targeting both non-statistician scientists in various fields and students of statistics. The main mode of presentation is via code examples with liberal commenting of the code and the output, from the computational as well as the statistical viewpoint. A supplementary R package can be downloaded and contains the data sets. In the second edition, the text and code have been updated to R version 2.6.2. The last two methodological chapters are new, as is a chapter on advanced data handling. The introductory chapter has been extended and reorganized as two chapters. Exercises have been revised and answers are now provided in an Appendix.

2008. Approx. 384 pp. (Statistics and Computing) Softcover
ISBN 978-0-387-79053-4

Easy Ways to Order▶ Call: Toll-Free 1-800-SPRINGER • E-mail: orders-ny@springer.com • Write: Springer, Dept. S8113, PO Box 2485, Secaucus, NJ 07096-2485 • Visit: Your local scientific bookstore or urge your librarian to order.

MIX
Papier aus verantwortungsvollen Quellen
Paper from responsible sources
FSC® C105338

If you have any concerns about our products,
you can contact us on
ProductSafety@springernature.com

In case Publisher is established outside the EU,
the EU authorized representative is:
Springer Nature Customer Service Center GmbH
Europaplatz 3, 69115 Heidelberg, Germany

Printed by Libri Plureos GmbH
in Hamburg, Germany